The Coiled Spring

HOW LIFE BEGINS

The Coiled Spring
HOW LIFE BEGINS

Ethan Bier
University of California, San Diego
La Jolla, California

COLD SPRING HARBOR LABORATORY PRESS
Cold Spring Harbor, New York

The Coiled Spring: How Life Begins

Acquisition/Developmental Editor Judy Cuddihy
Project Coordinator Inez Sialiano
Production Editor Pat Barker
Interior Designer Denise Weiss
Cover Designer Ed Atkeson/Berg Design

Front Cover Art: Painting by Judy Cuddihy

Library of Congress Cataloging-in-Publication Data

Bier, Ethan.
The coiled spring : how life begins / Ethan Bier.
p. cm.
Includes bibliographical references and index.
ISBN 0-87969-563-3 (pbk. : alk. paper)—ISBN 0-87969-562-5 (cloth: alk. paper)
1. Developmental genetics. I. Title.

QH453 .B53 2000
571.8′5—dc21 00-022975

All Cold Spring Harbor Laboratory Press publications may be ordered directly from Cold Spring Harbor Laboratory Press, 10 Skyline Drive, Plainview, New York 11803-2500. Phone: 1-800-843-4388 in Continental U.S. and Canada. All other locations: (516) 349-1930. FAX: (516) 349-1946. E-mail: cshpress@cshl.org. For a complete catalog of Cold Spring Harbor Laboratory Press publications, visit our World Wide Web Site http://www.cshl.org

This book is dedicated to
my parents and my son Benjamin

Contents

implications of these discoveries—how we think of ourselves as humans and how these discoveries can change our nature—are considered using topics such as the Human Genome Project, which is determining the complete genetic blueprint of humans; the implications of our newfound knowledge of the genetics of human disease and health; and genetic engineering of plants and animals. How the world of science fiction views these topics is also considered.

Bioboxes

Landmark progress in science is made by pioneering individuals. Those responsible for some of the key discoveries in developmental biology are highlighted in "Bioboxes" that appear throughout the text. These scientists represent the human side of the wonderful discoveries described in this book.

Preface

I undertook writing *The Coiled Spring* because now is an opportune time to provide the general science reader with an account of the rapidly unfolding field of developmental biology. Several factors contribute to this timeliness. First, the field is at the point where many of the general principles are well understood. This is by no means to say that we have answered all of the interesting questions. Quite to the contrary, many exciting discoveries remain to be made. But we do have a good idea about the outline of how development works, and this emerging story should be of significant interest to anyone curious to know how a fertilized egg smaller than the head of a pin makes a person, a fly, or a plant. One of the most unexpected and profound findings of the field has been the discovery that the basic mechanisms guiding development are the same in apparently disparate organisms such as flies and humans.

Another reason for bringing the field of development to the attention of a more general audience at this point is that our new understanding of developmental mechanisms is already beginning to have a great impact on the world in which we live. As a result, a basic knowledge of this field is important for all who are interested in shaping our common future. I hope this book will serve its intended purpose by familiarizing the reader with classic experiments in developmental biology, some of the cutting-edge research that explains these classic observations in simple mechanistic terms, and the implications of these discoveries for the future.

Many people have contributed to this book. First, there are all of the scientists in the field of developmental biology from the time of Goethe to the present. In addition to the many investigators working actively on topics covered in this book, there are yet greater numbers of impassioned scientists who work long into the night hours to unravel many other equally interesting mysteries about development. These latter topics have not been discussed in this book only because of space limitations. I am particularly indebted to colleagues whom I pestered mercilessly with questions about their fields, including Marty Yanofsky, Detlef Weigel, Kathy Barton, Laurie Smith, Phil Benfey, and David Kimelman. I also express my gratitude to those who kindly agreed to be featured in the short biographical sketches scattered throughout the text. The "biobox" subjects were all asked to respond to a set of similar questions, and, invariably, they gave insightful and heartfelt responses. For me, reading and then organizing the comments of these accomplished individuals was one of the most interesting and rewarding parts of writing this book. The single most obvious outcome of this query was that questions such as "What are the most important ingredients in scientific discovery?" evoked a wide range of opinions and commentary. The diversity of views on such topics underscores the fact that scientists are individuals and approach science from many different perspectives, employing a variety of distinct strategies and styles. There may be more ways to study embryos than there are ways for embryos to develop!

I also thank the scientific reviewers of this book who took the time to make valuable and critical comments on the first draft of the book. In addition, I thank my father Jesse Bier; my colleagues at the University of California, San Diego; Marty Yanofsky, Bill McGinnis, Randall Johnson, Georgiana Zimm, Larry Reiter, and Diane Ingles; and Detlef Weigel of the Salk Institute, for reading drafts of the book or various chapters and making insightful comments. These reviewers helped define the focus and made excellent suggestions about the organization of topics.

Thanks are also due to the people at Cold Spring Harbor Laboratory Press for their help, especially Inez Sialiano and Jan Argentine in the Development Department and Pat Barker and Denise Weiss in the Production Department. Likewise, I was fortunate to have the assistance of Meghan Scott, a dedicated UCSD undergraduate, who helped compile the glossary. I also thank Dan Ang, who put in many hours preparing the plates of original data, and members of my lab for putting up with this project. Most of all, I thank Judy Cuddihy, my tireless, good-natured editor at Cold Spring Harbor Laboratory Press, and friend, for all of her varied efforts and encouragement during the lengthy series of steps from start to finish on the book.

Finally, I am most grateful to my wife Kathryn Burton and close friend Marty Yanofsky for their constant encouragement and support during the course of conceiving and writing this book. I'm sure they are quite happy that the ordeal is over and that the coiled spring has sprung!

Foreword

Since the beginnings of conscious thought, human beings have looked with wonder at the world around them. Perhaps the most significant part of the development of consciousness was self-awareness, and with it came the profoundest of biological questions: "Where did I come from?" There are several levels to this question, including those of cosmic scope such as "Where did the universe come from?" or "Where did my species come from?" These are levels of origin that stretch back far beyond the life span of a single individual; hence, in ancient times all one could do was invoke the direct hand of a creator and say that it has been thus from the beginning.

The question "Where did I come from as an individual?" is an entirely different matter. Not only is individual origin within the experience of every person, but one can also find analogous origins in a myriad of other types of animals. These were indeed closely observed to understand their life patterns, first for the hunt, and subsequently to control their life histories in domestication. Thus, as mankind accumulated knowledge and moved from myth to philosophy and science, one of the first areas of serious scientific inquiry was the formation of new individuals. Indeed, embryological descriptions can be found in writings of the ancient Egyptians and Mesopotamians. Among the Greeks, Aristotle, often considered the first true embryologist, opened bird eggs at different stages of incubation and wrote careful descriptions of the process of ontogeny. Hippocrates, the father of modern medicine, also used animals to form the basis of a descriptive embryology.

Yet in the thousands of years since then, while the descriptions have gotten more and more accurate as the naked eye was aided by the magnifying lens, the microscope, and the electron microscope, they remained largely just that: descriptions. So, despite its early beginnings and profound interest, embryology largely stagnated as a descriptive science while chemistry, physics, and even other areas of biology, such as microbiology or the origin of species, flourished with theoretic advances and conceptual insights. Certainly, there were a few key experiments (like Spemann's organizer grafts, which caused duplications of the embryonic body axis resulting in conjoined twin salamanders) which provided some glimmer that the processes could be understood, but as any high school biology student of the 1950s to the 1970s knows, embryology meant memorizing the steps of meiosis and the stages of ontogeny (blastula, gastrula, neurula) and perhaps learning the anatomy of a dissected fetal pig—a profoundly interesting topic taught on a boring descriptive level because that was all that was available.

Within the last 20 years, however, there has been an absolute revolution in the biological sciences. With it, the moribund discipline of embryology has been reborn as the modern field of developmental biology. No longer a descriptive endeavor, developmental biology promises no less than an understanding of the genetic logic underlying the way in which an embryo is made. After more than 3000 years of description, within the careers of the current set of workers we will be able to answer that ancient question "Where did I come from?" on a very deep level. Embedded within that fascinating answer will be a wealth of benefits for mankind ranging from prevention of birth defects to regenerative repair of aged and damaged organs.

Although the biomedical rewards are still largely in the future, we are already at the stage where an understanding of the process of embryogenesis is in place, albeit painted in broad strokes. With such a profound area of scientific inquiry finally reaching maturity, there is a need for a popular science book to describe it to the lay reader. Ethan Bier, himself an important contributor to this rapidly progressing field, has risen to that challenge in providing this book.

The field of developmental biology has grown from two convergent lines of inquiry: the application of molecular approaches to the classic art of experimental embryology (cutting and grafting tissue in living embryos) and the focus of genetic studies on the problem of embryogenesis in simpler organisms such as the fly and worm. As Bier carefully explains, the basic genetic mechanisms that define the shape and structure of a developing animal and those that specify the various types of cells in its body are remarkably similar, whether one looks at an insect or a human being. This stunning revelation was totally unexpected by embryologists who worked on organisms with such different body forms and, seemingly, such different modes of embryogenesis. Yet in retrospect it should not have been so shocking. We did, after all, have a common ancestor with a fly. We are both animals. Yes, that ancient ancestor was simple in comparison to the modern forms, but it had a head end and a tail end and therefore needed a developmental mechanism for distinguishing the two. It had specialized organs for sensing light and needed a genetic program for specifying them. It had appendages for probing its environment and needed a way of forming them. It had a simple heart for distributing fluid within its body. It had a mouth and gut for ingesting and digesting food. And it had specialized cells that gave rise to sperm and eggs in order to propagate itself. All these had an underlying genetic basis, and as we (and flies) evolved from that ancient creature, we elaborated and built upon its genetic and developmental heritage.

Ethan Bier is a distinguished contributor to the genetic analysis of fruit flies, and he has the breadth of knowledge to be able to incorporate advances in vertebrate embryology into his thinking. This has given him the capacity to present the modern synthesis in developmental biology to the reader in clear and accessible language.

The Coiled Spring starts with the most fundamental of premises in modern developmental biology: All cells contain the same genetic information. He then logically progresses to ask, if this is so, why all cells in the body are not the same. The rest of the book (and indeed embryogenesis) answers this question.

In addition to highlighting the common basis for embryogenesis in flies and vertebrates, the book explores our understanding of parallel problems in plant development. Unlike flies and vertebrates, plants achieved multicellularity independently from animals. They therefore evolved embryogenic processes on their own. As the second great experiment in multicellular development, plants can be compared and contrasted with animal development to teach us the universal aspects of forming an organism and to provide examples of alternative solutions to achieving that end.

Bier establishes in the first pages a very successful style of drawing the reader in with an important idea, then describing the experimental evidence and the logic that led to that concept. The reader thus gets an appreciation for the way in which this particular science is advanced, along with an understanding of the way in which an embryo forms. In addition, the book includes boxes in which the contributions of some of the pioneers of the field are highlighted. These go beyond biographical data and descriptions of their work and focus on their personalities and motivations. These are only a small subset of the people who have had a major impact on the field, but including them gives the reader a personalized view of the field and of how science is done.

Finally, the book concludes with a chapter considering the ethical as well as the scientific implications of modern embryological research. With this chapter, Bier provides a glimpse of where we are going and what challenges lie ahead for the future as the embryological question, "Where did I come from?" is finally answered.

Cliff Tabin

Introduction

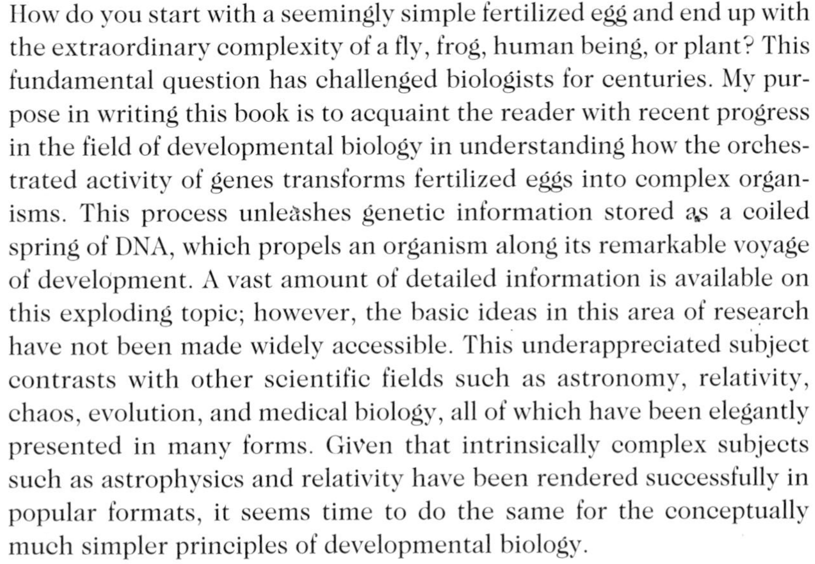

How do you start with a seemingly simple fertilized egg and end up with the extraordinary complexity of a fly, frog, human being, or plant? This fundamental question has challenged biologists for centuries. My purpose in writing this book is to acquaint the reader with recent progress in the field of developmental biology in understanding how the orchestrated activity of genes transforms fertilized eggs into complex organisms. This process unleashes genetic information stored as a coiled spring of DNA, which propels an organism along its remarkable voyage of development. A vast amount of detailed information is available on this exploding topic; however, the basic ideas in this area of research have not been made widely accessible. This underappreciated subject contrasts with other scientific fields such as astronomy, relativity, chaos, evolution, and medical biology, all of which have been elegantly presented in many forms. Given that intrinsically complex subjects such as astrophysics and relativity have been rendered successfully in popular formats, it seems time to do the same for the conceptually much simpler principles of developmental biology.

Frog eggs

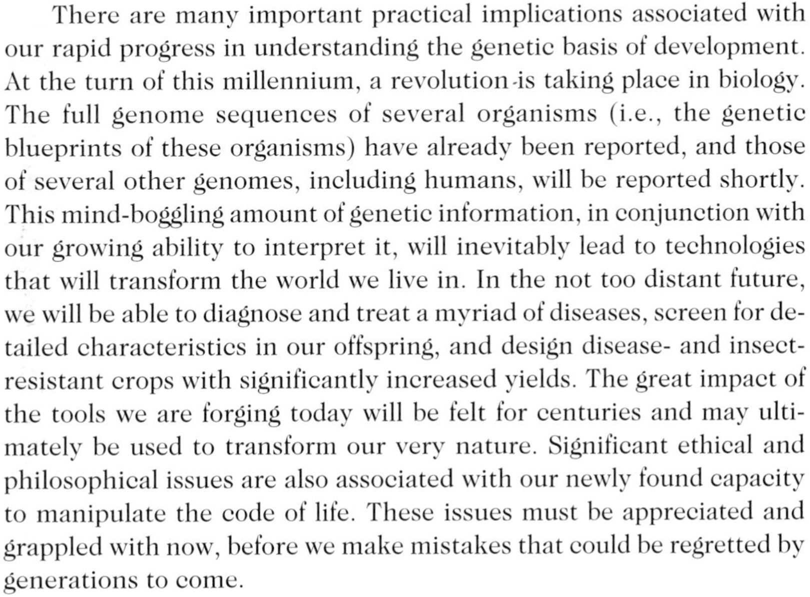

There are many important practical implications associated with our rapid progress in understanding the genetic basis of development. At the turn of this millennium, a revolution is taking place in biology. The full genome sequences of several organisms (i.e., the genetic blueprints of these organisms) have already been reported, and those of several other genomes, including humans, will be reported shortly. This mind-boggling amount of genetic information, in conjunction with our growing ability to interpret it, will inevitably lead to technologies that will transform the world we live in. In the not too distant future, we will be able to diagnose and treat a myriad of diseases, screen for detailed characteristics in our offspring, and design disease- and insect-resistant crops with significantly increased yields. The great impact of the tools we are forging today will be felt for centuries and may ultimately be used to transform our very nature. Significant ethical and philosophical issues are also associated with our newly found capacity to manipulate the code of life. These issues must be appreciated and grappled with now, before we make mistakes that could be regretted by generations to come.

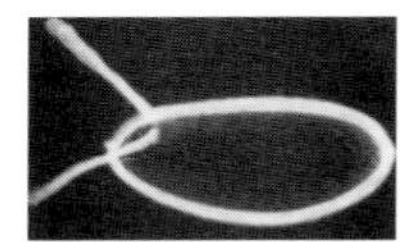

Fruit fly egg

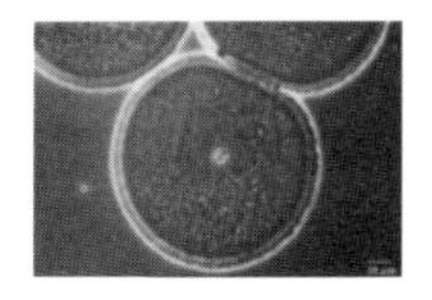

Mouse egg

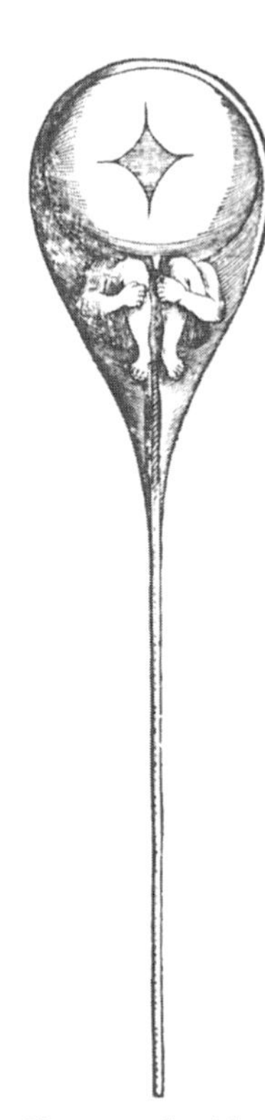

Homunculus idea

The major issue that fascinates developmental biologists is how a simple structure such as a single-cell egg can develop into a mature intricate adult organism, and with nearly infallible reliability. Two opposing models have been proposed over the years to explain how information stored in the egg might govern development. The first model, which represents one extreme end of the spectrum of possibilities, is that the egg is far more organized than it appears to be. The essence of this hypothesis is that the information for where to make various adult structures such as the head versus tail, the back versus belly, or where to place limbs, resides in the egg in a precise spatial code of some kind. This idea is similar to an old view of the sperm and embryo as a homunculus. The homunculus view of development can be likened to a series of Russian dolls within dolls, the very innermost doll being the fertilized egg. The inner doll in this model already is fully patterned, with well-formed head, tail, legs, and arms. According to this scenario, development consists of transferring information present in the fertilized egg to the next developmental stage, which differs from the fertilized egg in size but not in organizational complexity.

The alternative possibility, which could be called the "progressive patterning" hypothesis, is that the egg provides only a very crude sense of position (e.g., anterior versus posterior, or dorsal versus ventral) and that this modest amount of information initiates a series of simple patterning events in developing embryos. According to this view, developmental events at a given stage are determined by what has just happened; in other words, development consists of a series of simple small steps, each dependent on what has happened in the previous step.

The basic difference between these two polar views is that in the first, all positional information is intrinsic to the egg, whereas in the second, patterning information is created by the process of development itself. Although a consistent theme of this book is that the latter progressive patterning hypothesis is most in accord with what we know about development, it should be noted that there are examples of the Russian doll type of development exemplified by growth through molting (e.g., some insects and reptiles such as snakes).

The choice of the specific topics covered in this book was in many respects arbitrary. I state from the very outset that it is not meant to be a comprehensive survey of current topics in developmental biology. Rather, a very limited number of topics are examined in detail to illustrate major principles of development. The topics covered are not intrinsically more important than many interesting alternatives. The major reason for focusing on a narrow range of topics was to limit the book to a manageable length and to present classic examples that embody important generalizations about development as a whole.

The book is organized into two introductory chapters followed by three pairs of chapters on developmental topics. Each pair of chapters on development deals with one type of organism (i.e., invertebrates, vertebrates, and flowering plants). The first of each pair is devoted to establishment of the primary body axes (i.e., the anterior–posterior axis and the dorsal–ventral axis) during early embryonic development. The second of each pair of chapters describes the formation of a specific adult structure, such as an animal appendage or a plant flower. This organization is motivated in part by symmetry and in part because it reveals how very different developmental systems rely on similar progressive pattern-forming mechanisms.

The two introductory chapters provide the minimal essential facts about biology (Chapter 1) and modern recombinant DNA technology (Chapter 2) required to follow the six developmental vignettes. The object of these first two chapters is to familiarize the reader with the bipartite nature of genes (i.e., regulatory versus coding regions) and to describe the central premise of developmental biology, namely, that every cell in an organism contains a full complement of genetic information stored in the molecular form of DNA, but that different types of cells employ or "express" distinct subsets of this genetic information. Chapters 3 and 4 are dedicated to pattern formation in an invertebrate (the common fruit fly). In Chapter 3, the reader is taken on a hunt for genes involved in establishing the primary body axes of the embryo. In Chapter 4, the focus shifts to metamorphosis in flies, when adult structures such as appendages and eyes are formed. The next pair of chapters (Chapters 5 and 6) cover similar topics in vertebrate development. Chapter 5 deals with establishment of the primary body axes in vertebrate embryos, and Chapter 6 with formation of vertebrate limbs. The last duo, Chapters 7 and 8, are devoted to plant development. In Chapter 7, polarization of the plant embryo into root versus apex is featured, and in Chapter 8, the topic shifts to development of plant appendages such as leaves and flowers. A common theme uniting these apparently disparate forms of development (i.e., vertebrate versus invertebrate or animal versus plant) is that crude starting information is converted by a sequence of simple events into progressively more refined positional information.

One of the most provocative themes of contemporary developmental biology, and of this book, is the remarkable extent to which specific developmental mechanisms have been "conserved"—that is "kept intact"—during the course of evolution. When I began my graduate training in 1978, it was generally thought that the most recent common ancestor of vertebrates and invertebrates was a poorly defined blob of some sort. This view was based in part on the great superficial diversity of embryonic forms, which suggested that fundamentally different mechanisms must govern development of organisms with such apparently disparate body plans. On the basis of what we know now, however, it is clear that embryos of organisms as different as fruit flies and humans use virtually identical mechanisms to guide the early stages of development. For example, both invertebrate and vertebrate embryos rely on common sets of genes and mechanisms to subdivide the anterior–posterior axis into segmental units and the dorsal–ventral axis into basic tissue types such as skin, nerve, and muscle (discussed in Chapters 3 and 5). It also seems likely that there is a common set of genes governing development of eyes in all animals. One of the most remarkable findings regarding the similarities between vertebrate and invertebrate development is that genes directing many of these processes are functionally interchangeable between flies and man. One can take a gene from a human being, plug it into a fly, and see it work like the fly gene to make a fly! These deep similarities (or homologies, as we call them) have profound evolutionary implications because they indicate that the common ancestor of flies and humans was a highly organized creature that had invented the genetic machinery to create the primary body axes and basic tissue types. Although plants and animals are believed to have evolved independently into multicellular organisms from a single-cell ancestor, there are remarkable similarities in the general principles by which embryos of these two kingdoms of life develop into mature organisms.

This book also introduces the reader to the key mechanisms involved in establishing pattern during development. For example, a frequently used strategy for generating positional information in the developing embryo is to have a group of cells secrete a "signaling" factor that diffuses from its site of synthesis into regions where it is not produced. The concentration of a such a spreading signal will fall as a function of distance from its source if it is unstable or rapidly destroyed. If different concentrations of this signaling factor trigger distinct responses in cells, then the position of a cell relative to the source of the signal can be determined by the level of signal sensed by that responding cell. The term "morphogen" was coined to define such a hypothetical diffusible factor that elicits different cellular responses as a function of concentration. Several concrete examples of morphogens have been discovered during the last few years, and the principles by which these signaling molecules initiate pattern formation have been worked out. The importance of morphogens in all stages of development is a recurring theme throughout this book. The existence of morphogens bears directly on the question of whether development from a fertilized egg is driven by a homunculus-type map of the adult within the egg, or by a sequence of simple progressive patterning events. The problem now can be restated in simplified molecular terms. According to the Russian doll model described earlier, a large number of distinct cellular responses are evoked by slightly different concentrations of a spatially graded morphogen in the fertilized egg. The fine-grained information provided by such a morphogen is a molecular equivalent of the smallest Russian doll. According to the second progressive patterning model, only a limited number of distinct cellular responses (two or three) are elicited by broad ranges of morphogen concentration. Because the crude positional information represented by such a morphogen is too blurred to serve as a detailed blueprint for a complex organism, a series of subsequent patterning steps is required to create the final detailed product of development. As mentioned above, a major theme of this book is that the latter sequential-developmental strategy seems to be the rule. One rationalization for development proceeding as a series of small simple steps rather than taking place in one fell swoop is that a sequence of simple steps is more reliable and ultimately more accurate than a single giant leap.

In the final chapter of this book (Chapter 9), we consider some of the practical consequences of the ongoing biological revolution, as well as longer-term ethical and social implications. Although this biological revolution is based on progress in many areas of biology such as molecular biology, cell biology, and genomics, our understanding of how genes effect developmental transformation of eggs into organisms will allow us to interpret and exploit genetic information in remarkable ways. In addition to this final chapter, short biographies of various key figures in contemporary biology, along with some commentary by these investigators, accompany the relevant portions of text throughout the book. The point of the last chapter and of the biographical sketches is to provide a vivid sense of the thrill of discovery and to emphasize that scientific exploration, like all other human endeavors, is carried out by individuals who often ask questions in very different ways. The biographies, like the topics covered in this book, are necessarily very limited in scope. It is important to appreciate that many first-rate scientists have contributed to our vision of the coiled spring.

1 The Central Dogma of Biology

The central premise of developmental biology is that all cells in a given organism contain the same genetic material. The difference between various cell types, such as those making up the nervous system, muscle, or skin, is that they employ, or "express," overlapping but distinct subsets of the genetic information that all cells contain. This genetic information is stored in the molecular form of DNA (*d*eoxyribo*n*ucleic *a*cid), which is copied every time a cell divides, the result being that each daughter cell contains the same genetic information as its parent cell. This chapter provides basic facts about genes and DNA that are necessary for understanding how different cells end up accessing different genetic information during the course of development. As shown in the following chapters, transformation of a fertilized egg into an embryo and then into a complex adult organism relies heavily on mechanisms that regulate access to genetic information in different cells during the course of development.

ALL CELLS CONTAIN THE SAME GENETIC INFORMATION: THE GURDON EXPERIMENT

John Gurdon performed a classic experiment in 1970 in which he demonstrated that a specialized adult cell type such as a skin cell contains all the genetic information required to generate a complete organism (Fig. 1.1). He stuck a glass needle into a skin cell in the foot webbing of an adult frog and sucked out its nucleus. This would be the donor nucleus for his experiment. The nucleus is the information center of a cell and contains the genetic material (i.e., DNA). He also removed, and discarded, the nucleus from a fertilized egg cell. This "enucleated" egg, the host egg, no longer contained the genetic information that normally guides development of the frog embryo but contained all of the other ingredients of the egg. He then injected the skin cell nucleus into the enucleated egg. Following some additional manipulations (see sidebar), Gurdon could get his hybrid egg containing a skin nucleus to develop into a tadpole that then underwent metamorphosis to generate a complete adult frog.

Gurdon experiment frog

■ Cast of Characters ■

Activator A transcription factor that activates expression of a gene (turns the gene on).

Amino acids The 20 subunits from which proteins are built.

Chromosome A long string of DNA that contains 1,000–10,000 genes.

Cloning (an animal) The process of creating an exact genetic copy of an animal.

Coding region of a gene Gene region that encodes the amino acid sequence of a protein.

DNA The double helical molecule consisting of two complementary strands of bases that stores the genetic information of all living organisms.

DNA bases (A, C, G, T, and U) The four subunits of DNA (A, C, G, T) and RNA (A, C, G, U).

DNA polymerase An enzyme that carries out DNA replication.

DNA replication The copying of DNA, which is required for cell division.

Dolly A famous sheep that has been cloned.

Double helix The three-dimensional structure of a double-stranded DNA molecule.

Egg The cell that stores female genetic information.

Embryo A fertilized egg.

Gene The unit of heredity composed of DNA.

Genetic code The code relating the sequence of bases in RNA to the amino acid sequence in proteins.

Genetics The subfield of biology dealing with gene function.

Induction A change in the developmental course of a cell resulting from that cell's receiving a signal from another cell.

Morphogen A secreted signal that elicits different cellular responses at different concentrations.

Mutation An alteration in the base sequence of a gene.

Neural inducing factor A secreted signal liberated by the Spemann organizer that promotes neural over epidermal development.

Organizer A region of a developing organism that sends signals to neighboring cells to organize the formation of a morphological structure.

Protein A three-dimensional polymer constructed from amino acids.

Receptor A molecule on the surface of a cell that receives a secreted signal.

Regulatory region of gene The region of a gene that determines when and where the gene will be active (i.e., transcribed, or on) versus silent (or off).

Repressor A transcription factor that prevents transcription of a gene (i.e., turns a gene off).

RNA A single-stranded polymer of bases similar to single-stranded DNA and essential for protein synthesis.

RNA polymerase An enzyme that carries out RNA synthesis, called transcription.

Signal A molecule produced in one cell that alters the fate of a neighboring cell.

Spemann organizer The organizing region of a frog embryo that induces neighboring cells to form the central nervous system.

Transcription The synthesis of a single-stranded RNA copy of a DNA molecule.

Transcription factors A class of proteins that control the transcription of genes (i.e., that turn genes on or off).

Translation The conversion of a base sequence of DNA and RNA into a sequence of amino acids in protein.

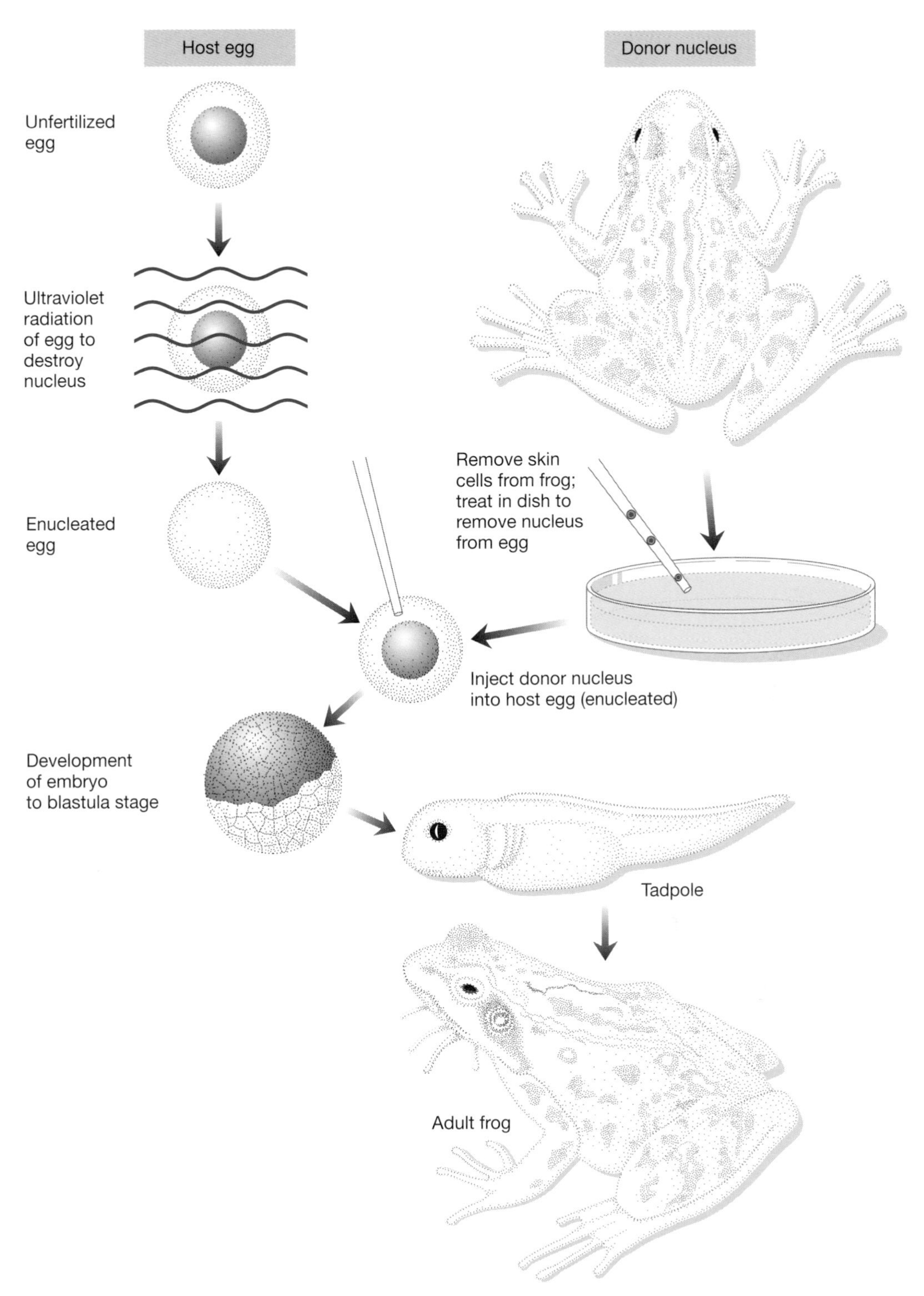

FIGURE 1.1. Every cell in an organism contains all the genetic information necessary to make a complete individual—the Gurdon experiment.

John Gurdon (1933–)

The 1997 report of the cloning of Dolly the sheep was greeted with great media hoopla. However, essentially the same experiment was done in the 1960s by John Gurdon, using the African clawed frog, with little notice outside the scientific community. Since these seminal experiments, nuclear transfer has been performed in many animal species, and several animal species have been cloned. However, along with the technical achievement of these experiments, we are dealing with their implications for aging and development, as well as the ethics of this work and what it means to be human.

Gurdon's seminal cloning experiments were encouraged by his thesis advisor Dr. Michail Fischberg at Oxford University. He thought that Gurdon should follow up on the 1952 nuclear transplantation experiments of Robert Briggs and Thomas King, who performed the first successful transplantation of living nuclei in animal cells; their work showed that cell nuclei lose their potentiality very early in development. Gurdon's graduate work from 1956 to 1960 (he received his D. Phil. in 1960) involved further developing the methodology for nuclear transplantation, including the identification of an appropriate genetic marker to distinguish host versus donor frogs. (Recall the importance of this technical detail in the experiments of Mangold and Spemann described in Chapter 1.) As Gurdon obtained results, he found that they did not agree with the conclusions of Briggs and King. However, he continued his work exploring the developmental potential of cells of the gut lineage.

In 1962, Gurdon obtained feeding tadpoles from the nuclei of intestinal epithelial cells. This was one of the most exciting moments in Gurdon's scientific career: "...seeing that the genetic marker used for nuclear transplantation was present in the excised tail tips of swimming larvae derived from intestinal epithelium cells." By combining results from initial as well as serial nuclear transplantations, Gurdon found that 7% of the intestinal epithelial cell nuclei contained the genetic information necessary to form all cell types of the feeding tadpoles. In his 1962 paper, he concluded "...the nucleus can promote the formation of a differentiated intestinal cell and at the same time contain the genetic information necessary for the formation of all other types of differentiated somatic cells in a normal feeding tadpole."

Gurdon successfully derived fertile adults from epithelial cell nuclei in 1966. Later, he confirmed these experiments with nuclei derived from skin cells isolated from the webbing of an adult frog to rule out the possibility that there were cells with germ-cell-like properties in the developing gut. According to Gurdon, these experiments showed that "...specialization of cells involves the differential activity of genes present in all cells, rather than the selective elimination of unwanted genes...."

These were milestone experiments in developmental biology and organismic cloning. Because many scientists at the time believed that mature differentiated cells could not give rise to an entire organism, Gurdon had to be exceedingly rigorous in his analysis. As he points out, "In the 1960s, it was natural that the scientific community preferred to believe the results of the highly respected Robert Briggs and his colleagues. It was therefore necessary that my experiments were totally convincing, and the use of a genetic marker was essential for this purpose."

Gurdon's approach to scientific investigation centers on choice of experimental material for answering relevant questions: "...it has probably been fortunate that I have chosen to use *Xenopus* (frog) eggs and oocytes for various questions that I consider to be important in developmental biology. It would not have been good for me if I had restricted my attention to one single question, having to learn to work with numerous different kinds of material. Thus, in my case, opportunism seems to have worked well...." But more than these approaches, Gurdon feels that "...it is important to be prepared to try out a new kind of experiment, even if a granting body would have refused such an experiment on the grounds that it might not work."

John Gurdon remained at Oxford until 1972 and then moved to the University of Cambridge. He has received more than 30 awards and honorary degrees for his achievements, as well as a knighthood. He serves on the boards of numerous organizations and is governor of the Wellcome Trust, one of the most important scientific funding organizations in Europe. Gurdon continues to be productive scientifically as he continues to probe the mechanisms by which cells communicate during development.

The Gurdon experiment was the first example of cloning an animal (i.e., the creation of an identical copy of an individual). More recently, there has been significant interest in the successful cloning of a sheep named Dolly at the Roslin Institute in Scotland, and subsequently of other animals. Dolly was voted the "Breakthrough of the Year" in 1997 by the esteemed journal *Science* and was featured on the cover of *Life* magazine. Although Dolly represents a significant technical step forward in manipulating mammalian embryos, this achievement was no different conceptually from the original Gurdon experiment. Instead of extracting the nucleus from an adult skin cell, a nucleus from a breast cell was transplanted into an enucleated sheep egg cell to create the clone of Dolly. Similar techniques most likely would permit the cloning of any animal, including humans. Obviously, there are serious ethical issues surrounding such manipulations; however, the scientific basis for this dramatic contemporary accomplishment is essentially just a repeat of Gurdon's classic experiment performed nearly 40 years ago.

As with many experimental procedures, the Gurdon experiment actually required an additional experimental twist to work. After injecting the donor nucleus into the enucleated host egg, Gurdon permitted the hybrid egg to go through a normal sequence of cell divisions to form a typical undifferentiated ball of cells known as a blastoderm embryo (see Fig. 1.1). He retrieved a nucleus from a cell in this blastoderm embryo and injected it back into a second enucleated fertilized egg. This last manipulation somehow "adapts" the original donor skin cell nucleus to the environment of the egg cell. After repeating this adaptation process several times, Gurdon found that the final fertilized egg carrying the adapted skin cell nucleus could go on to develop into a tadpole, which then metamorphosed into a complete adult frog.

Skillful manipulations were needed to perform the Gurdon cloning experiment and its recent mammalian counterpart, which revealed the full developmental capacity retained by specialized cells in an adult organism. In contrast, cloning turns out to be quite trivial in many plants. For example, one can grind up a carrot into a suspension of single cells and regenerate complete carrots from these isolated cells. The success of plant and animal cloning experiments demonstrates that cells in mature organisms carry a complete genetic blueprint for entire complex organisms.

ACCESS TO GENETIC INFORMATION IS REGULATED

If different types of cells contain the same genetic information, then what distinguishes them from one another? The answer is that each cell employs only a small fraction of the information carried in its DNA and that different cells access distinct subsets of their total genetic information. Thus, about 85% of the genetic information used in an adult cell of any given type (e.g., liver) is also used in other cell types (e.g., kidney). The remaining 15% of genetic information accessed by these two cell types, however, is different. Using computers as an analogy, different cell types could be likened to different programs such as word processing or drawing applications. When any program is running, some common computer functions supplied by the operating system (the 85% of genetic material) are used. Since word processing and drawing programs accomplish different tasks, however, additional subroutines (the other 15%) are accessed exclusively for either word processing or drawing.

The crux to understanding how cells progressively acquire distinct characteristics during development is to understand the mechanisms by which they access different subsets of genetic information. Because genetic information is stored in the molecular form of DNA, this question reduces to understanding how information carried in DNA is accessed, and how access to individual units of information called genes is regulated during development.

THE STRUCTURE OF DNA AND HOW IT IS COPIED

Before proceeding further, we need to know something about the structure of DNA and how this molecule stores genetic information. In addition to being one of the crowning accomplishments of science in the 20th century, the determination of the structure of DNA in 1953 by James Watson and Francis Crick provided a mechanism for how genetic information is propagated from one generation to the next. Genetic information also must be transmitted reliably during the many cell divisions required to build a complex organism. To create an organism containing billions and billions of cells from a single egg cell requires 20–30 cycles of cell division. A consequence of Gurdon's demonstration that cells in a mature organism contain the same genetic information as the egg is that DNA must be copied at each cell division. To assure that no genetic information is lost during cell division, a cell must duplicate all of its DNA to generate two copies of each DNA molecule prior to division so that one copy of each DNA molecule can be allotted to each of the two daughter cells. The result of cell division is the formation of two daughter cells that are genetically identical to each other and to their parent cell. As discussed below, it turns out that the mechanism for copying DNA (or "replicating" DNA, in the jargon of the field) is quite similar to the mechanism for accessing genetic information (or "transcribing" DNA).

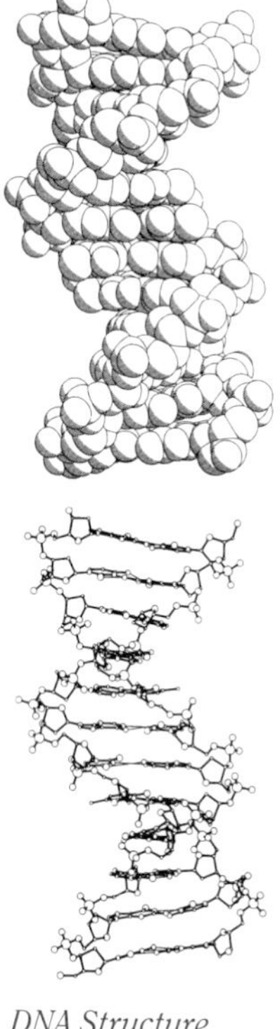

DNA Structure

A DNA molecule is a linear string of four different subunits called bases, which are abbreviated **A**, **C**, **G**, and **T** for adenine, cytosine, guanine, and thymine (Fig. 1.2). These bases are strung together like a strand of pearls on a necklace. DNA in all plant and animal cells comprises two strands of bases wrapped about each other in a configuration known as a double helix, which resembles a spiral staircase. The bases on the two strands of DNA always join together as defined couples to make the stairs of the DNA spiral staircase. If the base at a given position in one strand is an **A** then the other strand will have a **T** in that corresponding position, and vice versa. Similarly, if a **C** is present at a certain position in one DNA strand, then the other strand will contain a **G**. In DNA jargon we say that "**A** pairs with **T**" and "**C** pairs with **G**" or that "**A** is complementary to **T**" and "**C** is complementary to **G**." The

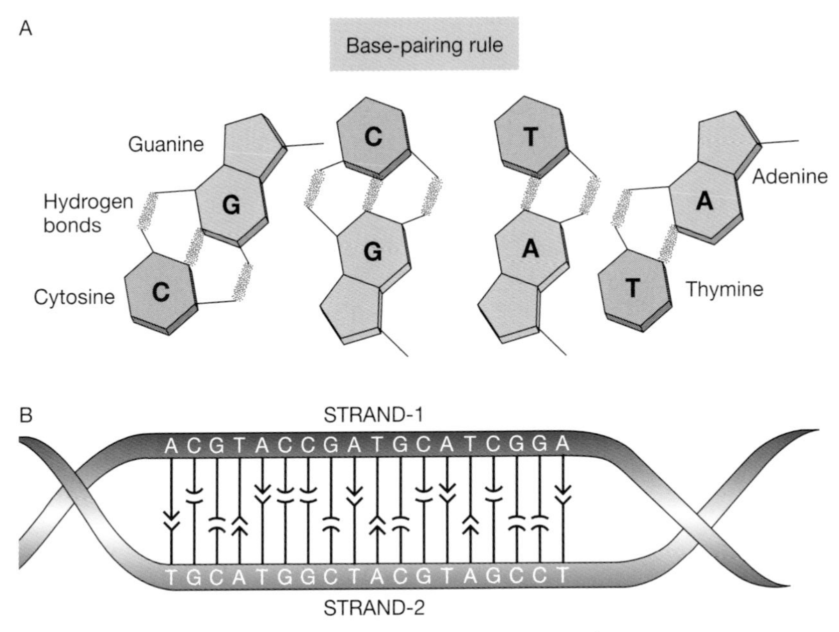

FIGURE 1.2. DNA comprises two complementary strands of bases.

reason that bases only pair as committed couples is that there are specific chemical and physical interactions between complementary bases. The shapes of the **A** and **T** bases fit together in a DNA double helix like interdigitating pieces of a puzzle. Similarly, the shapes of the **C** and **G** bases are complementary. Other combinations of bases, such as **A** and **C**, **A** and **G**, **T** and **C**, or **T** and **G** do not mesh, however, and therefore these potential base pair combinations do not form in double-stranded DNA.

An important implication of the strict complementarity between bases in double-stranded DNA is that if the sequence of bases on one strand is known, one can immediately deduce the sequence of bases on the opposite strand. In other words, each strand of DNA carries all the information necessary to recreate the other. This structural fact has deep ramifications for the way in which DNA molecules are replicated. In their famous paper describing the structure of DNA, Watson and Crick end with the dry understatement "It has not escaped our notice that the specific pairing we have postulated immediately suggests a possible copying mechanism for the genetic material."

The overall mechanism for replicating DNA is remarkably simple (Fig. 1.3). First, the two intertwined complementary strands of DNA separate from each other. Then, new strands of DNA are copied from each of the two existing single strands using the "A:T/C:G" rule. Because each separated single strand of DNA directs the synthesis of its

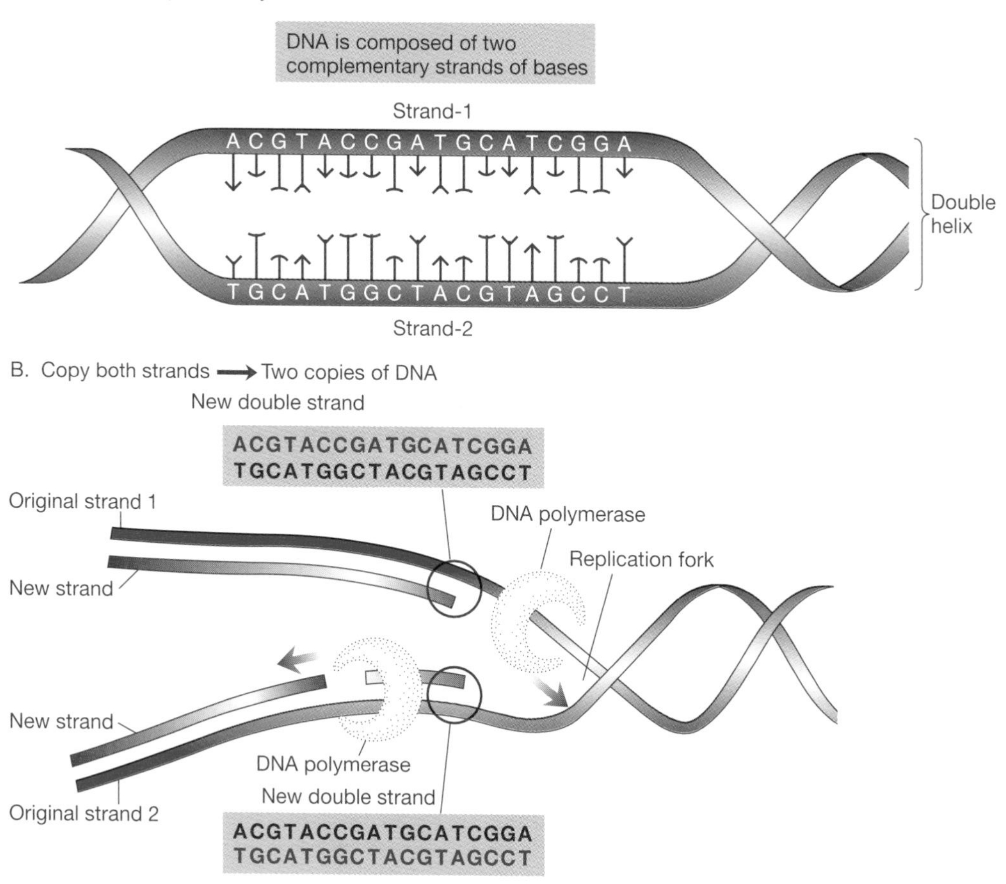

FIGURE 1.3. DNA replication.

counterpart, the result of replication is the synthesis of two identical double-stranded DNA molecules. A copying molecule known as DNA polymerase carries out DNA replication. DNA polymerase copies a single strand of DNA by hopping onto one end of the DNA and moving progressively along the DNA to the other end. At each point of its excursion along the single strand of DNA, it adds an appropriate complementary base to the newly synthesized growing strand (Fig. 1.3). Although we need not dwell further here on the details of DNA replication, it is worth noting that the key copying molecule involved in accessing the genetic information, called RNA polymerase, functions analogously to DNA polymerase. As discussed below, RNA polymerase copies a single strand of DNA into a very similar polymer of bases known as RNA. Let us examine how the genetic information stored in DNA is accessed by a cell.

GENES HAVE A BIPARTITE ORGANIZATION: CODING VERSUS REGULATORY REGIONS

The complete genetic blueprint of an organism is referred to as the organism's "genome." The genome can be imagined as a linear string of DNA base pairs (**A:T**, **C:G**, **G:C**, or **T:A**) 100 million to 3 billion base pairs long, depending on the organism. To give a sense of the immensity of this genetic information, if a single base were represented by a printed letter on a single-spaced typed page, the genome of a human being would be an encyclopedia one million pages long and standing 12 stories high. In reality, an organism's genome is split up into a small number of large strings of DNA called chromosomes (Fig. 1.4). In fruit flies, the genome is partitioned among four different chromosomes, whereas in humans the genome is subdivided into 23 discrete chromosomes. The information contained in each of these chromosomes is considerably longer than the *Iliad*. The elementary unit of genetic information carried on a chromosome is called a gene. A typical gene is composed of 1,000–10,000 base pairs of DNA. Complex multicellular organisms such as flies or ourselves contain 10,000–100,000 genes in their genomes. Genes exert their influence by directing the synthesis of another class of molecules called proteins. To a first approximation, each gene produces one protein as a product. Proteins are molecular "machines" that perform specific tasks inside and outside of cells. DNA polymerase and RNA polymerase mentioned above are examples of proteins having specialized functions. Thus, genes store the informa-

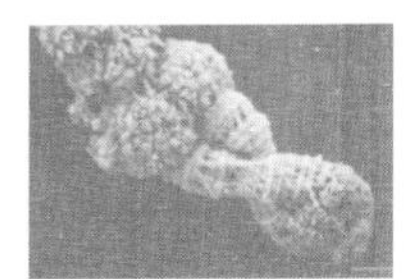

Scanning electron microscope image of chromosome

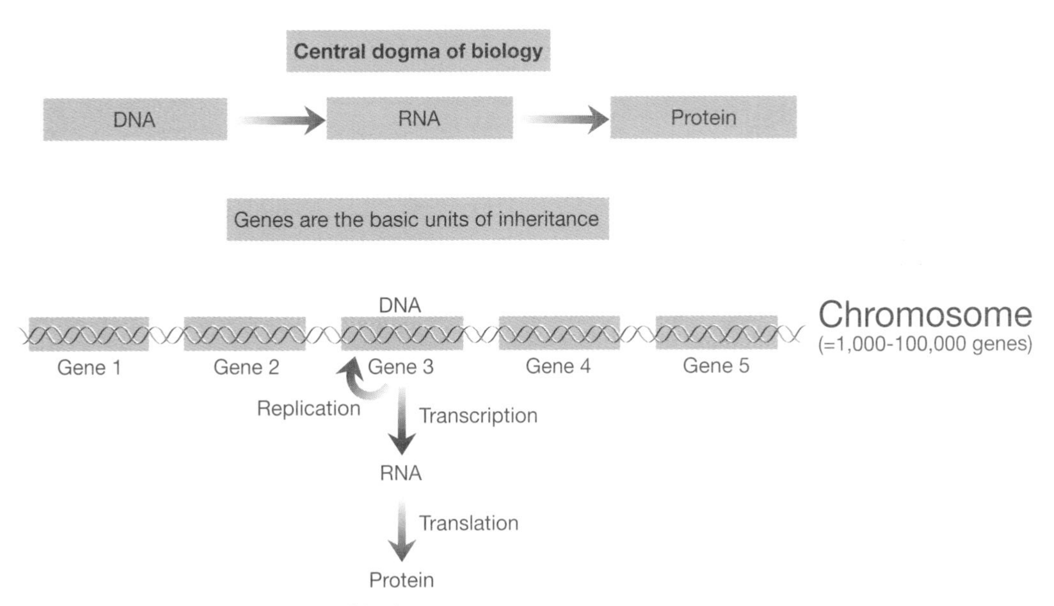

FIGURE 1.4. The central dogma of biology.

tion required to synthesize individual proteins, which do the actual work. As described in greater detail below, the simple graphic relationship between DNA, RNA, and protein, often referred to as the central dogma of biology, is: DNA → RNA → protein. A typical gene can be subdivided into two functionally independent components (Fig. 1.5). The first part, called the coding region of the gene, contains the information necessary for synthesizing a protein product, which, as mentioned above, executes the function of a gene by performing a particular cellular task. Proteins are linear chains composed of subunits called amino acids, which fold up to assume a myriad of complex three-

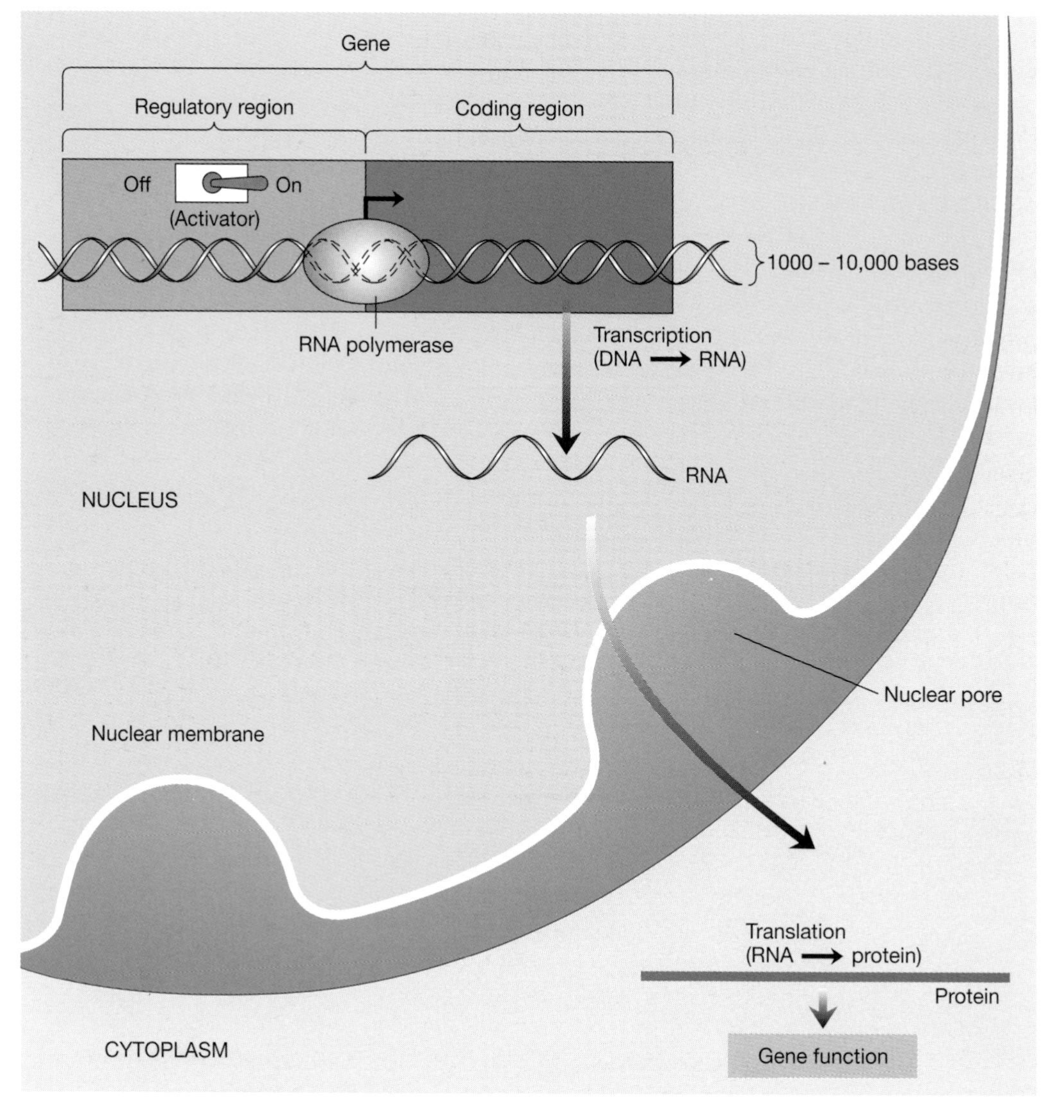

FIGURE 1.5. Genes have a bipartite structure.

■ A Gene Analogy ■

A gene can be likened to an electric appliance such as a lamp, TV, or toaster. The functional portion of the device is similar to the coding region of the gene (i.e., the light fixture of a lamp, the screen of the TV, or the heating coils of the toaster). The switches controlling these devices are analogous to the regulatory regions of genes. Each appliance has its own switch, which generally has two positions, on and off. In some cases, more sophisticated switches can adjust the output level continuously (e.g., a dimmer light switch or a volume control on a TV or stereo). Since each electronic device has its own switch, one can use appliances in various combinations to perform different tasks (e.g., cooking dinner versus cleaning up afterwards). To cook dinner, the stove is turned on but the dishwasher remains off, whereas the converse is true for my typical kitchen duty.

dimensional forms. The other functional part of the gene, called the regulatory or control region, is an on/off switch that determines whether or not the coding region of the gene will actively synthesize its protein product in a given cell type. A gene actively directing the synthesis of its protein product is said to be "expressed" or "switched on." When the protein is not being made, the gene is said to be inactive or "switched off."

As shown in more detail in subsequent chapters, development in many respects can be likened to a cascade of switches turning other switches on or off. If one considers all of the switches controlling expression of genes in two different cell types, most of the switches will be in the off position in both cells, 10–20% of the switches will be on in both cells, and 1–2% of the switches will be on in one cell but not in the other. The key to understanding development is to identify the genetic circuitry that turns on one small subset of switches in the first cell type and another subset of switches in a second cell type. Two cells become different during the course of development when, for reasons discussed below (e.g., because the cells are in different positions in the embryo), different switches are thrown in one cell versus the other. Development, like a hierarchical computer circuit, relies on a series of simple, often binary, decisions that result in specific combinations of switches being turned on or off. The end result of cells making a series of different switch-flipping decisions is that distinct subsets of switches are activated in different cells. Such cell-type-specific genetic circuits control different cellular programs, such as those required for the electrical behavior of nerve cells or the contractile properties of muscle cells.

GENES CODE FOR PROTEINS

The coding region of a gene directs the synthesis of its functional protein product by a two-step process (see Figs. 1.5 and 1.6). In the first step, known as transcription, genetic information stored in the form of DNA is copied into a molecule called RNA. RNA is a single-stranded

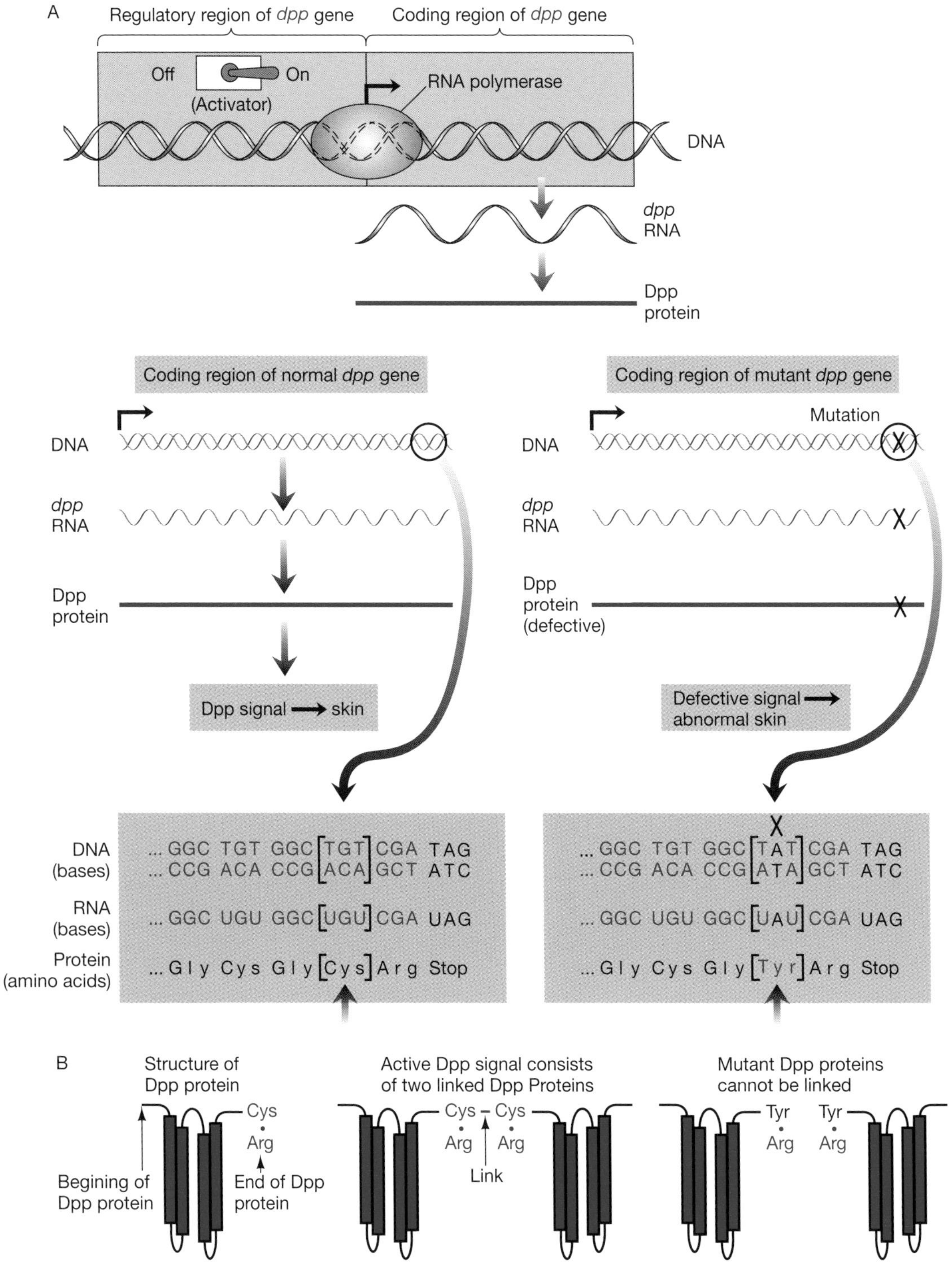

FIGURE 1.6. The coding region of a gene and mutants.

string of bases that is chemically very similar to a single strand of DNA. Three of the RNA bases are identical to those in DNA (i.e., **A**, **C**, and **G**), and the fourth base **U**, or uracil, is a minor structural variant of the base **T** present in DNA. A single strand of RNA bases can form a double helix with a complementary single strand of DNA to generate a double-stranded RNA/DNA hybrid molecule. As in the case of double-stranded DNA, the two strands of this hybrid molecule must have complementary sequences of bases. The pairing rule for a DNA/RNA hybrid molecule is the same as that for two DNA strands, with the modification that an **A** in the DNA strand pairs with a **U** in the RNA strand instead of a **T** in DNA (i.e., $\mathbf{T}_{DNA}$ pairs with $\mathbf{A}_{RNA}$, $\mathbf{A}_{DNA}$ pairs with $\mathbf{U}_{RNA}$, and $\mathbf{C}_{DNA\ or\ RNA}$ pairs with $\mathbf{G}_{DNA\ or\ RNA}$).

During transcription, the commandments written in the stone tablets of DNA are copied into a more portable parchment edition (i.e., RNA). DNA is transcribed into RNA by a copying protein called RNA polymerase (see Fig. 1.5). It is informative to compare how RNA polymerase and DNA polymerase make copies from single-stranded DNA. These two proteins are similar in that they both start at one end of the DNA molecule and progressively move down the DNA to the other end, inserting complementary bases into the growing strand as they go. DNA polymerase and RNA polymerase are very different, however, in two important respects. First, when DNA is replicated, as opposed to transcribed, both strands of the complete genome are copied. The result of replication is the synthesis of a duplicate copy of the entire genetic information of the organism (see Fig. 1.3). The process of transcription, on the other hand, copies only selected portions of the DNA into RNA. Only a small fraction of the genes carried by a given cell are actively transcribed (i.e., become switched on) in that cell. As discussed in more detail below, the control region of a gene determines whether or not that gene will be transcribed in a given cell. The second difference between DNA polymerase and RNA polymerase is that RNA polymerase only copies one of the two strands of DNA instead of both strands. The DNA strand that is copied to make RNA is called the coding strand. DNA/RNA hybrid molecules created by RNA synthesis, unlike stable double-stranded DNA, are formed only transiently. Once completed, the RNA copy of a gene is peeled away from the complementary DNA strand and is liberated as a single-stranded RNA molecule, which then directs the synthesis of the final protein product of the gene.

DNA is replicated and transcribed in the nucleus of the cell. The nucleus contains densely packed DNA and is surrounded by a porous nuclear membrane, which permits newly synthesized RNA molecules to exit the nucleus and enter the main cellular compartment called the cytoplasm. The cytoplasm is enclosed by a watertight cell membrane separating the inside of the cell from the outside world. The second step of reading the genetic information is called translation and takes place in the cytoplasm (see Fig. 1.5). The term translation reflects the fact that one molecular language is being translated into another. The first language, that of DNA or RNA, is represented by the order of the four bases. Since DNA and RNA obey essentially the same base complementarity rule, these two molecules can be considered to be closely related dialects of the same language. The language of protein, however, is very different from that of DNA or RNA and requires complex translation to convert the sequence of bases in DNA or RNA into a sequence of amino acids in a protein. The amino acid subunits of proteins are structurally unrelated to DNA or RNA bases. In addition, there are 20 different amino acids that make up proteins rather than only four bases comprising DNA or RNA. Given the great structural differences between DNA and protein, it is not surprising that a complex translating machine is required to convert a sequence of bases into the grammatically correct sequence of amino acids in the corresponding protein.

One important feature of translation merits particular mention. The basic question is, How does a sequence of only four different DNA bases in a gene direct the synthesis of a protein chain comprising 20 different amino acids? This paradoxical problem once led biologists to conclude that DNA could not be the genetic material because it was structurally too simple to encode chemically more complex proteins. The solution to this "coding" problem is that each amino acid is specified by a combination of three contiguous bases. The code relating the sequence of bases in RNA to the sequence of amino acids in a protein is referred to as the genetic code. The genetic code assigns each of the 20 amino acids to a contiguous group of three RNA bases (or triplets, in the lingo). There are 64 different combinations of bases grouped three at a time (i.e., $4^3 = 4 \times 4 \times 4 = 64$ different triplets). Because 64 is greater than 20, there is more than enough genetic information carried in base triplets to code for the 20 amino acids. In fact, because all but 3 of the possible 64 triplets of bases code for amino acids, more than one triplet codes for the same amino acid. The remaining 3 of the 64 triplets do not code for any amino acid at all, but rather terminate the process of translation. These special triplets are called stop codons.

The genetic code is universal to all known forms of life. One profound implication of the universality of the genetic code is that all existing life on earth derived from one ancestral life form, which had already evolved the genetic code still used today by all of its descendants. This incontrovertible evidence for a common ancestor of all life forms confirms one of Charles Darwin's most remarkable predictions. In *On the Origin of Species, by Means of Natural Selection,* Darwin concluded with this stunning flash of insight: "Therefore I should infer ... that probably all the organic beings which have ever lived on earth have descended from some primordial form, into which life was first breathed." Given how little was known in Darwin's time about the genetic or molecular basis of life, this inspired intuitive leap is surely one of the greatest intellectual achievements in biology.

One final point should be made about the structural difference between DNA and proteins. A double-stranded DNA helix can be pictured as a very long thin screw. The rod-like structure of DNA does not vary much depending on the sequence of bases. This is not to say that two DNA molecules with different sequences of bases are structurally identical. In fact, as shown below, it is important to consider subtle differences in the structure of DNA determined by the sequence of bases. In contrast to the minor impact that base sequence has on the physical structure of DNA, however, the sequence of amino acids has enormous consequences on the structure of proteins. Each protein folds into a unique, complex, three-dimensional structure. This great structural diversity permits different proteins to carry out a wide array of distinct cellular functions. For example, multistranded rope-like proteins give hair its linear shape and strength. In contrast, complex globular-shaped proteins called enzymes accelerate or "catalyze" chemical reactions by bringing together and promoting chemical interaction between two compounds. By controlling the rate of different chemical reactions, enzymes determine the chemical properties of different tissues such as the stomach, kidney, and liver. One could make an architectural analogy to the structural diversity of proteins encoded by structurally similar DNA molecules. The great variety of buildings with diverse shapes and sizes could be likened to proteins, whereas the uniform format of architectural plans that are drawn on similar sheets of flat paper could be compared to the relative structural monotony of DNA.

MUTATIONS CHANGE THE BASE SEQUENCE OF GENES

An important theme in this book is that much can be learned about the normal activity of a gene by studying the consequence of eliminating the function of that gene. The subfield of biology dealing with gene function is called genetics. Geneticists spend a great deal of time trying to find altered forms of genes known as mutants. The topic of mutants is considered in greater detail in Chapter 3, but this is a good place to introduce the principle of mutation (Fig. 1.6). A mutation is an alteration in the base sequence of a gene. In general, the base change lies in the coding region of the gene, although there are important examples of mutations that alter the regulatory region of genes. In fact, some aggressive forms of cancer are the result of regulatory mutations that inappropriately activate genes involved in promoting cell growth. For the time being, however, let us consider the most common form of mutation in which the sequence of bases in the coding region of a gene is changed.

Mutations alter the sequence of bases in a gene in various ways. The simplest type of mutation consists of a single base being changed to another (i.e., instead of base #947 in a gene being a G, the mutant gene contains an A in that position). This is called a point mutation. Mutations also can involve greater numbers of base changes, including deletion or insertion of multiple bases. Because the genetic code converts the sequence of bases in DNA into a sequence of amino acids in the protein product of a gene, point mutations often alter the sequence of amino acids in the protein. In some cases, a base change generates one of the three termination triplets. When this happens, translation of the protein encoded by the mutant gene terminates prematurely, leading to the production of a shortened protein. Mutant proteins that function poorly or not at all are referred to as loss-of-function mutations. The great majority of mutations result in loss of function. For example, in Figure 1.6A, a G → A base change is depicted in the fruit fly *decapentaplegic* gene (or *dpp* for short), which, as discussed in Chapter 3, is required for skin versus neural development. This mutation replaces an amino acid called cysteine (or Cys for short) with another amino acid called tyrosine (or Tyr). To form an active Dpp signaling molecule, two Dpp proteins must associate as a pair. The Cys amino acid at the end of the Dpp protein is critical for linking two Dpp proteins together (Fig. 1.6B). In the *dpp* mutant where the critical linking Cys is changed to a Tyr, the coupling of Dpp proteins is not possible. Because the solitary Dpp proteins in this mutant are unable to perform their normal function in promoting development of skin, nervous system cells are overproduced at the expense of skin.

Note that throughout this book gene names such as dpp are italicized and lowercase, whereas the protein products of genes, such as Dpp, *are not italicized. However, there are several other nomenclatures used in various fields of plant and animal genetics, and these are reflected in the text of this book.*

Mutations can arise spontaneously as a consequence of rare copying errors during DNA replication, or they can be induced at high frequencies by exposure to mutagenic compounds. Mutagenic chemicals (or mutagens) work by a variety of mechanisms. Most commonly, a mutagen reacts chemically with DNA bases and changes one base into

■ Mutations and Disease ■

Many genetic diseases in humans such as muscular dystrophy, β-thalassemia, and Tay Sachs disease result from loss-of-function mutations. Occasionally, however, mutant proteins are more active than normal. One serious example of such gain-of-function mutations is the hypervirulence of mutant strains of viruses such as the flu or AIDS viruses. The appearance in 1998 in Hong Kong of a new strain of chicken flu that could infect humans is a troubling and potentially devastating example of this type of gain-of-function mutation. Other medically important gain-of-function mutations are the infrequent, but rapidly spreading and pervasive, antibiotic-resistance mutations in bacteria. These mutations generally alter the structure of a bacterial protein that is the target of an antibiotic in such a fashion that the mutant protein can perform its normal function, but is no longer inhibited by the drug. With respect to development, however, most of the mutations we discuss in subsequent chapters decrease or eliminate the function of protein products.

another, or reduces the fidelity of DNA replication so that copying errors become common rather than rare. Not surprisingly, most mutagenic compounds also are highly carcinogenic (i.e., cancer causing). The fact that mutagens often are potent carcinogens is one of the key pieces of evidence in favor of the hypothesis that genetic alterations in cells cause cancer. Because mutagenic compounds greatly increase the incidence of mutation, they are used widely in generating mutants disrupting developmental processes. Methods by which mutations are generated and identified in developmentally important genes are discussed in Chapter 3.

HOW GENE EXPRESSION IS REGULATED

The second part of a gene, the regulatory region (see Fig. 1.5), determines where and when that gene will be copied from DNA into RNA (i.e., transcribed or expressed). Some genes must be expressed in all cells because their protein products are essential for general cellular functions (e.g., enzymes such as ATP synthetase involved in energy metabolism). Genes playing developmental roles, however, are frequently expressed only in certain regions of the embryo or in specific tissue types. For example, the dorsal portion of the early fruit fly embryo gives rise to skin, and *dpp,* the key gene required for the formation of skin, is expressed only in dorsal cells of the embryo. It often is important to restrict expression of patterning genes to particular regions of the embryo or to specific tissues. When such genes are incorrectly expressed in the wrong cells, those cells behave abnormally by giving rise to inappropriate structures or by dying if they are unable to respond to conflicting genetic instructions. For example, when the *dpp* gene is misexpressed (i.e., expressed inappropriately) in the lateral region of the embryo, it fights the neural genetic program active in those cells and transforms them into skin precursor cells (this example is discussed in more detail in Chapter 2 and Chapter 3; see, e.g., Figs. 2.4 and 3.10).

As mentioned above, genes come in two parts. The coding region of the gene is transcribed from DNA into RNA, which then is translated to make a protein product. In contrast, the regulatory region of the gene determines when and where the gene will be transcribed. The mechanism by which DNA in the regulatory region of a typical gene controls the transcription of the coding region of that gene into RNA is discussed in more detail in Chapter 3. For the moment, let us consider a simple experiment illustrating that the regulatory and coding regions of a gene provide independent and separable functions. The experiment is to make a hybrid gene consisting of the regulatory region of one gene and the coding region of another gene. For example, the regulatory region of the fruit fly *dpp* gene, which is expressed in dorsal skin cells of the embryo, is joined to the coding region of a bacterial gene called *lacZ* to create a hybrid *dpp:lacZ* gene. *lacZ* encodes an enzyme

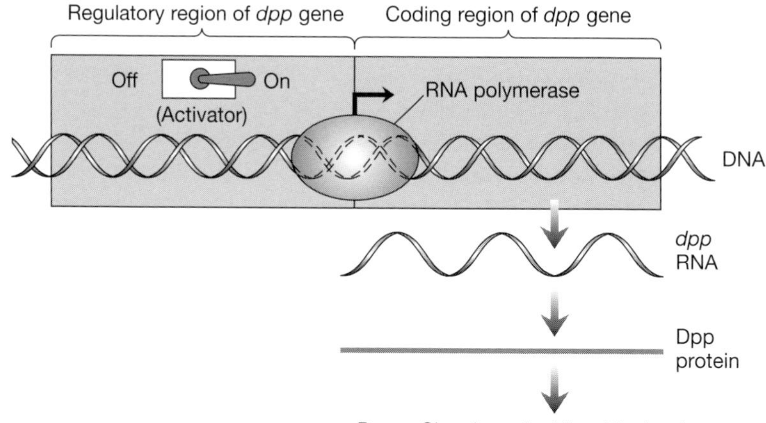

A. Fuse regulatory region of *dpp* gene to coding region of the *lacZ* gene

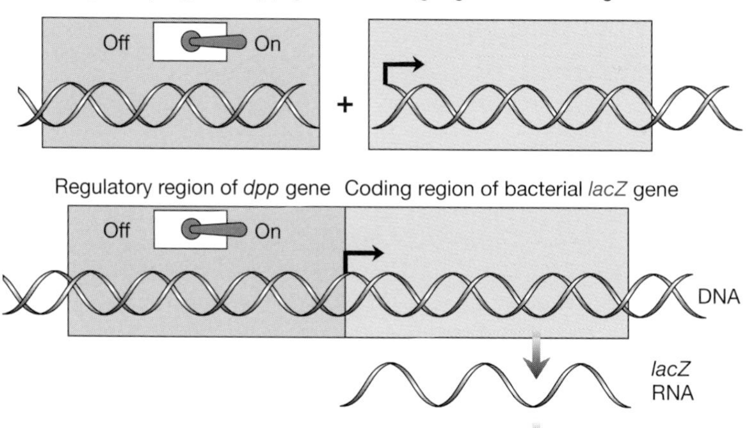

B. *dpp* regulatory region controls expression of the *lacZ* coding region in skin

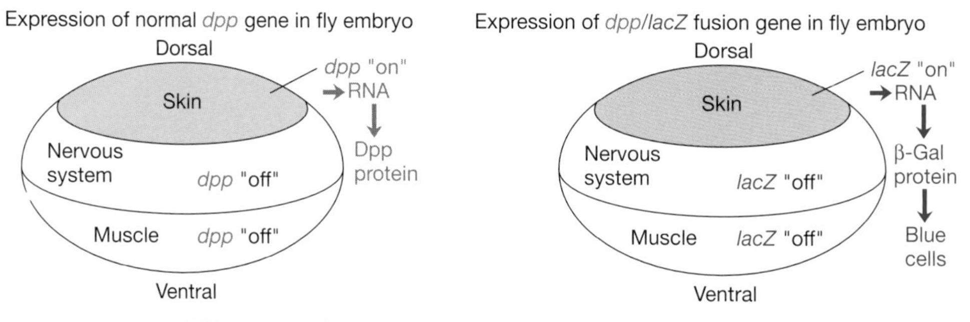

FIGURE 1.7. The regulatory region of a gene controls gene expression.

involved in sugar metabolism (Fig. 1.7A) that can cleave a sugar-like molecule to generate blue color. The hybrid *dpp:lacZ* gene is inserted into a fruit fly chromosome, where it directs the expression of the bacterial enzyme only in dorsal skin cells. When embryos derived from the flies carrying the *dpp:lacZ* hybrid gene are placed into a solution containing the sugar-like compound, the skin, but not internal tissues or organs, turns bright blue (Fig. 1.7B). This experiment demonstrates that the regulatory region of the *dpp* gene contains all of the information necessary to control the expression of the adjacent coding region in developing dorsal skin cells. Recalling our earlier example of switches in electrical appliances, it is possible to swap switches between two appliances. For example, one could attach the timer from a clothes dryer to a lamp and use it to turn on the light for adjustable periods of time.

TRANSCRIPTION FACTORS TURN GENES ON AND OFF

Many genes involved in development are on in some cells and off in others. An important question regarding the function of regulatory regions is, What molecular mechanisms underlie the selective activation of these genetic switches in specific cell types? The emerging answer to this question is that the regulatory region of a gene, like the coding region, functions by defining a code. In the case of the coding region, the universal genetic code converts a sequence of RNA bases into a sequence of amino acids during translation. In the case of the regulatory region, the code is mediated by a class of proteins called transcription factors. Transcription factors, which are present in the nucleus of the cell, are strongly attracted to DNA. As a consequence of this attraction, transcription factors stick to DNA, or bind to it. A key fact regarding transcription factors binding to DNA is that this binding is extremely dependent on the sequence of bases in DNA. A transcription factor can discriminate between very subtle differences in DNA base sequence and bind to its preferred site like a key fitting into a lock (i.e., a given transcription factor key will only fit a particular DNA-binding site). Because of this base sequence specificity, a given transcription factor will only bind to the regulatory regions of genes that have binding sites for it. Transcription factors interact with RNA polymerase as well as with DNA. As a result of this dual interaction, when a transcription factor binds to the regulatory region of a gene, it can greatly increase the ability of RNA polymerase to copy the coding region of the gene by leading the RNA polymerase to the position on the DNA where transcription begins.

A typical regulatory region of a gene (see Fig. 1.5) consists of a cluster of DNA-binding sites for several different transcription factors. Since each transcription factor binds to a specific preferred binding site, the set of transcription factors that can bind to a particular regulatory region is determined by the base sequence of that regulatory region. When a transcription factor binds to a regulatory region, it can function either by activating or repressing expression of the adjacent coding region. If the transcription factor activates expression of a gene, it is called an activator, whereas if it turns a gene off, it is referred to as a repressor. Some transcription factors activate one set of genes while repressing expression of another set of genes. The reason that transcription factors can exert opposite effects on different genes is that a given transcription factor activates or represses transcription by virtue of its interactions with other transcription factors. Because distinct groups of transcription factors bind to regulatory regions of different genes, a given transcription factor will be in different company when binding to one regulatory region versus another. In this sense, transcription factors are like people who behave differently in one social setting than in another (e.g., with parents versus friends). Thus, whether a particular transcription factor functions as an activator or a repressor of gene expression must be assessed on a gene-by-gene basis.

The pattern of gene expression driven by a regulatory region can be predicted if one knows the set of transcription factors that bind to that regulatory region and if one knows how each of these transcription factors functions (i.e., as an activator or a repressor). For a regulatory region to be active in a given cell, two conditions must be met: (1) All activating transcription factors that can bind to that regulatory region must be present in that cell and (2) all repressors capable of binding to that regulatory region must be absent. The "all activators and no repressors" condition for turning on a regulatory switch permits each gene to be controlled by an individually tailored regulatory code. The regulatory code is specified by the combination of optimal binding sites present in a particular regulatory region. The only set of transcription factors that can influence the activity of a regulatory region are those that can bind to that regulatory region. Since many different transcription factors are expressed in various patterns during development, it is not difficult to design a regulatory region that will activate gene expression in a given cell type at a particular time. In Chapter 3, we analyze the organization of several regulatory regions controlling expression of key embryonic patterning genes. In these model cases, the regulatory codes have been deciphered, and the sets of activating and repressing transcription factors acting on these regulatory regions are known.

CELLS COMMUNICATE WITH EACH OTHER THROUGH SIGNALS AND RECEPTORS

To reiterate the central point of this chapter, development can be viewed as a sequence of events in which initially equivalent cells acquire distinct patterns of gene expression. These differences in gene expression define different developmental potentials, such as whether a cell will become part of a muscle or part of the nervous system. As development proceeds, the potential of a given cell becomes progressively more limited until its identity is unambiguously established. This view of development as a progressively hardening plastic leads to two key questions: (1) How are differences in gene expression generated during development? (2) How do differences in gene expression alter the developmental potential of cells?

One important mechanism by which cells acquire and maintain distinct developmental potentials is by communicating with one another. Cellular communication may create a difference between initially equivalent cells, or it may exaggerate subtle preexisting differences between two cells. Communication between cells is mediated by two types of molecules referred to as "signals" and "receptors." When a cell sends a message, it liberates a signal, which is sensed by receptors present on neighboring cells (Fig. 1.8). A "receiving" cell senses a signal by virtue of the signal sticking to receptors on its surface. When a signal sticks to its receptor (or "binds" to its receptor, in the jargon), the receptor changes shape and becomes activated. Activation of the receptor alters gene expression in the responding cell, thereby defining the developmental potential of that cell. Signals and receptors are exquisitely monogamous molecules. A signal typically binds to only one receptor, and the receptor likewise is faithful to its signal.

Some signals are tethered to the surface of the signaling cell. Tethered signals can only be sensed by receptors on neighboring cells in direct contact with the signaling cell. This highly restricted form of communication is akin to carrying on a confidential conversation with a single person in a low voice. Other kinds of signals are secreted from cells, travel or "diffuse" some distance, and bind to receptors present on the surface of cells several cell diameters away. This more public form of communication is similar to broadcasting or widescale publication of a message. A developmentally important type of secreted sig-

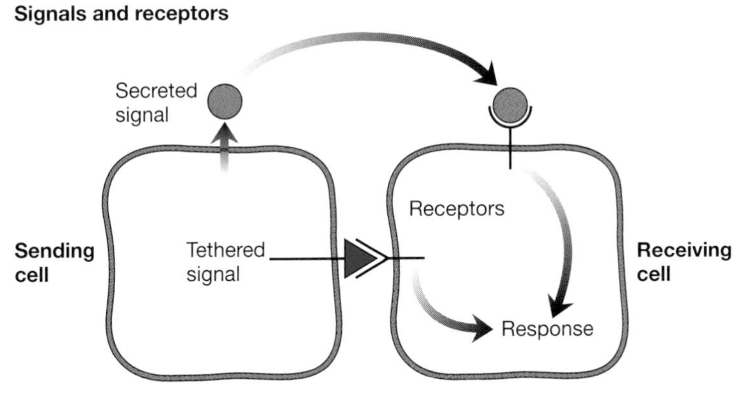

FIGURE 1.8. Cell-to-cell communication—signals and receptors.

nal is called a morphogen (Fig. 1.9). The term morphogen reflects the fact that such molecules determine the morphology or proportions of an organism or structure. The intriguing characteristic about morphogens is that they mean different things to different cells based on how much of the morphogen a cell receives. To be considered a morphogen, a molecule ideally should satisfy three criteria: (1) The molecule should be synthesized in some but not all cells, (2) the molecule should diffuse from its site of synthesis to become progressively less concentrated farther from the source of synthesis, and (3) cells should respond to different concentrations of the morphogen by activating expression of distinct sets of genes.

A typical morphogen signal is synthesized by a small group of cells. The morphogen is secreted from these cells and diffuses over several cell diameters to reach cells that do not synthesize the morphogen themselves. The combination of localized production of the morphogen signal and its subsequent diffusion away from its source creates a graded concentration of the morphogen (Fig. 1.9). To create and maintain a graded distribution of the morphogen, it also is necessary to destroy or inactivate the morphogen at a rate that balances its rate of synthesis. Cells synthesizing the morphogen, and their immediate neighbors, experience high concentrations of the morphogen; cells a small distance away sense intermediate levels of the morphogen; and cells farther from the morphogen source detect little if any signal.

Morphogen signals bind to and activate specific receptors, which then alter gene expression in receiving cells. The key defining characteristic of morphogens is that different concentrations of a morphogen activate expression of different subsets of genes. Cells close to the source of the morphogen receive high concentrations of the mor-

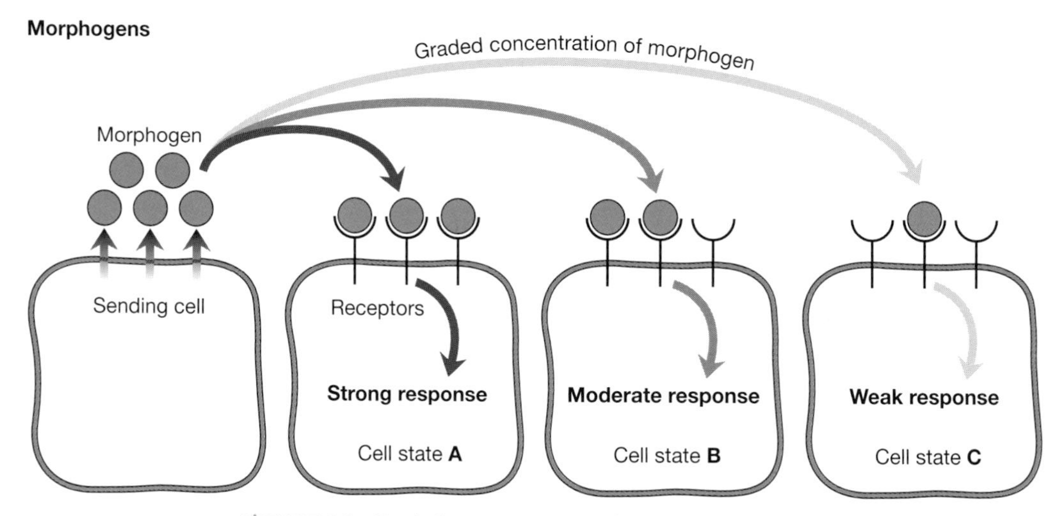

FIGURE 1.9. Graded concentrations of morphogens.

phogen and respond by activating expression of one set of genes to define cell state **A** (Fig. 1.9). Cells a short distance away from the morphogen source receive intermediate levels of the signal and respond by activating a different group of genes to define cell state **B**, and cells far away from the source activate yet another subset of genes to define cell state **C**. Because the response of cells to a morphogen is dosage dependent, the set of genes activated in a given cell is determined by the distance between that cell and the source of morphogen. There are examples where morphogens can elicit as many as five distinct responses depending on the concentration of morphogen. In such cases, distance from the morphogen source can be measured in five discrete increments. Although most examples of morphogens are secreted signals, other types of molecules can also satisfy the conditions for being morphogens. For example, in Chapter 3, we show that transcription factors can behave as morphogens under certain circumstances.

Another important type of cellular communication takes place when two equivalent cells communicate to determine which of two alternative identities each cell will adopt (i.e., cell type **A** versus cell type **B**). In these cases, it may be completely random whether a particular cell assumes the **A** or **B** identity. Often, however, one of the two communicating cells is biased toward one of the two identities. What is important in such binary decisions is that one **A** cell and one **B** cell always are produced. A common type of communication assuring this binary result is known as mutual inhibition (Fig. 1.10). In cases of mutual inhibition, one cell state (say **A**) is the default or preferred state. This means that in the absence of communication, both cells would become **A** cells (Fig. 1.10A). When the two cells communicate, however, both of them attempt to prevent the other from becoming the **A** cell type. This mutually inhibitory interaction is very unstable. As soon as one cell sends a stronger inhibitory signal to its neighbor than it receives, it assumes the default **A** state and forces the other cell to adopt the alternative **B** cell state (Fig. 1.10B).

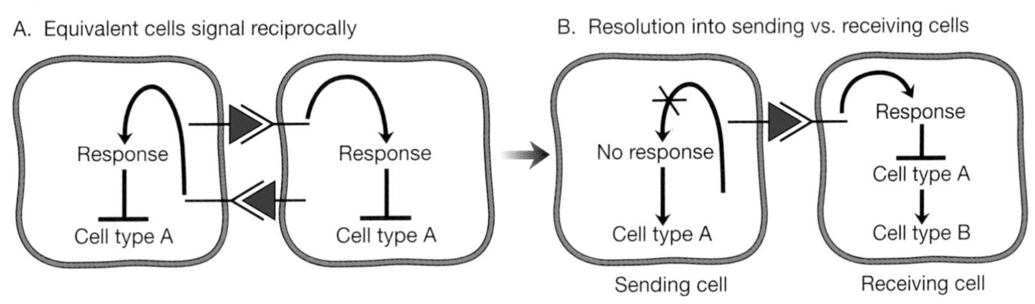

FIGURE 1.10. Cell signaling and mutual inhibition.

ORGANIZING CENTERS ORCHESTRATE EMBRYONIC DEVELOPMENT

In the early part of this century, Hilde Mangold, a talented graduate student of the prominent developmental biologist Hans Spemann, performed a very important experiment (Fig. 1.11). She cut out small regions from an early frog embryo (the donor embryo) and grafted them into various positions in a second embryo (the host embryo). The goal of these transplantation experiments was to identify regions of the embryo that might influence the developmental potential of neighboring regions. These tedious experiments paid off when Mangold transplanted a small piece of future dorsal tissue from a donor embryo into a ventral position in a host embryo. The dorsal region of vertebrate embryos normally gives rise to the brain, spinal cord, and backbone, whereas ventral regions give rise to nonneural structures such as skin, muscle, and blood. When Mangold grafted dorsal donor cells into a ventral position in a host embryo, she obtained a monstrous two-headed tadpole.

In these grafting experiments, two different species of amphibian embryos with morphologically distinguishable cells were used as donor and host. Because cells derived from these two different embryos could be told apart, it was possible to determine whether the second neural axis (brain and spinal cord) of a two-headed frog embryo was composed of donor or host cells. This analysis determined that the second neural axis was formed entirely from host cells. Because donor cells themselves did not contribute to the second neural axis, it could be inferred that they acted by organizing nearby host cells and "inducing" them to change developmental course by generating neural structures rather than skin. This inductive event, initiated by the dorsally derived donor "organizing" cells (now referred to as the Spemann organizer), was proposed to be mediated by secreted signals referred to as neural inducing factors. It was hypothesized that such neural inducing factors were liberated by the Spemann organizer and received by surrounding host cells. Cells exposed to sufficient concentrations of the neural inducer responded by developing as nervous system instead of skin. These experiments ultimately earned Hans Spemann a Nobel Prize for providing the first clear evidence for the existence of embryonic organizing centers that alter the developmental potential of neighboring cells.

Since the seminal experiments of Mangold and Spemann, organizing centers have been identified in various developing organisms. A common property of organizing centers is that they emit morphogen signals influencing the developmental potential of cells over considerable distances. This action at a distance permits a relatively small

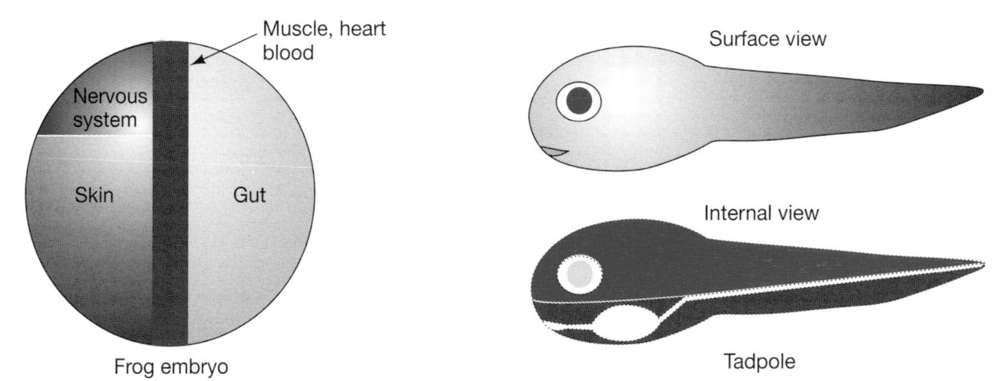

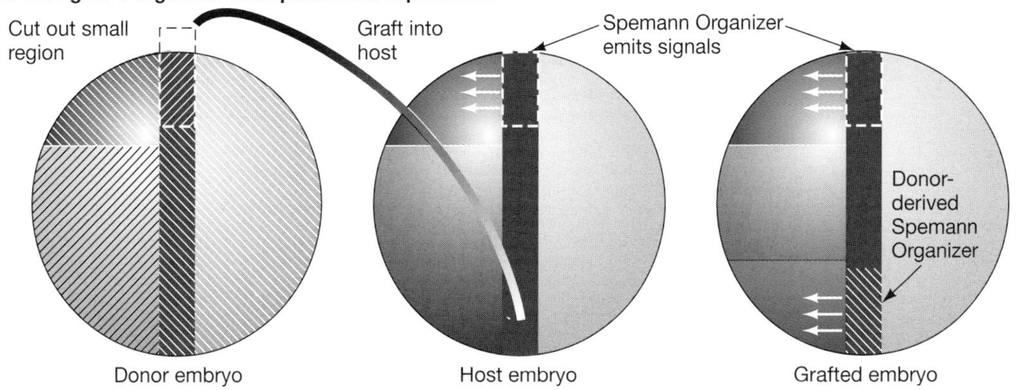

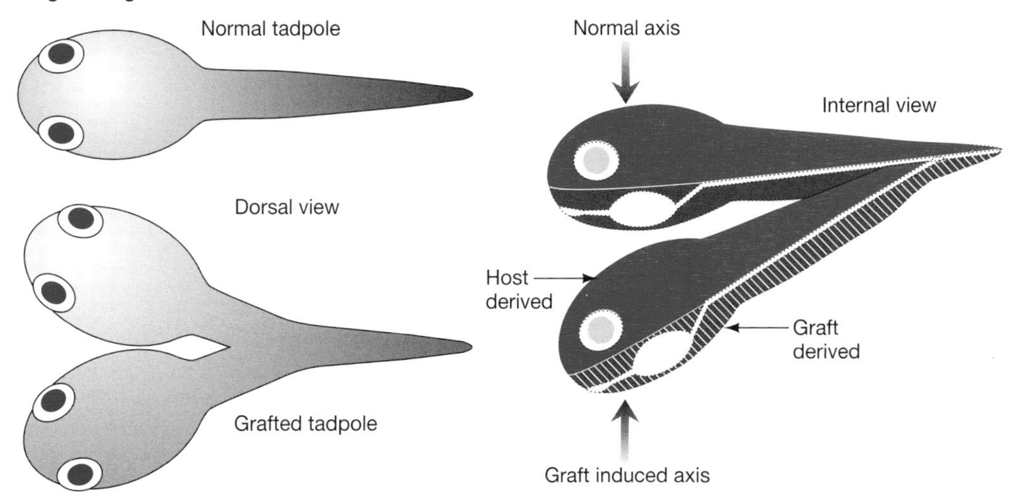

FIGURE 1.11. The presence of organizing centers—the Mangold–Spemann experiment.

group of organizing cells to orchestrate the developmental paths of large numbers of surrounding cells. The molecular identities of signals mediating the activity of several organizing centers, including the Spemann organizer, have been determined in recent years and are discussed further in subsequent chapters.

Hilde Proescholdt Mangold (*ca.* 1898–1924)

Hilde Proescholdt Mangold's doctoral dissertation won a Nobel Prize for Hans Spemann in 1935 and spawned a search for the inductive, or organizer, factor—often dubbed the "mystery of the century" in embryology and developmental biology—that continues to this day. Mangold's studies with Spemann derived from a line of research focusing on induction that began in the early 19th century with the work of Karl Ernst von Baer and continued with Caspar Friedrich Wolff and others during the 19th century. The term "Auslosung," which could be translated as "permissive induction," was used in 1901 by Curt Herbst to describe his evidence for interactions between embryonic parts in the sea urchin—a stimulus triggering an already-present response.

In this atmosphere of "the first golden age of developmental biology," Spemann performed his groundbreaking eye lens ablation and induction experiments published in 1901. In 1918, he focused on experiments in which small clumps of cells were removed from gastrula-stage amphibian embryos and grafted into different locations in other amphibian embryos at the same stage. He found that these transplanted cells followed the same development as the cells already in the new location.

Hilde Proescholdt came to Hans Spemann's lab at the Zoological Institute in Freiburg in the spring of 1920 from the University of Frankfurt. Her benchmates in the lab included the developmental biologist Johannes (Hans) Hoftfreter and Viktor Hamburger. Hamburger described her as "...open, frank, and cheerful. She had a penetrating intellect...." Spemann encouraged Mangold to transplant the various parts of an early unpigmented newt gastrula into different positions in a pigmented newt gastrula. These were technically very difficult experiments performed using a low-power binocular microscope, glass needles, and hair loops. Almost immediately in May, 1921, Proescholdt obtained an embryo that had a large secondary neural tube. This pigmented embryo had received a graft from the upper lip of the blastopore of an unpigmented donor embryo. Because Proescholdt could distinguish between pigmentation of the donor and host cells, she was able to make the critical observation that the secondary axis was formed entirely of host cells (i.e., pigmented). This experiment therefore provided strong evidence for a secreted factor produced in the host cells which redirected the developmental course of host cells surrounding the implant. Over two breeding seasons, she had only six embryos that she thought could be presented in the famous 1924 organizer paper. In all, she made 275 chimeras, of which 55 survived, and 28 had prominent secondary neural axes. Spemann proposed that the donor material had induced the ectoderm of the recipient embryo to become neural tissue. She and Spemann submitted their paper to Roux's Archiv in June 1923, and it was published in 1924. They described the organizer: "A piece of the upper blastoporal lip of an amphibian embryo undergoing gastrulation exerts an organizing effect on its environment in such a way that, if transplanted to an indifferent region of another embryo, it causes there the formation of a secondary embryonic anlage. Such a piece can there for be designated as an organizer."

Possibly indicative of women's status in the world of science in the early 20th century, Spemann included his name as author on Mangold's thesis, a procedure not followed with the male students in his lab. Viktor Hamburger, in his remembrance of Mangold, says that Spemann was correct in doing this because "...she apparently did not fully realize the significance of her results." Coincidentally, Ethel Browne Harvey, a graduate student in another lab, performed analogous experiments on hydra 12 years earlier in 1909. She showed induction by transplanted tissue of a secondary axis of polarity in the host hydra. Spemann seemed to know of her work, but he never cited it. It was said that Harvey, too, did not understand the significance of her findings. However, in later years, Harvey told a colleague, "You know that it was I who first discovered the organizer."

While finishing up her doctoral studies, Proescholdt married Otto Mangold, who was also in Spemann's lab, and they moved to Dahlem-Berlin in 1924, where Otto Mangold took up a position at the then Kaiser Wilhelm Institute for Biology. Tragically, Hilde died from severe burns in an explosion of a gasoline heater in her kitchen in September, 1924, at the age of 26. Otto Mangold wrote up his wife's findings under her name and published them in 1929 in a Festschrift for Spemann. Sadly, in Hamburger's words, "It was not granted for her to live to see the great impact her experiment had on the course of experimental biology."

Subsequent work on the inductive factor included discoveries in the 1930s that dead organizer tissue could induce neural plates, and the organizer elicited inherent capabilities of cells but did not provide detailed instructions. The organizer was linked to growth factors in the 1950s, and the inducer was found to be diffusible in the 1960s. Organizer molecules could finally be identified by the newly developing technology of the 1980s, and in the 1990s, noggin, chordin, and follistatin were identified as three likely neural inducers. Thus, the inducing principle, or organizer, is likely to be several factors and not just one. And "the mystery of the century" finally appears to be reaching some resolution.

■ Summary ■

In this chapter, we have seen that the units of genetic information, genes, which consist of DNA, comprise two parts, a coding region and a regulatory region. All cells in an organism contain the same DNA, which is a double-stranded molecule made up of subunits called bases (A,C,G,T). Different genes are active or expressed, however, in different cell types. When a gene is expressed, the coding region of the gene directs the synthesis of a single-stranded RNA copy of itself. This RNA in turn directs the synthesis of a protein according to the universal genetic code, which converts the sequence of triplets of RNA bases into a sequence of amino acids comprising protein. Proteins are molecular machines that carry out the myriad of functions required for life. The hierarchical relationship of DNA→RNA→protein is known as the central dogma of biology. Mutations are changes in the coding region of a gene that lead to the production of abnormal proteins. Most mutations reduce or eliminate the function of the protein product of a gene. In some rare cases, however, mutant forms of proteins have new activities that can render them dangerous and capable of causing diseases such as cancer.

The regulatory region of a gene functions as an on/off switch to determine where and when a gene will be expressed as RNA and ultimately as protein (i.e., according to the central dogma). Proteins known as transcription factors bind to DNA in the regulatory region of a gene in a sequence-specific fashion and interact with RNA polymerase either to increase the rate of transcription (activators) or to block transcription (repressors). The combination of activator and repressor transcription-factor-binding sites in the regulatory region of a given gene determines which transcription factors can bind to that regulatory region and influence expression of that gene. This combination of activator- and repressor-binding sites thereby defines a regulatory code for when and where the gene can be expressed.

During development, cells communicate with one another to determine which genes will be activated in different regions of the embryo. Cell–cell communication is based on cells sending secreted signals that are received by receptors on other cells. Signals can act over varying distances. Some signals are tethered to the cell surface of the producing cell and therefore only activate receptors in immediate neighboring cells. Other signals travel (or diffuse) varying distances to activate receptors on cells that are separated from those producing the signal. When signals bind to their receptors, they initiate a chain of events that lead to changes in transcription factor activity in the receiving cell, which in turn alters the pattern of gene expression in that cell.

A special type of signal known as a morphogen can cause cells to activate different subsets of genes, depending on the level of that morphogen. Morphogens are often produced in a restricted region of a developing embryo and diffuse into surrounding regions. Because cells close to the source of morphogen experience high levels of the signal, whereas cells farther way sense lower levels, morphogens can activate different patterns of gene expression at different distances from their source. In this way, morphogens can organize the expression of genes and hence the development of large territories in the embryo. Morphogens that can mediate the effect of organizing tissues defined by classic cell transplantation experiments, such as those of Mangold and Spemann, have been identified. In subsequent chapters, we show how development relies on the interplay between signaling morphogens and transcription factors.

2 Molecular Methods for Analyzing Development

This chapter provides an overview of standard methods of molecular biology and recombinant DNA technology that are needed to study development (outlined in Fig. 2.1). Gene cloning, or the isolation of DNA corresponding to a single gene, which is an essential first step in assessing the function of a gene, is described first. Next, a method for determining pattern of gene expression during development known as in situ hybridization is considered. As emphasized in the previous chapter, the central premise of developmental biology is that many genes involved in pattern formation are expressed in restricted spatial and temporal patterns. Knowing where and when a gene is expressed provides the first hint about the function of the gene. For example, a gene that is expressed only in the developing nervous system is likely to have a function in neural development. Finally, methods are discussed for eliminating the function of genes or expressing genes in inappropriate patterns or at incorrect times during development. These experimental alterations of gene activity are critical for analyzing the function of a gene. In general, we expect that too little versus too much gene activity will have opposite effects on development.

■ Cast of Characters ■

Bacterial clones Isolated colonies of genetically identical bacteria containing plasmids.

Gene clone = DNA clone A single species of plasmid DNA containing an inserted piece of DNA from another organism.

Gene cloning The isolation of a gene clone from a collection of clones, or library.

In situ hybridization A method for determining the pattern of gene expression in an organism or tissue.

Library A collection of bacterial clones containing fragments of the entire genome of an organism.

Misexpression The inappropriate expression of a gene in time or space.

Plasmid Small, circular, independently replicating DNA molecule carried by bacteria.

Recombinant DNA The joining of distinct DNA molecules into a single molecule in a test tube.

Transgene A cloned or recombinant gene which has been inserted into the genome of an organism.

Transgenic organism An organism containing a recombinant gene or transgene.

OVERVIEW OF RECOMBINANT DNA TECHNOLOGY

In Chapter 1, we discussed the structure of a typical gene. To recapitulate briefly, the functional or "coding region" of a gene consists of a string of 1,000–10,000 DNA bases. This DNA is "expressed" when one of the two DNA strands is transcribed (i.e., copied) into a complementary string of RNA bases, which then is translated via the genetic code into a linear sequence of amino acids comprising the protein product of the gene. The protein product of a gene provides a specific cellular function in the same way an electric appliance performs a particular task in the home.

To study the function of a gene in detail, it is necessary to obtain purified DNA corresponding to the coding region of that gene. Isolating the DNA corresponding to a particular gene is accomplished by cloning the gene (Fig. 2.1). As shown in this and subsequent chapters, there are several reasons to clone a gene. One great benefit of cloning a gene is that one can determine its DNA base sequence. Knowing the DNA sequence of a gene, in turn, permits deduction of the amino acid sequence of the protein product of that gene using the universal genetic

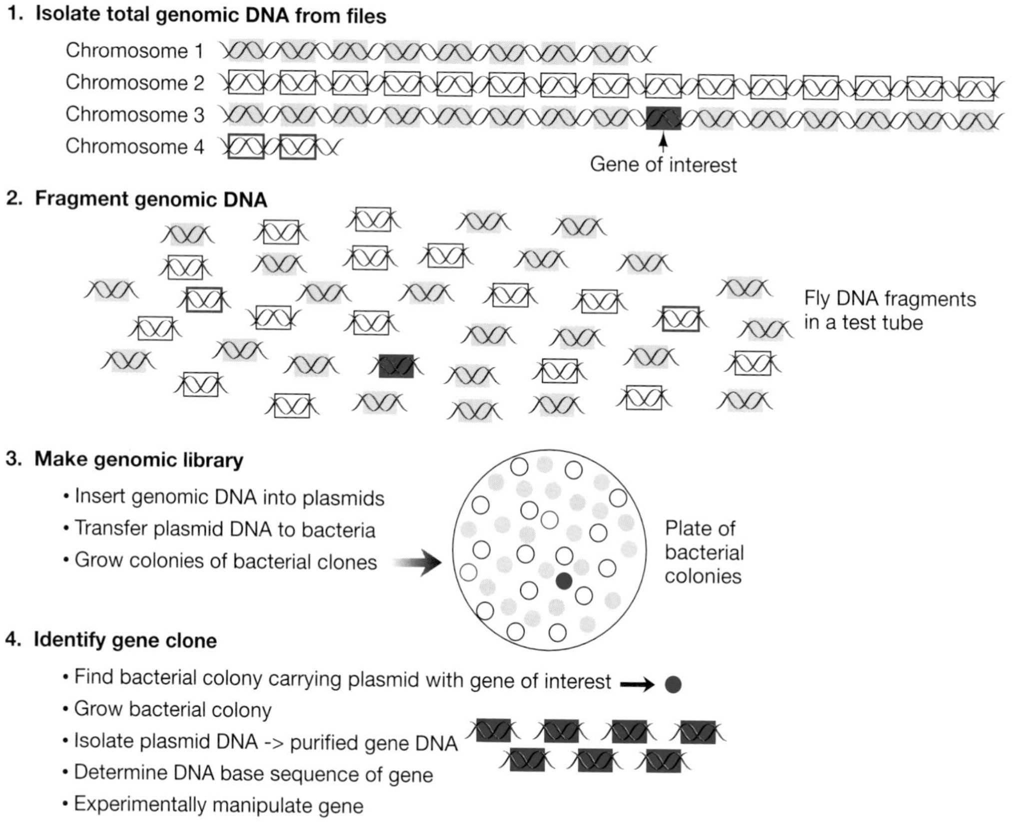

1. Isolate total genomic DNA from files

Chromosome 1
Chromosome 2
Chromosome 3
Chromosome 4
Gene of interest

2. Fragment genomic DNA

Fly DNA fragments in a test tube

3. Make genomic library

- Insert genomic DNA into plasmids
- Transfer plasmid DNA to bacteria
- Grow colonies of bacterial clones

Plate of bacterial colonies

4. Identify gene clone

- Find bacterial colony carrying plasmid with gene of interest
- Grow bacterial colony
- Isolate plasmid DNA -> purified gene DNA
- Determine DNA base sequence of gene
- Experimentally manipulate gene

FIGURE 2.1. Cloning a gene.

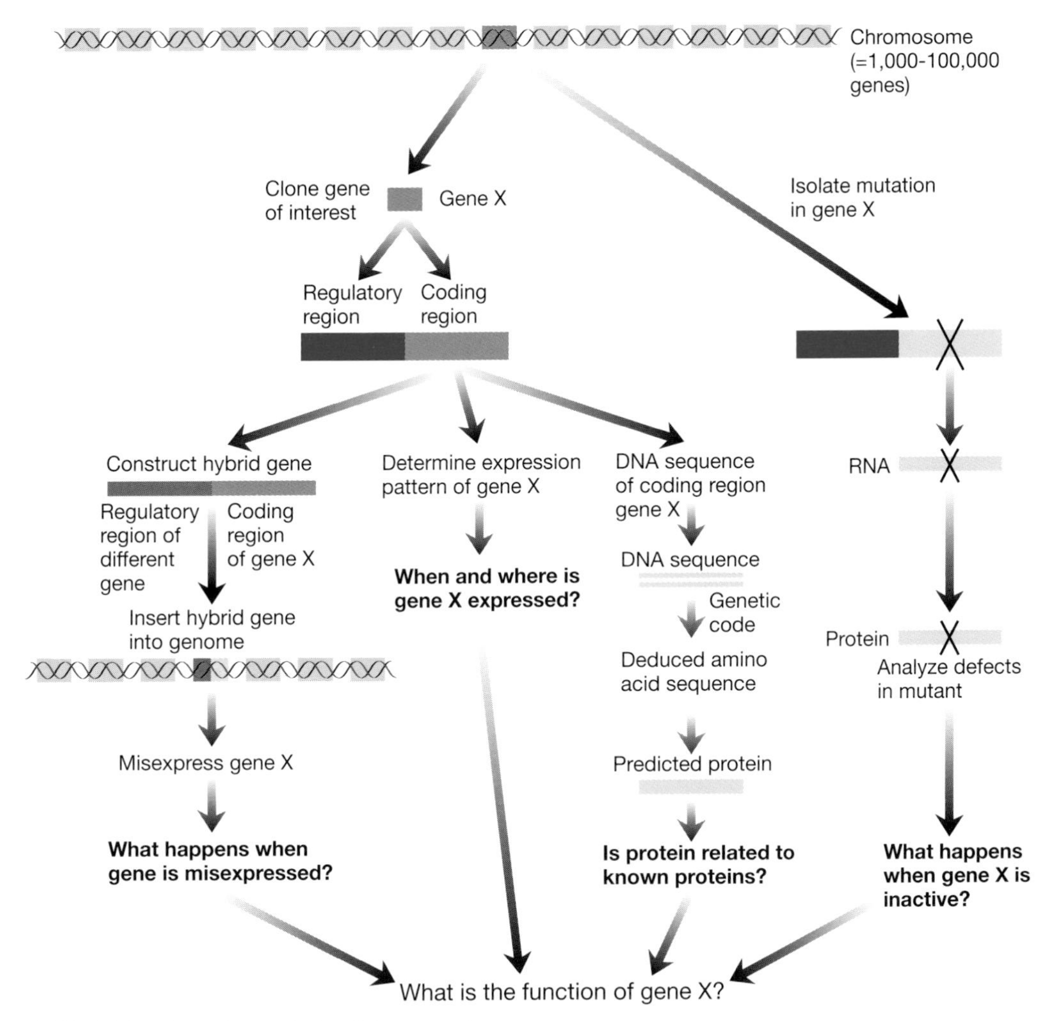

FIGURE 2.2. Molecular analysis of gene function.

code to translate. The sequence of amino acids often provides important clues regarding the function of a protein. For example, if a protein is an enzyme involved in sugar metabolism, it is likely to have an amino acid sequence similar to that of a known enzyme that carries out a related function. In addition, many transcription factors and molecules involved in signaling (e.g., signals and receptors) are members of families of genes that share amino acid sequences.

The vast amount of valuable information stored in DNA has promoted a major national initiative known as the Human Genome Project. The goal of the genome project is to determine the complete DNA sequence of the human genome as well as of genomes from several experimental organisms including bacteria, yeast, worms, flies, mice, and the mustard plant. Recently, the entire genome sequences of several bacteria have been completed. Since approximately 65% of all identified bacterial genes are related to known genes, it is possible to guess the function of most of these genes based on their predicted amino acid sequences.

If a gene of interest is found to encode a novel protein that is unrelated to any known protein, the deduced amino acid sequence of that protein is still valuable for manipulating that gene experimentally. For example, on the basis of amino acid sequence information, one can design altered versions of a protein to determine which parts of it are functionally important. Before we discuss how one goes about manipulating gene function, we must begin with gene cloning to isolate the gene in the first place.

GENE CLONING

A typical animal or plant genome contains 10,000–100,000 genes, which are strung together on chromosomes like beads on a string. To understand the function of a particular gene, it is necessary to manipulate it (e.g., determine its base sequence, determine its expression pattern, and alter its function; see Fig. 2.2). To perform any of these important manipulations, one must first isolate that gene as a single molecular entity. The problem seems daunting at first glance. A single gene represents only 1 part in 10,000 to 1 part in 100,000 of the genome, depending on the organism. How can one locate such a small needle from such a large haystack?

The method used to isolate a single gene from the genome is known as gene cloning (Fig. 2.1). Gene cloning, which should not be confused with cloning a complex organism such as Dolly the sheep, is based on work performed by Herb Boyer and colleagues at the University of California, San Francisco and by Stanley Cohen, Paul Berg, and colleagues at Stanford University. A well-appreciated fact at the time that gene cloning was developed was that bacteria carry small independently replicating DNA molecules called plasmids. Plasmids contain several genes necessary for their own replication and can carry a few additional hitchhiking genes.

Boyer, Cohen, and Berg reasoned that they might exploit the ability of plasmids to carry foreign genes if they could insert small pieces

■ Plasmids and Antibiotic Resistance ■

A medically important example of plasmids harboring these hitchhiking genes is the phenomenon of antibiotic resistance. Genes conferring resistance to antibiotics are often stowed away on plasmids. If one ingests a bacterium carrying a drug-resistance plasmid, it is possible for that plasmid DNA to be released from the bacterium and transferred to other bacteria such as those naturally lining the gut. When this happens, subsequently infecting bacteria can acquire the drug-resistance gene from those residing in the gut. These infectious bacteria rapidly become insensitive to the antibiotic in question and can make it difficult to treat serious diseases such as tuberculosis.

of DNA from an organism such as a fly, mouse, human, or plant into a plasmid. They worked out a strategy for generating a collection of recombinant plasmids (from which the term "recombinant DNA" derives) each containing a different small fragment of an organism's genome (Fig. 2.1). Bacteria harboring recombinant plasmids are grown on plates at low densities so that the descendants of single founding bacteria form isolated colonies or "bacterial clones." Because the founding bacterium of a colony will have taken up only a single DNA molecule, all bacteria within a clone carry the same recombinant plasmid. The term "clone" reflects the fact that bacteria within a single colony are genetically identical, which is similar, on a much reduced scale, to cloning a sheep such as Dolly. It is possible to generate sufficiently large numbers of colonies carrying different recombinant plasmids so that some bacterial clone within that collection will carry any gene of interest. Such a complete collection of bacterial clones is called a library. In a library, each bacterial colony contains a different fragment of genomic DNA and can be likened to a book. The base sequence of the genomic DNA carried on a particular plasmid would correspond to the sequence of words in a book. Once one has found and checked out a book of interest, one can read it to discover the story inside. The seminal work on generating recombinant plasmid molecules eventually won Berg the Nobel Prize in Chemistry in 1980.

The trick in cloning a particular gene is to identify which of the 10,000–100,000 different bacterial clones in a library contains the plasmid harboring the gene of interest. Because each bacterial clone carries only a single inserted fragment of genomic DNA in its plasmid, one must generate a collection of 10,000–100,000 different clones to ensure that every fragment of an organism's genome is represented somewhere in that library. There are a variety of tricks for sorting through such enormous libraries of bacterial clones to find a gene of interest. In the future, this problem will be greatly simplified through the efforts of the genome project, which soon will provide the DNA sequence of all genes in libraries of flies, mice, humans, and plants. This detailed information will permit bacterial clones in libraries to be organized according to the order of genes on chromosomes. For instance, a bacterial clone carrying gene number 223 of human chromosome 6 will be placed between clones carrying genes 222 and 224 on that same chromosome. In contrast to such an ideally ordered set of clones, currently available libraries of plasmid clones are randomly organized collections and can be likened to a library in which all of the books have been dumped onto the floor and scattered about. Obviously, it will be much easier and more efficient to clone genes when the books are placed on shelves in alphabetical order. Although discussion of gene-hunting methods falls beyond the scope of this book, one of various existing approaches generally can be used to find a particular gene of interest within the scattered mess of a random library. Once a bacterial clone carrying a gene of interest is identified, one can grow cultures of that bacterial strain and purify large quantities of the plasmid DNA. This single species of plasmid DNA is referred to as a "DNA clone" or a cloned gene, which is propagated in the bacterial clone carrying the plasmid.

If one has a cloned gene in hand, it is straightforward to determine the base sequence of that cloned DNA. The amino acid sequence of the protein encoded by that gene then can be deduced according to the universal genetic code (see Chapter 1). Cloned DNA also can be used to determine the expression pattern of the gene during development and to manipulate the function of the gene (Fig. 2.2). We now turn to determining the expression pattern of a cloned gene.

VISUALIZING GENE ACTIVITY IN DEVELOPING EMBRYOS

The most direct method for determining the expression patterns of cloned genes during development is in situ hybridization (Fig. 2.3). In situ hybridization permits the experimenter to determine visually which cells in a developing organism express a gene of interest as RNA (i.e., transcribe the gene from DNA to RNA). Because transcription (i.e., DNA→RNA) rather than translation (i.e., RNA→protein) is the most commonly regulated step in gene activation, in situ hybridization generally provides an accurate measure of gene activity (e.g., DNA→RNA→protein). It also is possible to determine the distribution of the protein products of many genes using other methods. In such cases, there usually is very good agreement between the patterns of RNA and protein expression.

The principle of in situ hybridization derives from the structure of double-stranded DNA. As described in Chapter 1, DNA is composed of two complementary strands of bases, which obey the "**A:T**," "**C:G**" base-pairing rule. Similar DNA:RNA double-stranded hybrid molecules are formed during the process of transcription. In situ hybridization can detect the presence of a single type of RNA as a result of base-pairing between DNA and RNA strands in DNA:RNA hybrids, which can only occur when the sequence of bases on one strand is exactly complementary to that on the other strand. This stringent base-pairing condition provides a means of monitoring the presence of a single RNA species in the vast sea of different RNAs contained in any cell type.

The first step of in situ hybridization is to synthesize a single-stranded DNA copy of the cloned gene of interest (Fig. 2.3, step 1). This synthesized DNA, which is referred to as a probe, must be marked in some way so that it can be detected later. In the early days of in situ hybridization (i.e., the 1980s), DNA probes were marked with radioactive bases. The presence of the radioactive probe DNA was ultimately detected with an emulsion of photographic film. We now generally use a chemical means of marking the bases in probes which can be de-

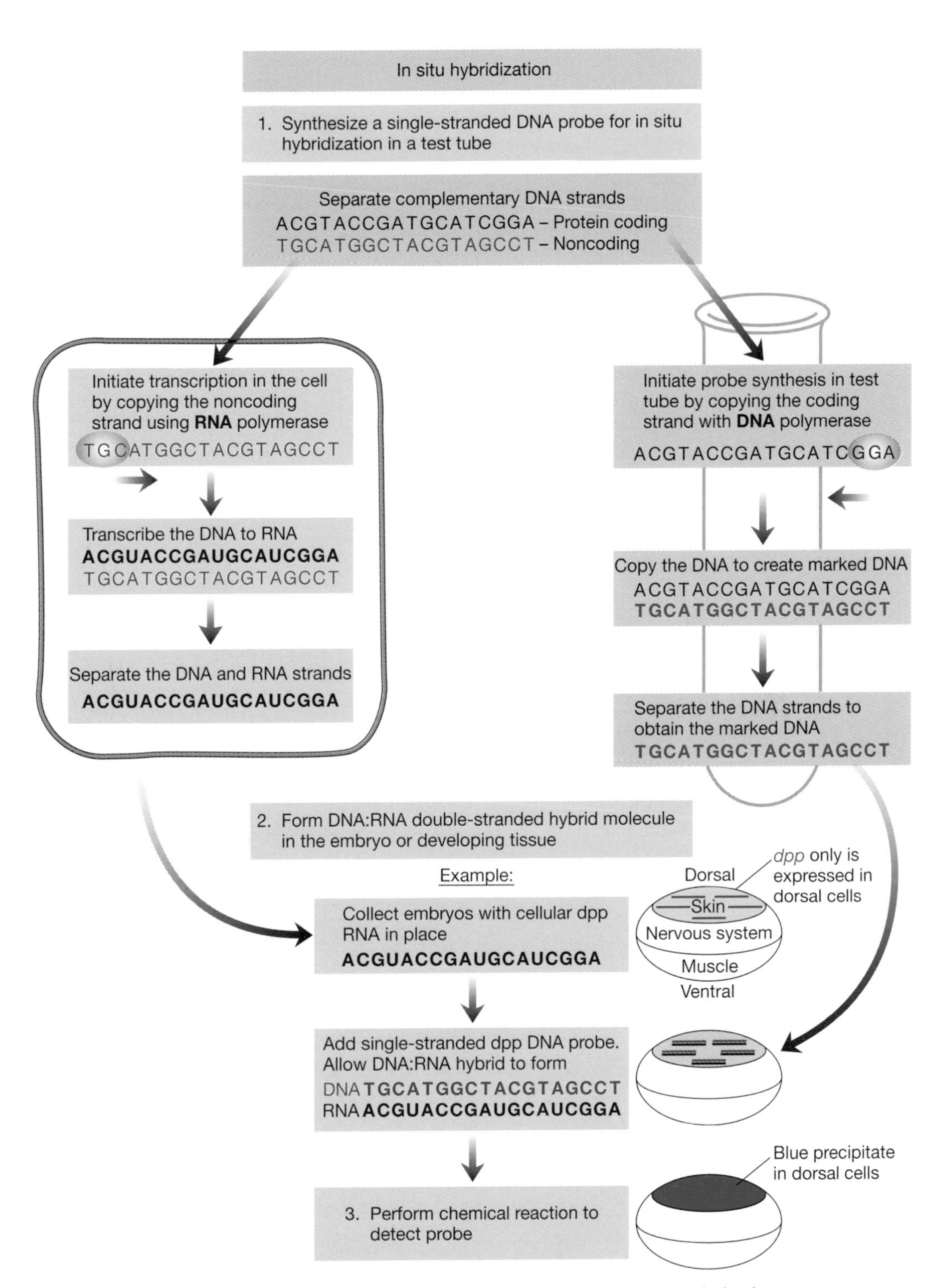

FIGURE 2.3. Determining the expression pattern of a gene by in situ hybridization.

tected by the production of blue or brown color reactions. To make a probe, a single strand of probe DNA is synthesized in a test tube using DNA polymerase, the enzyme that functions during DNA replication to synthesize double-stranded DNA from a single strand of DNA. Because DNA polymerase copies a single strand of DNA into its complementary strand, one can arrange matters so that the marked probe strand is complementary to the expressed RNA strand. Because the single-stranded DNA probe and the single-stranded RNA transcribed from the gene are complementary, they can form a stable DNA:RNA double helix (Fig. 2.3, step 2).

In the next step of in situ hybridization (Fig. 2.3, step 2), the single-stranded DNA probe is added to embryos or a developing tissue under conditions favoring the formation of double-stranded DNA:RNA hybrid molecules. Because the probe is complementary to the RNA product of only one gene, it can only form a hybrid with that specific RNA. As a consequence, DNA:RNA hybrids only will form in cells that were expressing the gene of interest at the time the embryos were prepared for in situ hybridization. Following the hybridization step, any excess probe that did not form a DNA:RNA hybrid is washed away. In the final step of the in situ hybridization procedure (Fig. 2.3, step 3), a chemical reaction is carried out to reveal which cells contain the chemically marked probe DNA. Only cells containing the DNA:RNA hybrid molecules will turn color (e.g., blue) at this point. Thus, the blue cells are those expressing the gene of interest. Although uncolored cells also contain DNA corresponding to the gene of interest (see the Gurdon experiment in Chapter 1), they do not actively transcribe this gene into RNA. The ability to determine gene expression patterns at the resolution of single cells has contributed significantly to the rapid progress in developmental biology over the past 15 years.

HOW MUTANTS REVEAL THE FUNCTION OF GENES

The classic genetic approach to studying a biological problem is to identify mutant versions of genes involved in the process of interest. As mentioned in Chapter 1, a mutant gene is abnormal because its DNA base sequence is altered (see Fig. 1.6). Changes in base sequence generally cause mutant genes to make defective protein products or no protein at all. A geneticist infers the function of a given gene from what goes wrong when that gene is not active. To make another analogy to electronics, it is as though someone interested in learning about how a TV works went about removing individual components from the circuitry and then tried to figure out what each circuit element did in a functioning TV. For example, if one removed a circuit involved in color balance, the image might become too red, but would remain sharp. On the other hand, if a circuit involved in achieving maximal resolution were taken out, the color should be fine but the image would get fuzzy.

In the following chapters, we show how geneticists have generated large numbers of mutants affecting the development of fruit flies, fish, mice, and plants. The most critical step in determining the function of a gene of interest is to generate a loss-of-function mutant in that gene. As an illustration, we consider two well-studied mutants in patterning genes, one from fruit flies and the other from the mustard plant (Fig. 2.4). The example from fruit flies is a gene we have already mentioned called *decapentaplegic*, or *dpp* for short. The name *decapentaplegic* means that many (*decapenta* = 50) defects (*plegic*) arise when this gene is not functioning properly. The multiplicity of defects associated with *dpp* mutants reflects the fact that the *dpp* gene is involved in patterning several different facets of the embryo and adult. During early embryonic development, in situ hybridization reveals that the *dpp* gene is expressed only in cells in the dorsal region of the embryo. This dorsal part of the embryo normally gives rise to skin but not nervous system (Fig. 2.4A, left). In mutant embryos lacking function of the *dpp* gene, dorsal cells behave abnormally by giving rise to large numbers of neural precursor cells (Fig. 2.4A, middle). From this observation, one can conclude that the *dpp* gene normally suppresses early neural development in the dorsal region of the embryo.

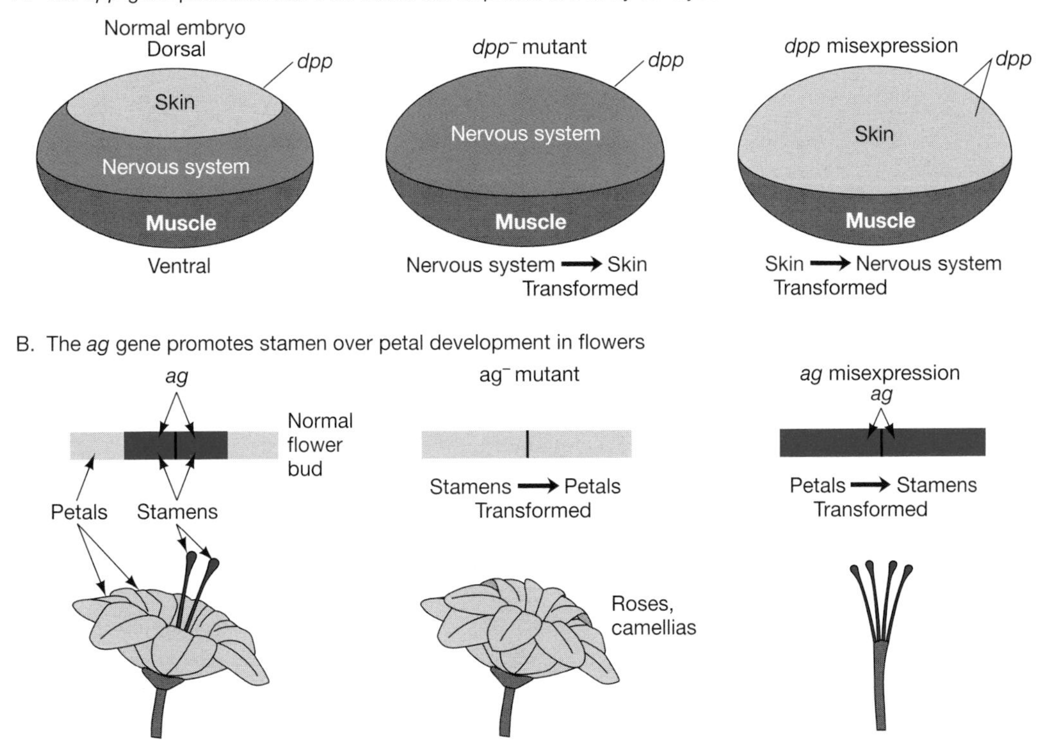

FIGURE 2.4. (*A*) Analysis of gene function in fruit flies. The *dpp* gene promotes skin over neural development in embryos. (*B*) Analysis of gene function in the mustard plant. The *ag* gene promotes stamen over petal development in flowers.

Another example of a patterning gene is the *agamous* gene (or *ag* for short) from the mustard plant, which plays an important role in flower development (Fig. 2.4B). The name *agamous* (meaning no gametes or germ cells) reflects the fact that plants mutant for this gene are infertile due to the absence of sperm or eggs. Mutants in the *ag* gene were known as far back as the third century B.C. when the Greek philosopher and naturalist Theophrastus (~ 371–287 B.C.) described double roses in a botanical treatise entitled *Inquiry into Plants*. Similar observations were made nearly a thousand years later in China, and then again in the 16th century in lengthy comparative studies of abnormal floral development. Transformation of one part of a flower into another (metamorphosis of flower parts) was extensively studied by the polymath Johann Wolfgang von Goethe (1749–1832), who is best known as a novelist and poet. In 1790, he wrote "From our acquaintance with this abnormal metamorphosis (meaning floral mutants), we are enabled to unveil the secrets that normal metamorphosis conceals from us, and to see distinctly what, from the regular course of development, we can only infer." Many attractive modern cultivated flowers such as roses and camellias, which have whorls of concentric petals, are mutants that lack function of the *ag* gene (Fig. 2.4B, middle). The increased number of petals in *ag* mutants results from stamens, which are male reproductive organs in the flower, developing as petals. In situ hybridization demonstrates that the *ag* gene is expressed in stamens but not in petals. Because *ag* is normally expressed in stamens (Fig. 2.4B, left), and since mutants lacking *ag* function form petals in place of stamens, one can deduce that this gene suppresses petal development in stamens.

Johann Wolfgang von Goethe (1749–1832)

Mention the name Goethe and most people immediately think of the great German romantic poet and author of *The Sorrows of Young Werther* and *Faust*. Johann Wolfgang von Goethe was also a scientist who made contributions in anatomy, geology, physics, and botany. In fact, he felt that the most important of all his works was *Theory of Colors* (*Zur Farbenlehre*, 1810), his anti-Newtonian analysis of the theory of light. In many respects, poetry and science were closely linked for Goethe—science appeared in his poems ("Entopic Colors" and "The Metamorphosis of Plants" are examples) and poetry appeared in his scientific papers. He characterized many of his scientific views and beliefs in his *Spruche in Prosa* (*Prose Aphorisms*).

Goethe, who was born in Frankfurt, studied law at the University of Leipzig as well as in Strasbourg. Rather than being inspired by the law, in Strasbourg especially, Goethe was inspired by the Gothic style and the birth of a new type of German literature. He returned to Frankfurt in 1771 to practice law, but promptly unleashed the Shakespeare mania for which the Sturm and Drang movement is famous. He arrived at the court of Duke Karl August in

Wiemar in 1775, and it was here that his scientific interests asserted themselves. In 1784, he demonstrated the existence of the intermaxillary (premaxillary) bone in humans, thus establishing the continuity of anatomy across species. Previous to this, it was believed that the absence of this bone in humans separated them from all other animal species. Unfortunately, and not known to Goethe, this discovery was also made in France in 1780. His influential theory regarding metamorphosis in plants was devised in 1789 and published as the book *The Metamorphosis of Plants* in 1790. Goethe resigned his position at Court in 1817 and turned to writing scientific essays, as well as continuing his prodigious literary pursuits.

A true romantic, Goethe believed that the manner in which his passions and emotions found form in his poems followed the same laws that make flowers bloom. He did not agree with the Linnean method of categorizing species of plants—one first had to understand how the forms and species developed as a metamorphosis of each other. Goethe thought that there was an "Urpflanze," or primal plant, and that all plant forms were a transformation of the leaf. He examined plants by observing and describing their sameness of form and their stages of growth. In studying malformations in plants, Goethe made comparative observations of the formation of flowers and studied the interconnections that he observed. One of his major intellectual achievements, which is now a central tenet of modern genetics, was understanding that it is possible to learn about normal processes by studying abnormal variants (now known as mutants) in which the process is disrupted. In Goethe's words, "From our acquaintance with this abnormal metamorphosis, we are enabled to unveil the secrets that normal metamorphosis conceals from us, and to see distinctly what, from the regular course of development, we can only infer."

"Type" was the key to understanding the development of forms—single variants of form could be understood from this general idea of type, an ideal primal organism. One had to train his or her faculties rigorously for observation and thinking, and then from this contemplative relationship with nature one could obtain the deepest knowledge of phenomena. "My attention has always been directed exclusively towards objects that surrounded me in the earthly realm and which could be directly perceived through the senses."

Goethe believed in the sensory nature of science and nature and in its unity; he was opposed to reductionist techniques that looked behind the scenes, and he dissected what he saw with his own eyes. "Seek nothing behind the phenomena, they are themselves the theory." He felt that scientific instruments such as microscopes and prisms could distort reality; one must trust the immediate truth of sensory perception.

The fusion of science and poetry that formed the fabric of Goethe's work is not something seen very often in 20th century science. However, his work, in addition to making its mark on plant genetics, has influenced such scientists as the German physicist Werner Heisenberg (1901–1976), who formulated the famous uncertainty principle, as well as modern-day environmentalists.

The strategy of inferring the function of a gene by analyzing the consequence of eliminating the function of that gene is very powerful. In the next chapter, the genetic analysis of fruit fly embryonic development is described. In these studies, nearly all genes involved in patterning the anterior–posterior and dorsal–ventral axes of the embryo were identified by isolating mutants that disrupt the function of these patterning genes. Analysis of early-acting fruit fly patterning mutants revealed that a hierarchy of gene action guides embryonic development. The subsequent cloning and mechanistic analyses of these fruit fly patterning genes over the last 15 years have borne out the major predictions proposed initially by geneticists. This triumph of the genetic method will serve as the gold standard for generations of developmental biologists to come.

EXPERIMENTAL MANIPULATION OF GENE ACTIVITY

Another way to analyze gene function is to express a gene at an inappropriate time or place. If one has cloned a gene of interest, it can be misexpressed by exploiting the bipartite structure of genes. One constructs a hybrid gene composed of the coding region from the gene to be misexpressed fused to the regulatory region from another gene, which is expressed in a different pattern from the test gene. In Chapter 1, we considered such an experiment in which the regulatory region of the *dpp* gene was joined to the coding region of the bacterial *lacZ* gene (see Fig. 1.7). The bacterial *lacZ* gene has no function in a fly embryo, but rather serves as a marker gene. When the hybrid *dpp:lacZ* gene is inserted into a fly chromosome, embryos laid by flies carrying this hybrid gene express the *lacZ* gene in skin cells. Because the *lacZ* gene encodes an enzyme that turns cells blue, one can generate embryos with blue skin. The same strategy can be applied to driving expression of a cloned gene of interest. In this latter case, instead of expressing a passive marker gene (e.g., *lacZ*) in a certain pattern, one misexpresses a developmentally active gene in inappropriate cells.

Many different known regulatory regions can be used to control the expression of a test gene in virtually any desired pattern. Some regulatory regions direct gene expression in restricted patterns such as in broad sectors, narrow stripes, or specific tissues (e.g., skin, nerve, muscle, or gut). With such an array of available regulatory regions, one can drive expression of a gene of interest in cells that normally do not express that gene. There also are regulatory regions that activate gene expression in all cells. A very important type of regulatory region is one that is conditionally active. For example, there are regulatory regions that can be activated by increasing temperature. Genes controlled by such regulatory regions normally are involved in combating heat stress. It is necessary to employ conditionally active control regions in situations where misexpression of a gene of interest at an early stage would kill the organism before it reached the developmental stage one wished to study.

Once a hybrid gene consisting of the coding region of a gene of interest and the regulatory region of another gene has been constructed, it can be inserted into the genome of a living organism. An individual containing such a recombinant gene is called a "transgenic" organism, and the hybrid gene is referred to as a "transgene." Although there is not sufficient space here to discuss methods by which transgenic organisms are generated, it should be noted that it is routine to generate transgenic individuals in species as diverse as bacteria, yeast, flies, worms, mice, and plants. Once one has generated a transgenic organism containing a hybrid transgene of interest, it is possible to examine the effect of misexpressing that gene during development.

■ The Fly *dpp* Gene Suppresses Neural Development ■

For an example of a misexpression experiment, we return to the fruit fly patterning gene *dpp* mentioned above. These particular experiments were actually performed in my own laboratory. Ron Blackman, a collaborator of ours, fused the coding region of the *dpp* gene to a temperature-dependent control region that could be activated in all cells by raising the temperature from 20°C to 37°C. He then inserted this hybrid *dpp* gene into a fruit fly chromosome to generate a strain of transgenic flies. Transgenic fruit flies carrying this hybrid *dpp* transgene survive at room temperature because the *dpp* gene is not expressed under these conditions. To determine the consequence of misexpressing the *dpp* gene in all embryonic cells, we collected embryos carrying the hybrid *dpp* transgene and then shifted them to 37°C (98°F). This manipulation activates expression of *dpp* in all cells. We then examined these experimental embryos for abnormalities. Normally the *dpp* gene is expressed only in the dorsal region of the embryo, which gives rise to skin, but is not expressed in lateral or ventral regions of the embryo, which give rise to the nervous system and muscle, respectively. The result of forcing expression of *dpp* in all cells is that cells which normally would become neural are prevented from doing so (Fig. 2.4A, right). This defect is opposite to that observed in mutant embryos lacking *dpp* function (i.e., dorsal cells generate ectopic neural tissue; Fig. 2.4A, middle). Both experimental results can be explained by hypothesizing that the *dpp* gene suppresses neural development.

Another example of a misexpression experiment involves the *ag* gene, which controls floral development in plants. Many rose-like flowering plants lack the function of the *ag* gene and form petals in place of stamens (Fig. 2.4B, middle). Because *ag* is expressed in developing stamens, but not in petals, it is likely that *ag* normally functions in stamens to suppress petal formation. Consistent with this hypothesis, misexpression of *ag* in petal primordia results in flowers having extra stamens and fewer petals than normal (see box and Fig. 2.4B, right).

Although the *ag* and *dpp* genes are as unrelated as plants and animals, they nonetheless function by similar mechanisms; namely, by suppressing alternative developmental programs. The function of *dpp* is considered in more depth in Chapters 3 and 5, since this gene plays a central role in limiting neural development in both invertebrate and vertebrate embryos. The *ag* gene also is revisited in Chapter 8, when we consider how this and other genes determine the identity of flower organs during plant development.

■ The *ag* Gene in Plants Promotes Stamen Over Petal Development ■

A leading plant molecular biologist, Martin Yanofsky, constructed a transgenic plant that contained a hybrid gene consisting of the coding region from the *ag* gene fused to the regulatory region from a gene active in both petals and stamens. When transgenic plants containing this hybrid gene attempt to flower, they form structures having many stamens, but no petals (Fig. 2.4B, right). This bizarre flower is the result of petals being transformed into stamens. There is a naturally occurring mutant form of the snapdragon called *macho,* which similarly has an excess of stamens (hence the mutant name) at the expense of petals. As in the case of the *ag* misexpression experiment, the defect in *macho* mutants results from misexpression of *ag* in the primordia of petals. The consequence of misexpressing *ag* conforms with defects observed in the loss-of-function *ag* mutant and can be explained in part by *ag* suppressing petal development in stamens.

■ Summary ■

In this chapter, we considered several methods that are needed for isolating, or "cloning," individual genes and for analyzing the function of genes. The starting point in analysis of gene function is cloning a gene of interest from an organism such as a mouse. One creates a large collection of bacteria (a library) that carry DNA fragments from that organism inserted into a small replicating form of DNA called a plasmid. If the plasmid library is complete (i.e., contains all fragments of the mouse genome), it is generally possible to use various strategies to isolate a colony of genetically identical bacteria (or a bacterial clone), which contains a plasmid carrying the gene of interest. All bacteria in this colony contain the same plasmid, which is referred to as a gene clone or DNA clone. This cloned DNA can be subjected to DNA sequence analysis to determine the order of base pairs in the gene of interest. Knowing the sequence of bases in DNA, one can deduce the amino acid sequence of the protein product encoded by the gene using the universal genetic code to translate.

Once one has cloned a gene of interest, it is possible to study the function of that gene by asking what happens to the organism if the gene function is eliminated or expressed in inappropriate cells. To misexpress the gene, one must first figure out where and when the gene is normally expressed in the developing organism, which can be determined by the in situ hybridization method. When the spatial and temporal expression patterns of a gene are known, it is possible to misexpress the gene experimentally in cells that would not ordinarily do so. By comparing the consequences of eliminating the function of a gene with those resulting from expressing the gene in the wrong place or time, one can learn a great deal about the normal function of that gene.

In the next chapter, the methods of genetic analysis described in this chapter will be applied in a comprehensive search for genes involved in creating the primary body axes of fruit fly embryos. Similar types of analysis form the basis for material covered in Chapters 4–6, in which remarkable parallels have been observed in the development of vertebrate and invertebrate embryos and appendages. The analysis of plant embryonic and floral development (Chapters 7 and 8) also relies on the same genetic approaches and reveals striking similarities in the general principles of development operating in organisms as different as plants and animals.

3 Establishing the Primary Axes of Fruit Fly Embryos

In 1995, Christiane Nüsslein-Volhard, Eric Wieschaus, and Ed Lewis shared the Nobel Prize in medicine in recognition of their visionary genetic analysis of embryonic development in the fruit fly. This chapter focuses on the impact of their genetic experiments and the subsequent detailed analyses conducted in many other laboratories. Discoveries made in the course of these studies revealed the fundamental mechanisms by which the primary body axes are established during early embryonic development and provided an essential starting point for a mechanistic analysis of how genes control developmental processes. To start, we summarize the development of the fruit fly embryo. Then, we describe the large-scale mutant hunt from which emerged an outline of how genes control embryonic development. Finally, we gather these various strands of the story to discuss specific mechanisms by which genes that have been identified in the mutant hunt can generate and maintain differences between cell types during development.

The fruit fly embryo is particularly well suited to a comprehensive genetic analysis of development because the two major perpendicular body axes, the anterior–posterior (A/P) axis and the dorsal–ventral (D/V) axis, are established through independent molecular mechanisms. Despite these fundamental differences, however, the A/P and D/V axes are created by strikingly similar principles. In both cases, crude information specifying position already exists prior to fertilization. The mother creates this positional information by depositing molecules called morphogens (see Chapter 1) asymmetrically within the egg. Maternally provided morphogens are most highly concentrated in one extreme pole of the egg and progressively diminish in concentration toward the opposite pole. Different concentrations of a morphogen are required to activate different genes in one of a few adjacent nonoverlapping domains of the embryo. These domains represent coarse primary territories of gene expression and serve as sources of secondary morphogen signals, which then subdivide the embryo on a finer scale. Ultimately, the embryo is partitioned into basic units of organization. In the fly, the fundamental units of organization along the A/P axis are segments comprising defined intervals of the head, thorax, and abdomen, whereas along the D/V axis, the units of subdivision are

■ Cast of Characters ■

Terms

Autoactivation The ability of a gene to activate its own expression.

Blastoderm embryo An early hollow cylindrical embryo prior to gastrulation.

Body axes The primary perpendicular axes of the fruit fly (anterior–posterior and dorsal–ventral), which are established during early embryonic development.

Carrier An individual having one mutant and one good copy of a given gene.

Cuticle The tough outer covering of the larva.

Denticles Hairs arrayed in rows on the ventral cuticle of the larva.

Diploid A cell having two copies of every gene.

Ectoderm The outer embryonic germ layer of cells that gives rise to skin and nervous system.

Embryonic gene A gene whose function is required in the embryo.

Embryonic mutant A mutant lacking the function of an embryonic gene.

Endoderm The inner embryonic germ layer of cells that gives rise to gut.

Essential gene A gene required for the survival of the fly.

Gap gene An embryonic gene that functions to define a large block of cells along the A/P axis.

Gap mutant A mutant lacking the function of a gap gene in which a large section of the cuticle is typically missing in one restricted region of the A/P axis.

Gastrulation The organized movement of cells during midembryonic development that creates a laminated embryo with distinct tissue layers.

Genetic screen = mutant screen A systematic hunt for mutations.

Homeobox The DNA-binding region of a subtype of a transcription factor such as those encoded by the homeotic genes.

Homeotic gene A gene required to define segmental identity.

Homeotic mutant A mutant lacking the function of a homeotic gene in which the identity of a specific segment is transformed into that of an adjacent segment.

Invagination An internalizing cell movement in which sheets of cells fold into a developing structure.

Larva A hatched embryo.

Maternal genes Genes that are active only in the mother and/or egg and are required for development of the fertilized embryo.

Maternal mutant A mutant in a maternal gene that lacks a function in the egg supplied solely by the mother.

Mesoderm An embryonic germ layer that gives rise to muscle, heart, and fat (the ventral region of a fly embryo).

Metamorphosis The transformation of an embryo into an adult.

Morphogen A secreted signal that elicits different cellular responses at different concentrations.

Morphogenesis The process by which a developing organism attains its final shape.

Mutagen A chemical compound that causes mutations.

Mutant An organism with a mutated gene causing an identifiable defect.

Mutation An alteration in the base sequence of a gene.

Neuroectoderm The portion of the ectoderm (outer germ layer) from which the nervous system forms (the lateral region of a fly embryo).

Pair-rule gene A patterning gene required for the formation of structures in every other segment.

Pair-rule mutant A mutant in a pair-rule gene that lacks cuticle derived from every other segment.

Patterning The process by which cells acquire distinct identities during development.

Progressive refinement A sequence of simple patterning events during development.

Segment polarity gene A patterning gene required for the formation of segmentally repeated structures.

Segment polarity mutant A mutant in a segment polarity gene that exhibits defects within every segment such as deletions and/or duplications.

Transcription factor A protein that controls the transcription of other genes (i.e., turns genes on or off).

Genes

achaete-scute genes (AS-C) Encode transcription factors required for development of the nervous system.

Antennapedia A gene encoding a homeobox transcription factor that defines the second thoracic segment (T2), which has wings and legs.

bicoid Encodes the maternal morphogen (Bicoid) which is provided by the mother to pattern the A/P axis.

Bithorax A homeotic gene encoding a homeobox transcription factor which is active in the third thoracic segment (T3).

decapentaplegic (dpp) Encodes a secreted signal required for formation of the dorsal ectoderm.

dorsal Encodes the maternal morphogen (Dorsal), which is a transcription factor responsible for patterning the D/V axis.

engrailed A segment polarity gene required for the formation of the posterior portion of each segment.

even-skipped A pair-rule gene encoding a transcription factor that is required for the formation of even-numbered segments.

hedgehog A segment polarity gene encoding a secreted signal that is required for the formation of posterior structures in the segment.

hunchback A gap gene encoding a transcription factor that is required for formation of the most anterior region of the embryo.

knirps A gap gene encoding a transcription factor that is required for formation of middle-posterior (abdominal) regions of the embryo.

Krüppel A gap gene encoding a transcription factor that is required for formation of the thoracic region of the embryo.

naked A segment polarity gene required for the formation of anterior structures in the segment.

odd-skipped A pair-rule gene encoding a transcription factor that is required for the formation of odd-numbered segments.

rhomboid (rho) A D/V patterning gene required for the development of the neuroectoderm.

short gastrulation (sog) A D/V patterning gene encoding a secreted factor that opposes the action of *dpp* in the neuroectoderm.

snail A D/V patterning gene encoding a transcription factor that represses neuroectodermal gene expression in the mesoderm.

twist A D/V patterning gene encoding a transcription factor which activates mesodermal gene expression in the mesoderm.

primary tissue types such as skin, nervous system, and muscle. Cells in the embryo that collectively give rise to a particular specified region of the larva or adult are often referred to as a primordium for that structure. This process by which cells acquire distinct identities during development is known as patterning.

Our rapid progress in understanding how morphogens initiate patterning in embryos derives principally from genetic studies. The primary goal of this chapter is to link the function of individual genes to the patterning of the A/P and D/V axes. We begin with a summary of early fruit fly development to put the genetic studies in context.

AN OVERVIEW OF EMBRYO DEVELOPMENT IN THE FRUIT FLY

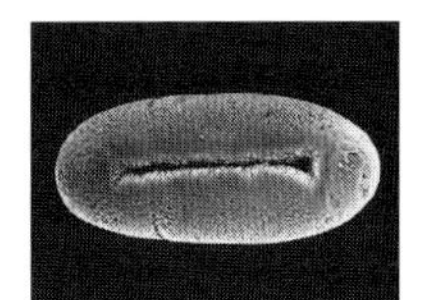

Early gastrulating fruit fly embryo.

The fertilized fruit fly embryo, like all other embryos, starts off with a single diploid nucleus (i.e., a nucleus having two copies of every gene). This diploid nucleus is the product of a fusion of the nuclei from the egg and the sperm, which each supply a single copy of every gene. This initial diploid nucleus undergoes a series of 14 rounds of cell division to generate approximately 5,000 cells, forming what is known as the blastoderm embryo. Cells in a blastoderm embryo are arranged in a single layer, or monolayer, on the surface of the egg and enclose a mass of yolk in the interior. At this stage, the embryo resembles a miniature football with all the cells lying at the surface like a skin. The first 13 divisions take place very rapidly (i.e., every 8–10 minutes). Only during the 14th round of division does the process slow down to the typical embryonic cell division period of approximately 1 hour. It is during this hour-long 14th cell cycle that most of the early pattern formation events discussed below are initiated. During each round of division, the DNA is fully replicated so that all cells have an identical content of DNA. As discussed in Chapter 1, distinctions between cells emerge during development as a result of distinct groups of genes being transcribed from DNA to RNA in different cells. Because all cells in a developing organism contain the same DNA, however, they retain the potential to express the various patterning genes as RNA during this period.

At the end of the 14th cell cycle (about 4 hours after fertilization), cell movement (or morphogenesis) begins. The period of embryonic development during which cells move as groups or migrate as individual cells is referred to as gastrulation. Distinguished developmental biologist Lewis Wolpert once said that "It is not birth, marriage, or death, but gastrulation that is truly the most important time in your life." The point of this unusual philosophical perspective is that formation of a normal-looking and functional individual is determined by complex cellular movements and interactions between cells during gastrulation. This highly choreographed dance places cells that are fated to form a given tissue in the correct position at the appropriate time to receive precisely timed and delivered signals from other cells. The exquisite ballet of gastrulation must be executed very accurately if the ballerina is not to be dropped. If cells are not where they belong when they should be, the embryo falls to an ugly mangled death, and a dead maggot is not a pretty sight!

Although a full description of gastrulation is beyond the scope of this book, there are three major types of cellular movements that should be discussed (see Fig. 3.1). The first important morphogenic event involves the internalization of the ventral-most third of the blastoderm embryo to generate an embryo with two-cell layers (Fig. 3.1, A–F). The internalized cells, referred to as mesoderm, give rise to muscle, heart, and fat. Mesodermal cells move inside the embryo by virtue of a shallow furrow that forms along the ventral midline of the embryo (Fig. 3.1B). This ventral furrow increases in depth as cells fold into a horseshoe shape (Fig. 3.1C,D). This very common form of internalizing cell movement is called invagination. When all of the future mesodermal cells have folded into the ventral furrow, the two lateral sides of the embryo are brought together along the future ventral midline (Fig. 3.1E). Next, mesodermal primordium pinches off from the rest of the embryo to form a closed tube (instead of an open horseshoe), and the outer cells (called ectoderm) fuse along the ventral midline to suture the embryo closed again (Fig. 3.1F). The tube of mesoderm then flattens into a sheet (as when one steps on a spent cardboard toilet paper

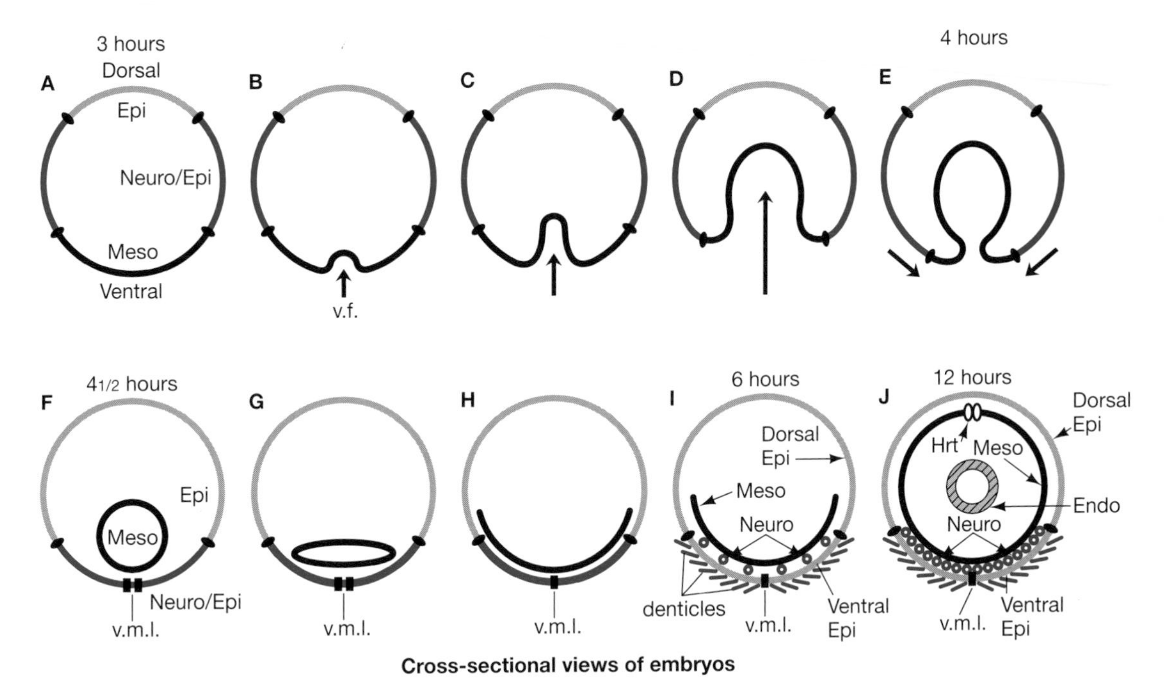

FIGURE 3.1. Gastrulation of fruit fly embryo. Epi = epidermis; Meso = mesoderm; Endo = endoderm; v.m.l. = ventral midline; v.f. = ventral furrow; Neuro = neural precursor cells or neuroblasts.

tube) (Fig. 3.1G). The resulting monolayer of mesoderm cells adheres tightly to the overlying monolayer of ectodermal cells to form a bilayered embryo (Fig. 3.1H).

In the second phase of embryonic tissue stratification, 5–6 hours after fertilization, the nervous system begins to separate from the outer ectoderm (Fig. 3.1I). Specialized neural precursor cells called neuroblasts become enlarged and segregate themselves from the ventral portion of the ectoderm, which comprised the lateral portion of the blastoderm-stage embryo prior to mesoderm invagination (the blue region in Fig. 3.1A). As described in greater detail below, this so-called neuroectoderm generates both neuronal and epidermal cells. Individual neuroblasts break their contacts with neighboring ectodermal cells and squeeze themselves between the ectoderm and mesoderm. This segregation of individual neuroblasts from the ectoderm, referred to as delamination, differs from the coherent invagination of mesoderm cells, which move as a single sheet of cells. Neuroblast delamination creates a three-layered embryo with mesoderm on the inside, neuroblasts in the middle, and skin-forming ectoderm on the outside. The process of neuroblast delamination is complete by approximately 6 1/2 hours after fertilization. Cells in each of these three primordial layers divide two to three more times and then give rise to the stereotyped pattern of muscle, heart, central nervous system, and skin of the fully formed larva (Fig. 3.1J).

Between 6 and 12 hours after fertilization, the inner layer of the embryo comprising the gut primordium (or endoderm = internal cell layer) forms by the separate infolding of cells at the anterior and posterior poles of the embryo. The openings of these two invaginations ultimately become the mouth and anus of the larva, respectively. The invaginating tubes of gut cells issuing from these two orifices grow progressively inward until they meet and join at the center of the embryo to create an uninterrupted tract connecting the mouth to the anus. By 12 hours after fertilization, the embryo is a four-layered structure consisting of endoderm, mesoderm, nervous system, and skin (Fig. 3.1J). By this time, the embryo consists of approximately 50,000 cells, cell division has almost ceased, and cells of the nervous system have sent out processes to contact maturing muscles. Four hours later (i.e., 16 hours after fertilization), the embryo is nearly fully formed and can twitch as a result of functional connections between nerve and muscle.

The final touches on embryonic development, such as the formation of an intricate pattern of hairs called denticles in the tough outer covering of the larvae known as the cuticle, are in place by 20–22 hours after fertilization. At 22 hours after fertilization, the embryo hatches from its egg casing and emerges as a larva or maggot. The larva feeds for a day and then molts to generate a larger larva, which feeds for another day, molts, and gives rise to a ravenous final-stage larva, which grows rapidly to achieve its full size. At the end of larval life (4 days after fertilization), the larva stops moving, attaches itself to a nearby surface, and begins metamorphosis. Metamorphosis, which is described in greater detail in Chapter 4, lasts 4–5 days. During this period, nearly the entire larval body is destroyed and rebuilt according to a completely new blueprint for constructing an adult fly. The fly hatches at the end of metamorphosis and is capable of flying and mating within 1 day. The entire life cycle of the fly takes about 10 days from embryo to sexually mature adult—a pretty fast-paced life even judged by modern-day standards.

THE GREAT DEVELOPMENT MUTANT HUNT: A GENE SAFARI

Results of the comprehensive genetic analysis of fruit fly embryonic development carried out by Christiane Nüsslein-Volhard, Eric Wieschaus, and Gerd Jürgens were first published in summary form in 1980, and then as a full report in 1984. These classic papers, which served as the basis for the Nobel Prize being awarded to the co-heads of the project, are landmarks because they described the identification of mutations in nearly all genes involved in establishing the embryonic body plan. In this section, this mutant hunt (or screen, as geneticists call it) and the question of how it was possible to mutate nearly all genes involved in early embryonic development are discussed.

■ First, a Refresher on Mutations ■

Recall from Chapter 1 that a mutation is a change in the sequence of DNA bases within a gene which reduces or alters the activity of that gene (revisit Fig. 1.6). The altered DNA of the mutated gene leads to the synthesis of a nonfunctional or aberrant protein. The defective protein may cause a specific developmental abnormality or may lead to more general cellular defects such as disruption of energy production. An organism having a mutated gene that causes some type of identifiable defect is called a mutant. The goal of a search or "screen" for patterning mutants is to generate a collection of many random mutants and then pick through them to identify those having recognizable patterning defects. As discussed in Chapter 1, a typical gene is a string of about 1,000 DNA bases, which are represented as **A**, **C**, **G**, and **T**. A mutation generally is a single change in one of the 1,000 bases of a gene (e.g., base #947 of the *dpp* gene, normally a **G**, is changed to an **A**). Because many different base changes can disrupt the function of a gene, two independently generated mutations affecting a given gene almost always disrupt that gene in different positions (i.e., one mutation in the *dpp* gene is a **G** to **A** change in base #947 and another *dpp* mutation is a **C** to **T** change in base #481).

■ Two Similar Mutants May Affect the Same Gene or Different Genes ■

Most mutations reduce or abolish the function of a gene. Generally, mutations have no effect if the organism has one good copy of the gene. An individual having one mutated copy (*m*) of a given gene and one good copy (+) of that gene is called a carrier. In genetic shorthand, such an individual is often denoted as *m*/+. A carrier appears normal because one functional copy of a typical gene is all that is needed. The second copy of the gene serves as a backup. An *m*/+ parent passes on one of its two genes (i.e., either the *m* or + version of the gene) to its offspring. The mutated gene, therefore, is inherited by half of its progeny. When a male carrier for a given mutation (i.e., *m*/+) mates with a female carrying that same mutation (i.e., *m*/+), on average one quarter of their progeny will inherit the mutated gene from both of their parents (i.e., $1/2 \times 1/2 = 1/4$ *m/m* progeny). These *m/m* individuals are mutant, since they lack a normal copy of the gene. Many human genetic disorders such as phenylketonuria (PKU) or Tay Sachs disease are propagated by carriers. Thus, two parents who are carriers for Tay Sachs (i.e., +/*ts*) have a 25% chance during each pregnancy of producing a child inheriting the mutant gene from both of them (i.e., *ts/ts*). Such an unfortunate child will be born with Tay Sachs disease.

One-quarter of the offspring from two carrier parents will be affected by a genetic disorder:

Two types of mother's eggs (+ or *ts*)

Two types of father's sperm (+ or *ts*)

	+	*ts*
+	+/+	*ts*/+
ts	+/*ts*	*ts/ts*

+/+ = 1/4 offspring] normal
+/*ts* or *ts*/+ = 1/2 offspring] normal
ts/ts = 1/4 offspring] mutant

If two independently generated mutants exhibit similar defects, it is possible they disrupt the same gene (Fig. 3.2). Although independently generated mutations are unlikely to change the same base in a gene, they may disrupt the function of that gene to similar extents (e.g., you can cut a string in many places to sever it). Alternatively, the two mutations may affect two different genes functioning in a common process. For example, suppose there are two genes involved in cell–cell signaling, where gene #1 codes for a protein signal that is received by a receptor coded for by gene #2 (Fig. 3.2A). Mutations crippling either gene #1 or gene #2 would have similar consequences because they both disrupt response to the signal. It is easy to distinguish between these two possibilities by crossing the two mutants to each other. If the progeny of two mutants have the same defect as observed in their parents, one can conclude that the parents carry two different mutations in the same gene. Consider such a case in which two mutations disrupt the function of gene #1. These two distinct mutations in gene #1 could be represented by *m1* and *m*1* (Fig. 3.2B). When an *m1/m1* mutant parent is mated to an *m*/m*1* mutant parent, the progeny will inherit the *m1* mutant gene from the first parent and the *m*1* mutation from the other parent. Such progeny, which are designated *m1/m*1,* have no functional copies of gene #1 and consequently, like their parents, will be unable to produce the signal. On the other hand, if the progeny of two similar mutant parents are normal, the two mutations must affect two different genes (Fig. 3.2C). In the case of our example, the first parent could be mutant in gene #1 (*m1*) but be normal in gene #2 (+). Such a parent, which cannot produce a signal, but could receive it, is denoted as *m1/m1* and +/+. The second parent might be normal in gene #1 (+) but be mutant in gene #2 (*m2*). This parent, which could produce a signal but would be unable to receive it, is denoted as +/+ and *m2/m2.* Progeny of these two mutant parents, which are denoted *m1*/+ and *m2*/+, have one functional copy of gene #1 and one functional copy of gene #2. As these normal-looking progeny have one functional copy of both genes, they are only carriers for the *m1* and *m2* mutations. Because cells in these carriers can both produce the signal and receive it, cells can communicate and respond normally.

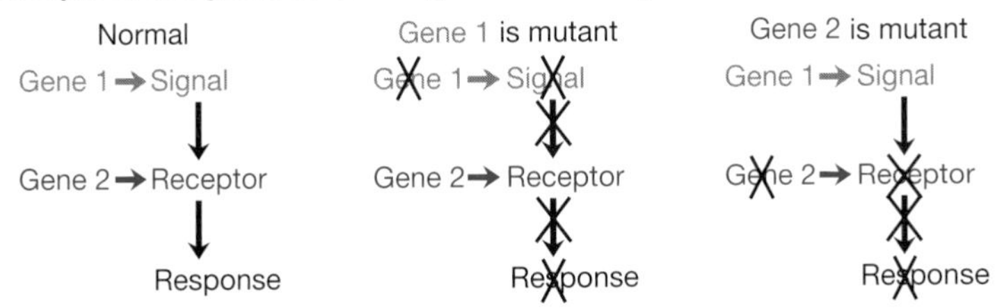

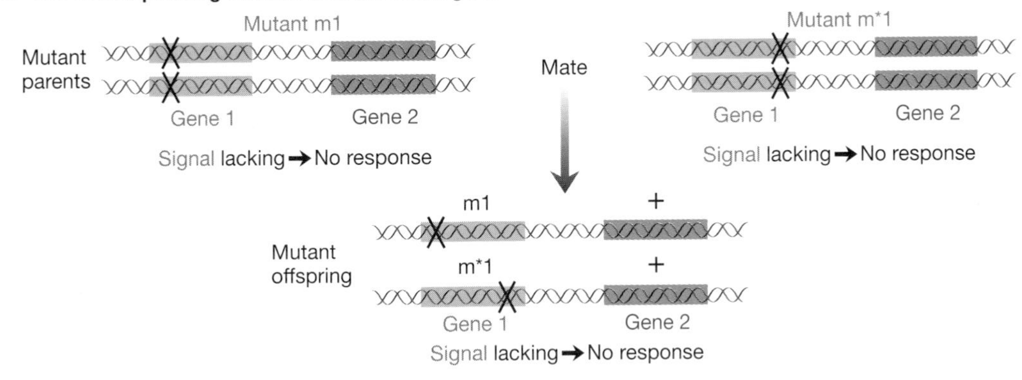

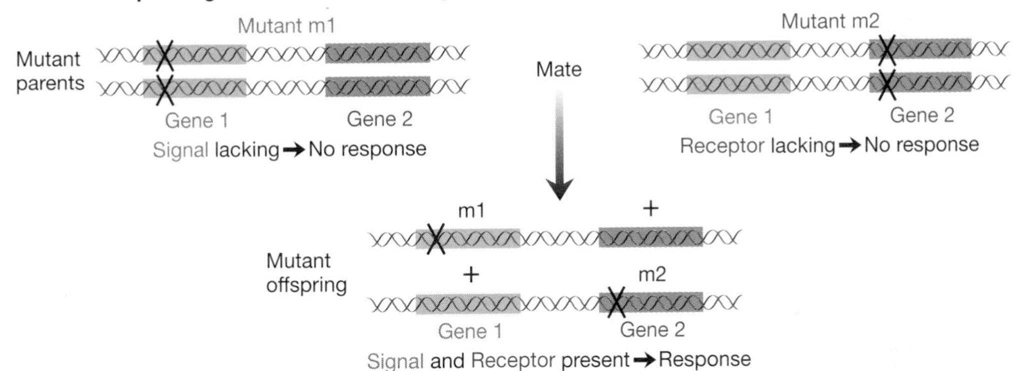

FIGURE 3.2. Two similar mutants may disrupt the same or different genes. (*A*) Two genes functioning in a common process. (*B*) Two nonresponding mutations in the same gene. (*C*) Two nonresponding mutations in different genes.

How Geneticists Stalk Genes

Geneticists approach biological problems from a somewhat odd angle. Instead of physically dissecting a biological specimen as an anatomist does to see how the organism is constructed, or separating a cell into molecular components as a biochemist would be prone to do, a geneticist dissects a biological process by asking, What can I do to disrupt the system? They address this question by first generating mutations in genes that control the process of interest. They then deduce the function of a particular gene by examining the nature of defects resulting from the absence of that gene activity. This strategy is similar to that used by electricians or mechanics when trying to diagnose a problem with a malfunctioning piece of electronics or a broken-down car.

The goal of the massive mutant screen carried out by the Nüsslein-Volhard, Wieschaus, and Jürgens "gene team" was to generate a large collection of random mutants (about 30,000 in all). Each of these mutants was analyzed to determine whether mutant embryos exhibited morphological defects along the A/P or D/V axes. The screen identified about 750 mutants that had interpretable patterning defects. The investigators then asked how many different genes had been disrupted in this collection of 750 mutants and found that about 150 genes had been mutated. They reached this conclusion by crossing mutants with similar defects to each other and asking whether the progeny had defects similar to their parents. As discussed above, such crosses can resolve whether two mutants with similar appearances have disruptions within the same gene or within two different genes involved in the same biological process. The ratio of 5 mutants per patterning gene obtained in the screen meant that each of the 150 patterning genes had been mutated 5 times on average (i.e., 750 patterning mutants divided by 150 patterning genes = 5 mutants per gene).

One can think of a gene as a target in a mutant hunt. The goal of a comprehensive mutant hunt is to hit every target at least once. If you fire enough rounds at the fly's DNA to hit an average target gene five times, you will have missed very few genes entirely. Because only 150 genes were found that could be mutated to give defects in embryonic pattern formation, their collection of 750 patterning mutants contained an average of five different lesions in each patterning gene. If the gene team had doubled their heroic efforts by screening 60,000 initial mutants instead a mere 30,000, they would have recovered twice as many total patterning mutants (i.e., 1,500 versus 750). However, these additional mutants would have included few additional disrupted genes beyond the 150 genes identified in the first group of 750 mutants. Thus, in this larger screen they would have ended up with an average of 10 distinct lesions per gene instead of 5. Because extending a mutant hunt beyond the point of recovering 5 independent mutations in an average gene predominantly generates more hits in the same small set of genes, the screen is said to be saturating. The degree of saturation can be quantified statistically. When you have examined a sufficient number of mutants to have identified an average of 5 mutations in each gene within your collection, you will have found more than 95% of all genes that could be identified using screening methods. In other words, the odds are better than 20 to 1 that any given gene important for pattern formation will have been hit at least once in such a large collection of mutants.

Trophies of the Hunt: Mutants Fall into Clear Categories

Through arduous but standard genetic methods, the gene team generated over 30,000 independent lines of flies that each carried an average of one mutation in an essential gene (an essential gene is one required for the survival of the fly). When a mutagenic chemical (or mutagen) is fed to adult flies, they pass on mutant genes to their offspring. The dose of the mutagen determines the average number of mutations that a parent passes on to its progeny. It is possible to adjust the mutagen dose so that on average one mutated essential gene will be inherited by the progeny of a chemically treated parent. There are approximately 6,000 essential genes in the fruit fly. Therefore, in a collection of 30,000 independent mutants, each carrying a mutation in one essential gene, a typical gene is hit five times on average (i.e., 30,000/6,000 = 5). The investigators established strains of flies in which both males and females were carriers for the same mutation. It took more than 3 years to generate and analyze these 30,000 independent strains of flies.

As a first step in analyzing the large number of mutants they generated, the investigators collected eggs from each strain of mutant flies and determined whether one-quarter of the eggs failed to hatch (i.e., were dead). Those strains having one-quarter dead embryos, the *m/m* class of embryos, were assumed to be mutant in a gene required for embryonic survival. Approximately 25% of all mutant strains tested (i.e., ~8,000 mutants) were lacking the function of a gene essential for embryonic survival. These mutants, referred to as embryonic mutants, were saved for further analysis. Most of the remaining 22,000 mutants disrupted genes that were not essential for embryonic development (i.e., *m/m* individuals hatched and survived as larvae), but were required in subsequent developmental stages such as metamorphosis.

Embryos were collected from each of the 8,000 embryonic mutants and examined under a microscope to determine whether they were morphologically normal or exhibited consistent patterning defects of some kind. Screening through such a large number of mutant lines required a rapid method for examining embryos. A simple technical innovation permitted the gene team to identify nearly all mutants with defects in A/P or D/V axis formation. This was accomplished by treating the unhatched mutant embryos (*m/m* individuals) with harsh dissolving agents that digested everything but the hard outer covering of the embryo called the cuticle. The cuticle, which is similar in texture to fingernails, provides an imprint of the underlying embryonic skin. Specialized structures such as hairs and distinctive sensory organs form in stereotyped positions of the cuticle. For example, the segmented nature of the embryo along the A/P axis is readily apparent from the reiterated arrays of hairs known as denticles, which are arranged in a trapezoidal configuration in the ventral anterior half of

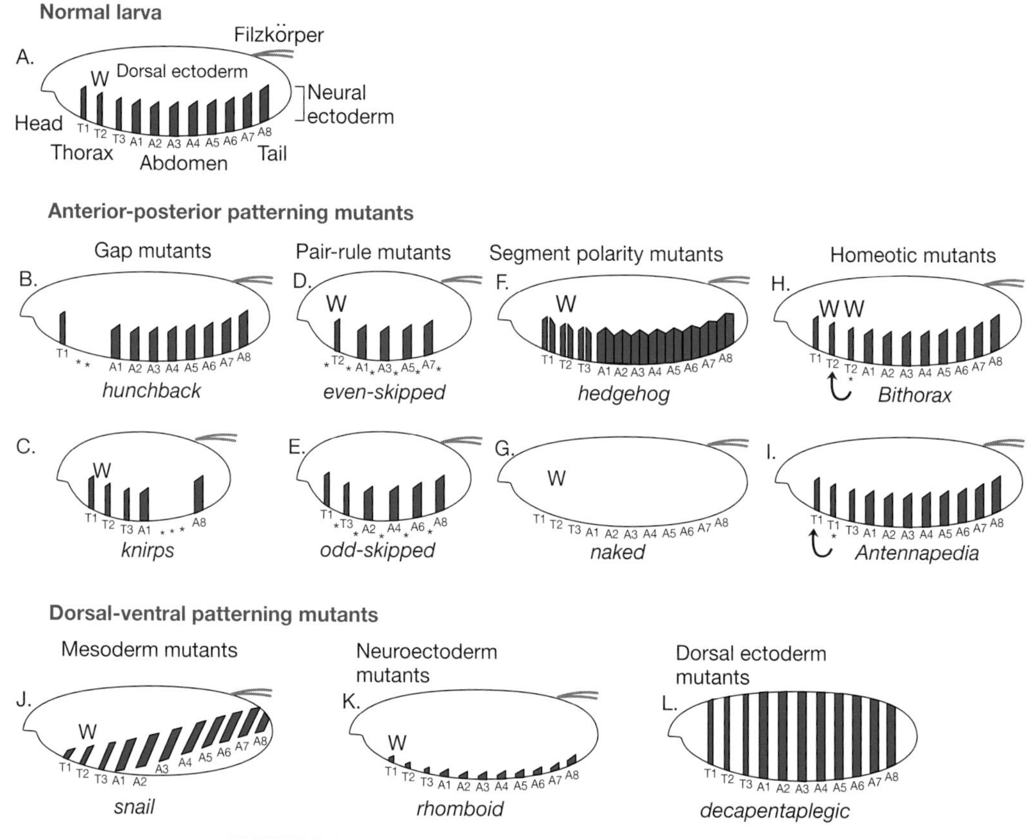

FIGURE 3.3. Patterning mutants in *Drosophila*.

each segment (drawn schematically in Fig. 3.3A; see also Plate 11). Because denticles are restricted to the ventral region of the cuticle, they also distinguish ventral from dorsal positions. Other cuticular structures known as Filzkörper are found only in dorsal regions (Fig. 3.3A). The easy preparation of cuticles from unhatched mutant embryos and their rapid examination made it possible for these three determined scientists to examine embryos collected from each of the 8,000 embryonic mutant strains. As mentioned above, the mutant screen ultimately identified approximately 750 A/P and D/V patterning mutants, corresponding to 150 different genes, that cause interpretable external morphological aberrations. As it turns out, several mutations that lead to gross malformation of internal tissues, such as muscle, were also identified in this screen since such defects generate misshapen embryos and correspondingly twisted cuticles.

Christiane Nüsslein-Volhard (1942–)

One of only eight women to receive a Nobel Prize, Nüsslein-Volhard's monumental experiments over several years creating over 30,000 strains of fly mutants and then screening them for defects in embryonic development of the fruit fly laid the foundation on which other geneticists could build to find similar genes in more complex organisms.

Christiane Nüsslein-Volhard was born in Magdeburg, Germany. She grew up with a great deal of freedom in a cultured and closely knit family, and by the age of 12, she had made up her mind to become a biologist. She entered Frankfurt University and, after finding the biology curriculum limited there, moved to Tübingen, where she entered a program in biochemistry. After completing her Diploma in 1969, she joined the laboratory of Heinz Schaller at the Max-Planck-Institute as a graduate student. There were many interesting people in Tübingen at that time, including Hans Meinhard, who was developing theoretical equations for how gradients might form and lead to subdivision of embryos into segments. For her thesis, Nüsslein-Volhard worked on a molecular biology project in which she purified and studied the DNA-binding properties of a viral RNA polymerase, which was quite a feat since, at that time, DNA sequencing methods had not yet been developed.

After giving a great deal of consideration to what to do next, Nüsslein-Volhard moved to Basel, Switzerland in 1975 to join the laboratory of Walter Gehring as a postdoctoral fellow. One of the most important people she met in the Gehring lab was Eric Wieschaus, who had just finished his Ph.D. thesis. She became friends with Eric, and although he soon left to go to Zurich for a postdoc, she maintained regular contact with him over the next several years. In Gehring's lab, she began studying the few embryonic mutants known at that time. She also began developing the methods necessary to conduct a large-scale mutant hunt, such as the quick method they used to collect cuticles from multiple mutant lines and a simple technique to clear them so that it was possible to see detailed features of the cuticle pattern along the anterior and posterior axes (the segmentally repeated pattern of ventral denticles). She tried to understand the underlying developmental problem in an embryonic mutant that had two tails instead of a head and tail, but became frustrated by the high variability of the defects. She also conducted the first small-scale screen for patterning mutants in which she found the critical D/V patterning mutant *dorsal*.

In 1977, Nüsslein-Volhard moved to the laboratory of the renowned embryologist Klaus Sander, who had performed various physical manipulations on embryos showing that they contained polarizing factors which could be reshuffled (e.g., by centrifugation) to repolarize the embryo. There, in collaboration with Margit Schardin, she created a fate map of the blastoderm-stage fly embryo (a map of where cells in the blastoderm embryo would end up in the hatched larva). These experiments proved to be very important for subsequent molecular analysis of patterning genes, since this detailed map made it possible to tell what larval structure would be generated by sets of blastoderm cells that expressed a particular patterning gene.

Nüsslein-Volhard's next move was perhaps the most critical in her career as she joined up with Eric Wieschaus to share a laboratory as independent investigators. Gary Struhl, then a graduate student with Peter Lawrence, showed them the segmental patterning defects in embryos from Antennapedia mutants, which encouraged them soon after to initiate the first major component of their comprehensive screen for embryonic patterning mutants. They worked together closely to invent additional clever tricks to speed up the screening process and, with the help of two technicians, they screened through 4,200 mutants mapping to the second chromosome (the second chromosome corresponds to approximately 40% of the fly genome) in 3 short months. They immediately realized that their collection contained distinct well defined classes of mutants (gap genes, pair-rule genes, and segment polarity genes) and that the genes disrupted in these mutants were likely to function hierarchically. The results from this first component of their collaborative effort were written up and published in a paper in 1980, which ultimately became the basis for awarding the Nobel Prize to Nüsslein-Volhard and Wieschaus in 1995. With Gerd Jürgens, patterning mutants mapping to the remaining chromosomes were generated and mapped. This work was written up in three classic papers in 1984. The modern era of developmental biology had been launched.

Soon after the first component of their screen was completed, Eric Wieschaus moved to take a faculty position at Princeton and Nüsslein-Volhard took a position at the Max-Planck-Institute in Tübingen, where she now heads her own division of the Institute. She and members of her laboratory went on to do additional screens to identify maternally acting genes required for embryonic patterning (e.g., bicoid) and performed many critical experiments such as cytoplasmic transplantation experiments to identify mutants likely to encode morphogens. Transplantation experiments of this kind suggested that *bicoid* was likely to encode a morphogen produced in the anterior end of the embryo since cytoplasm taken from the anterior end of normal embryos could rescue the defects of *bicoid* mutant embryos, whereas cytoplasm derived from the posterior end of the embryo could not. Her group then cloned the *bicoid* gene and demonstrated that Bicoid was indeed an anteriorly restricted morphogen.

After discovering and then characterizing genes controlling early axis formation in fly embryos, Nüsslein-Volhard's laboratory began a similar screen on a much greater logistical scale for patterning mutants in the vertebrate zebrafish. The results of this tour de force screen in conjunction with that of a parallel screen done in the laboratory of Wolfgang Driever (a former graduate student of Nüsslein-Volhard) were published in an entire volume of the journal *Development* in 1995. In this remarkable series of papers, many mutants affecting the formation of organs and tissue types were described, as well as those affecting axis formation (e.g., the zebrafish *chordin* gene). Her group continues to lead the field of zebrafish development, and this work has spawned yet another generation of bright young investigators.

Eric Wieschaus (1947–)

Eric Wieschaus shared the 1995 Nobel Prize in medicine with Christiane Nüsslein-Volhard and Ed Lewis in recognition of his work on the great genetic screen of fruit fly developmental mutants. However, he did not foresee a career in science when he was young. Rather, he spent much of his time drawing, painting, reading, and playing the piano. He remembers, "I dreamed of becoming an artist when I grew up."

Wieschaus was born in South Bend, Indiana, and then moved with his family to Birmingham, Alabama, when he was six. Following his junior year in high school, Wieschaus enrolled in a summer science program sponsored by the National Science Foundation in Lawrence, Kansas. He greatly enjoyed the experience as well as the intellectual company of other students in this program. This experience convinced Wieschaus that he wanted to become a scientist, and when he entered college at Notre Dame that next fall he was sure that he wanted to major in biology.

In his sophomore year at Notre Dame, Wieschaus, in need of money, took a job making fly food (a congealed combination of gelatin and molasses) in the laboratory of Harvey Bender where he learned the basics of fly genetics. Although he enjoyed his work with flies, he much preferred the embryology course he was taking. He vividly recalls the blitz of questions that poured into his mind. "I will never forget the thrill of seeing cleavage and gastrulation for the first time in living frog embryos. I immediately wanted to understand why cells in particular regions of the developing embryo behaved the way they did. What were the mechanisms that made them different from each other? What forces drove such dramatic rearrangements in the cytoplasm and the shape of cells?"

Wieschaus went to Yale for his graduate work, where he first joined the laboratory of Donald Poulson, from whom he learned about embryonic mutants such as Notch. He then switched labs to work with Walter Gehring, then at Yale, so he could learn techniques to study living embryos. Gehring's lab was very small at this time and Wieschaus was his only student. Wieschaus learned a great deal about doing experiments from working directly with Gehring during this formative period. For his thesis work, he developed a version of mutant clone analysis to determine what adult tissue cells in particular positions in the embryo could give rise to. This method was similar to that pioneered by García-Bellido but was applied to an earlier stage of development (i.e., early embryos versus larvae). From these experiments he learned that cells in the embryo did not give rise to progeny that moved to another segment, but they could generate cells that would contribute to both a leg and a wing. This meant that segmental borders are formed before distinctions between appendages are made.

As Wieschaus was finishing up his experiments in Gehring's lab (now moved to Basel, Switzerland), Christiane Nüsslein-Volhard joined the lab. They rapidly become friends and kept in regular contact after Wieschaus moved to Zurich to work with Rolf Nothiger, where he continued to work on methods for mapping the fate of cells. Nüsslein-Volhard was also working on making such maps in Gehring's lab, and their combined studies resulted in a detailed map of where cells in the blastoderm embryo end up in the fully formed larva. This map turned out to be very valuable for their subsequent analysis of patterning mutants because this information made it possible to establish the relationship between patterns of gene expression and defects in mutants lacking the function of genes. Wieschaus conducted some of these experiments with Trudi Schüpbach, which proved to be a particularly successful collaboration as it culminated in marriage.

In 1978, Wieschaus took a position as an independent investigator at the European Molecular Biology Lab (EMBL) in Heidelberg. One reason he took this position was that Nüsslein-Volhard had also been offered a position at the EMBL. They both looked forward to having an opportunity to work together on projects they had discussed at length over dinners they had together in Basel. Wieschaus recalls "a broad consensus that understanding embryonic development would require identifying the relevant genes, but a great uncertainty about whether this was actually feasible in a multicellular organism as complex as a fly. The major worry was that the number of genes could be very large, and that the phenotypes would be so pleiotropic (i.e., affect so many different structures) that it would be impossible to come up with simple pathways or models for function. Our decision to attempt saturation mutagenesis depended on genetic techniques developed in *Drosophila* over the preceding 50 years. Two experiments in the 1970s were particularly important: Judd's genetic analysis of the white region, which told us that the number of vital (i.e., essential) genes was finite, probably less than 5,000, and the Lindsley and Sandler synthetic deficiency experiments, which told us that most loss of function mutations would be recessive. Those two observations argued that saturation mutagenesis was feasible; without that knowledge we might not have attempted the screen."

Undaunted by potential pitfalls, Wieschaus and Nüsslein-Volhard soon began what was to be the great collaborative effort of their careers in which they systematically screened for mutants in all genes that were required to establish the primary body axes of the embryo. Wieschaus remembers this as the most intellectually exciting period of his scientific career and that "almost every day we could expect to encounter a new phenotype (developmental defect) that would force us to re-evaluate some long held assumption about embryonic development. Both Janni and I had read the Turing-type models of Gierer and Meinhardt and the experimental embryology of Klaus Sander. Initially we tried to think about *Drosophila* and our results in those terms. I was also heavily influenced by the compartmental model of Antonio García-Bellido, at the time the best and most thoroughly thought out hierarchical models for development. Ultimately none of these models really worked for our data, although parts of them clearly found their way into our understanding of the phenotypes."

After completing the first component of the comprehensive genetic screen in collaboration with Nüsslein-Volhard, Wieschaus moved to Princeton University in 1981, where he has remained ever since. His laboratory has made several important discoveries about genes functioning in the early embryo to pattern the A/P axis and has identified mechanisms by which cells signal to one another during D/V patterning of the wing. His own interest, however, continues to be in understanding how cells change shape to move and thereby result in the wondrous dance of cells during gastrulation.

Unlike most established investigators or Nobel laureates (Ed Lewis excepted), Wieschaus remains committed to doing his own experiments. He describes himself as "a hands on person—what is important to me are the individual experiments I do in the lab. These have always brought me more pleasure than any of the big ideas or the final refined understanding." He characterizes his scientific approach as being driven by two different but not opposing currents, namely "...a strong almost aggressive desire for logical structure and an attraction to problems that have a strong visual component. As a child I was good in math, but wanted to be an artist. In a certain sense, those two aspects of my personality have found some balance in my chosen research area, developmental genetics. Because embryos are beautiful and because cells do remarkable things, I still go into the lab every day with great enthusiasm." He also notes introspectively that "I do experiments on questions that interest me. These are not necessarily the ones that are recognized as the most important by the scientific community at a given time. I often work alone. This gives me freedom but it means that I don't benefit from constant intellectual feedback and criticism characteristic of most competitive fields." In pointing to the key elements of discovery, Wieschaus acknowledges the need for hard work as well as keeping an open mind "Hard work... Accuracy at a level relevant to the particular question being asked.... Willingness to push an experiment slightly farther than initially intended in its design.... Willingness to re-think what the results are actually telling you...."

PATTERNING GENES FUNCTION HIERARCHICALLY

The most important insight derived from the gene screen was that many of the patterning mutants fall into a small number of obvious subgroups based on shared sets of defects. Along the A/P axis, four general groups of mutants affecting the segmental organization of the embryo (segmentation mutants) can be distinguished, which are named for their characteristic defects. *Gap* mutants exhibit large gaps in the cuticle in one restricted region of the A/P axis, but are largely normal elsewhere; *pair-rule* mutants lack cuticle in every other segment; and *segment-polarity* mutants exhibit defects such as deletions and duplications of cuticular structure within every segment. Finally, in *homeotic* mutants, the identities of specific segments are transformed into those of adjacent segments.

■ Naming Genes in the Fruit Fly ■

Genes in fruit flies are generally named after the type of defect observed in mutants lacking the function of that gene. For example, the first fruit fly mutant discovered by the legendary Thomas Hunt Morgan was named *white* because flies lacking function of the *white* gene have white instead of red eyes (the normal eye color for fruit flies). Similarly, patterning mutants were typically named for the types of defects observed in larval cuticles. Thus, gap gene mutants that have small abnormally shaped cuticles were sensitively named *hunchback, Krüppel* (cripple in German), and *knirps* (dwarf in German). Similarly, pair-rule mutants such as *even-skipped* or *odd-skipped* were named on the basis of which segments were missing in these mutants, and names of segment-polarity mutants such as *hedgehog* or *naked* derived from the appearance of cuticles in embryos deficient for these genes.

Analysis of A/P Patterning Mutants Suggests That Segmentation Genes Act Hierarchically

The nature of A/P patterning defects defining the four major groups of segmentation mutants (i.e., gap, pair-rule, segment polarity, and homeotic) suggested to the gene team that patterning genes act in sequence during development. They reasoned that mutants exhibiting the broadest defects lacked the activity of genes functioning earliest in development and that those with more restricted defects lacked the function of genes acting during later stages. Thus, gap genes should act first because mutants lacking function of these genes exhibit defects across several contiguous segments. Among the five gap mutants, *hunchback* mutants lack cuticle derived from the anterior-most region of the embryo (Fig. 3.3B), *Krüppel* mutants lack cuticle in the thoracic region, and *knirps* mutants lack posterior cuticular structures (Fig. 3.3C, see also Plate 1J for actual data). Pair-rule mutants, which were predicted to function next in the patterning hierarchy, lack units one segment wide (i.e., every other segment is missing). Within the group of eight pair-rule mutants, *even-skipped* mutants lack even-numbered segments (Fig. 3.3D, see also Plate 1K for actual data) and *odd-skipped* mutants lack odd-numbered segments (Fig. 3.3E, see also plate 1L for actual data). Segment-polarity and homeotic mutants (discussed in next section), which have defects in specific subsets of structures within segments, were proposed to function last in the cascade of gene action. The *engrailed, hedgehog,* and *naked* mutants are examples of the more than 20 segment-polarity mutants. In *engrailed* and *hedgehog* mutants, the posterior half of each segment is transformed into a mirror symmetric copy of the anterior portion of the segment. As mentioned previously, denticles are normally restricted to the anterior-ventral region of each segment (Fig. 3.3A). The posterior-to-anterior segmental transformation associated with *engrailed* and *hedgehog* mutants results in cuticles having a solid pattern of denticles resembling a prickly hedgehog (Fig. 3.3F, see also Plate 1M for actual data). Conversely, in *naked* mutants, the anterior region of each segment develops as a mirror duplication of the denticle-free posterior portion of the segment, leading to a cuticle devoid of denticles (Fig. 3.3G). As shown below, the simple hierarchical ordering of gene action inferred from mutant appearance has been borne out remarkably well by subsequent detailed molecular studies.

Homeotic Genes Determine Segment Identity

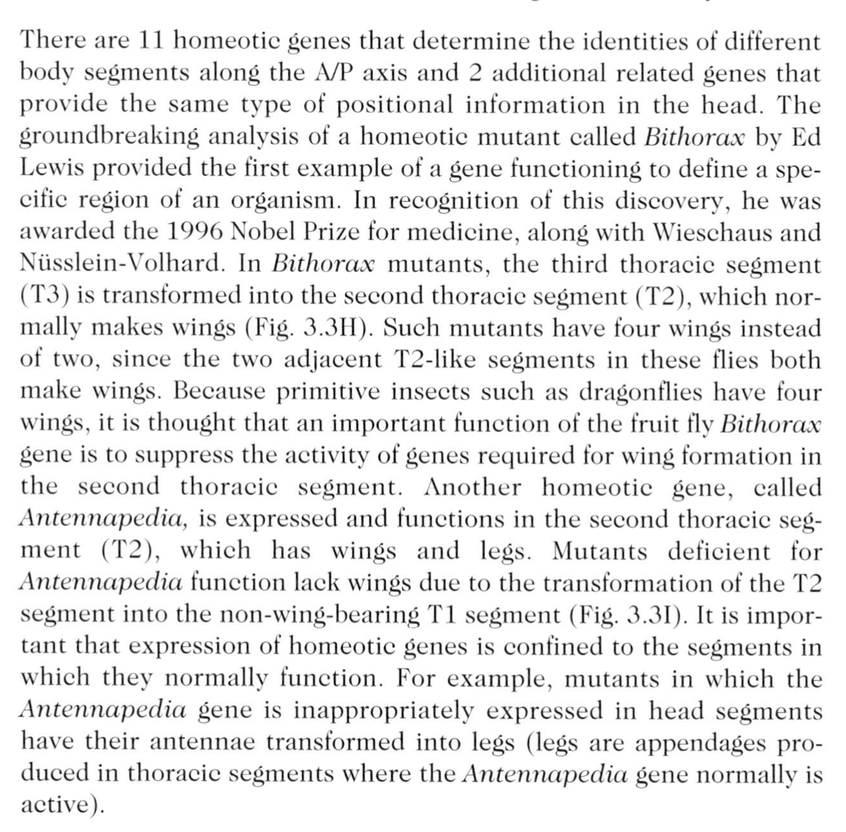

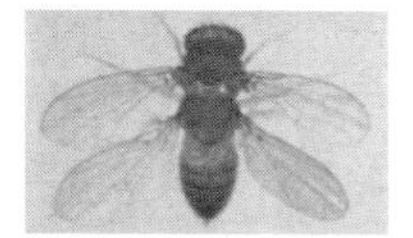

Bithorax *mutant*

There are 11 homeotic genes that determine the identities of different body segments along the A/P axis and 2 additional related genes that provide the same type of positional information in the head. The groundbreaking analysis of a homeotic mutant called *Bithorax* by Ed Lewis provided the first example of a gene functioning to define a specific region of an organism. In recognition of this discovery, he was awarded the 1996 Nobel Prize for medicine, along with Wieschaus and Nüsslein-Volhard. In *Bithorax* mutants, the third thoracic segment (T3) is transformed into the second thoracic segment (T2), which normally makes wings (Fig. 3.3H). Such mutants have four wings instead of two, since the two adjacent T2-like segments in these flies both make wings. Because primitive insects such as dragonflies have four wings, it is thought that an important function of the fruit fly *Bithorax* gene is to suppress the activity of genes required for wing formation in the second thoracic segment. Another homeotic gene, called *Antennapedia,* is expressed and functions in the second thoracic segment (T2), which has wings and legs. Mutants deficient for *Antennapedia* function lack wings due to the transformation of the T2 segment into the non-wing-bearing T1 segment (Fig. 3.3I). It is important that expression of homeotic genes is confined to the segments in which they normally function. For example, mutants in which the *Antennapedia* gene is inappropriately expressed in head segments have their antennae transformed into legs (legs are appendages produced in thoracic segments where the *Antennapedia* gene normally is active).

Edward B. *Lewis* (1918–)

In 1933, the primary founder of the field of fly genetics, Thomas Hunt Morgan, received the Nobel Prize in Medicine "for his discoveries concerning the role played by the chromosome in heredity." Morgan was unable to attend the Nobel ceremony in Stockholm, Sweden. In his place, F. Henschen from the Royal Caroline Institute wrote an eloquent presentation speech in which he said of Morgan's work, "Who could dream some ten years ago that science would be able to penetrate the problems of heredity in that way, and find the mechanism that lies behind the crossing results of plants and animals; that it would be possible to localize in these chromosomes, which are so small that they must be measured by the millesimal millimetre, hundreds of hereditary factors, which we must imagine as corresponding to infinitesimal corpuscular elements (e.g., genes). And this localization Morgan had found in a statistic way! A German scientist has appropriately compared this to the astronomical calculation of celestial bodies still unseen but later on found by the tube—but he adds: Morgan's predictions exceed this by far, because they mean something principally new, something that has not been observed before." These same words of praise could be equally well applied to the scientific grandson of Morgan, Ed Lewis, for his brilliant analysis of homeotic mutants and his remarkable prediction that homeotic genes would be members of a gene family that arose through a series of gene-duplication events during the course of evolution.

Ed Lewis was born in Wilkes-Barre, Pennsylvania. He graduated from the University of Minnesota with a B.A. in biostatistics and entered graduate school at Caltech in 1939 just as World War II broke out. He quickly made important genetic observations known as the *cis-trans* effect, which he wrote up for his Ph.D. thesis in 1942, just in time to enlist in the armed forces. Following the war, Lewis accepted a position at Caltech as an instructor. He then went to Cambridge University as a Rockefeller Foundation Fellow for a year and returned to Caltech as a faculty member, where he has remained ever since.

As a graduate student with one of the founding pioneers of fly genetics, Allan H. Sturtevant, Lewis had complete freedom to explore a novel genetic phenomenon in fruit flies called the *cis-trans* effect. Lewis's primary observation was that two mutations, *a* and *b*, had no effect when combined onto a single chromosome (i.e., mutations carried in the form *a b*/++ resulted in normal flies), but caused very severe defects when the mutations were present on opposite chromosomes (i.e., the mutations were carried in the form *a* +/+ *b*). From today's vantage point, the most likely explanation for this unusual genetic phenomenon is that the *a* and *b* mutations disrupted the same gene, but in different positions (see Chapter 2). Although very difficult, it was possible for Lewis to put these two distinct mutations together so that the same gene was simultaneously disrupted in two different places. Animals carrying this doubly disrupted gene and one good copy of the gene on the other chromosome (i.e., *a b*/+ individuals) were normal, as is typical for a mutant carrier (see Chapter 3), whereas flies having one of the mutations (*a*) on one chromosome and the second mutation (*b*) on the other chromosome had no good copies of the gene and therefore exhibited mutant defects. Lewis remembers this discovery as the most exciting in his career: "It required looking at tens of thousands of flies to get *a* and *b* on the same chromosome and predicting that it would be different from *a* +/+ *b*, and ran contrary to all genetic theory at the time."

In describing his most famous work analyzing homeotic genes, Lewis recalls, "I was testing the idea that new genes arise by tandem duplication of a gene followed by one of the duplicates diverging to carry out a new BUT RELATED function. And that ultimately is what the HOX complex turns out to be, but it took molecular genetics to establish that they must have come from a common ancestor by repeated tandem duplication." We now know, in addition, based on the subsequent molecular analyses of these genes (see bioboxes on McGinnis, Levine, and Scott), that this duplication of homeotic genes occurred before the split of vertebrate and invertebrate lineages and thus that segment identities are defined by the same fundamental mechanisms in all segmented animals.

Lewis isolated and studied many mutations in the Bithorax gene and closely related homeotic genes. He, much like a contemporary Charles Darwin, devoted incredible undistracted focus to his work over many years before publishing in 1978 (at the age of 60!) his most important paper where he put forth his ideas on how different homeotic genes acted in a spatially restricted fashion (e.g., only in particular segments) to define segment identity. Because Lewis has not strayed from his singular passion to understand the function of homeotic genes, it is not surprising that he sees focus as a key ingredient to his scientific success in commenting "INTENSE depth of focus is the only way in my opinion, but one can then turn to other problems and use that approach." Regarding the general elements for discovery he continues that what is most important is "having a testable hypothesis even if wrong; usually persistence is required; usually it has to be the right time, and one has to be in the right place where there is an atmosphere or freedom to pursue new ideas and methods. Of course, sometimes it is simply serendipity, but that obviously is not what a scientist should count on."

Ed Lewis is still performing experiments at Caltech. In addition to receiving the 1995 Nobel Prize in medicine for his pioneering work on homeotic genes, Lewis has won many other awards and distinctions, including the National Medal of Science (USA, 1990) and the Albert Lasker Basic Medical Research Award (1991).

Many Patterning Genes Encode Transcription Factors

Nearly all gap genes, pair-rule genes, and homeotic genes encode transcription factors and are expressed in spatially and temporally restricted patterns. That is to say, transcription of the patterning gene from DNA into RNA occurs only in cells in certain positions in the embryo and not in others. The fact that many key patterning genes encode transcription factors is easy to rationalize given that these genes function by controlling the expression of other genes. Gap genes are expressed in single broad domains of cells, pair-rule genes are expressed in stripes of cells having a periodicity of every other segment, and homeotic genes are expressed in single stripes corresponding to cells in the segments they define (Fig. 3.4; see also Plate 1B, C for actual data). In general, there is an excellent match between the region in which a patterning gene is expressed as RNA and the location of defects in mutants lacking the function of that patterning gene. Some

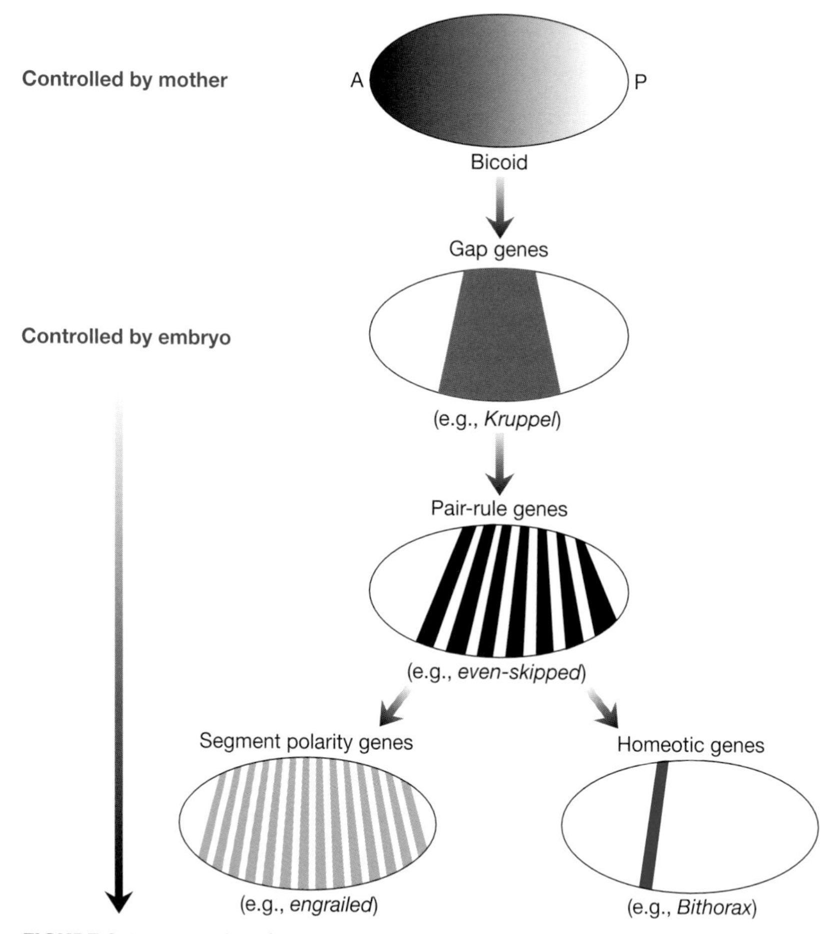

FIGURE 3.4. Hierarchy of gene action in A/P patterning.

genes of the segment-polarity class also encode transcription factors (e.g., *engrailed*), whereas others (e.g., *hedgehog*) encode proteins involved in cell–cell communication. The *engrailed* and *hedgehog* genes are expressed in stripes of cells comprising the posterior half of each segment and function to define posterior versus anterior cell identities (or cell fates). Cell–cell signaling molecules, such as Hedgehog, modify or stabilize the expression of transcription factors (e.g., *engrailed*) in subdomains within each segment. We consider the function of the *engrailed* and *hedgehog* genes further in Chapter 4, when we discuss pattern formation within a segment. As shown below, the sequence of patterning gene action during development (e.g., gap genes → pair-rule genes → segment-polarity genes) is consistent with the progressive refinement in their expression patterns (e.g., from broad single domains to sharply defined periodic stripes).

Confirmation That A/P Patterning Genes Function Hierarchically

One confirmation of the idea that A/P axis-forming genes function in hierarchical sequence is that a given patterning gene is expressed abnormally in embryos mutant for a gene acting earlier in the genetic hierarchy, but is expressed normally in embryos mutant for a gene functioning later in the genetic cascade. Thus, gap genes, which function at the top of the genetic hierarchy, are expressed normally in pair-rule mutants, segment-polarity mutants, and homeotic mutants. Pair-rule genes, functioning next in the developmental sequence, like gap genes, are expressed normally in segment-polarity mutants and homeotic mutants, but are misexpressed in the earlier-acting gap mutants. The segment-polarity genes and homeotic genes act last in the A/P patterning hierarchy and are expressed abnormally in both gap mutants and pair-rule mutants. For example, in the case of the segment-polarity gene *engrailed,* which normally is expressed in segmentally repeated stripes, the even-numbered stripes are missing in the pair-rule mutant *even-skipped,* whereas the odd-numbered stripes are deleted in *odd-skipped* mutants.

HOW NOSE-TO-TAIL PATTERNING OCCURS

Different Genes Are Required to Pattern the D/V Axis Than the A/P Axis

Embryonic mutants identified in the great gene screen disrupt patterning either along the A/P axis or the D/V axis, but not along both axes. This selectivity indicates that the two body axes are established by independent genetic mechanisms. Like A/P mutants, D/V mutants fall into obvious subgroups. The first group, typified by the *snail* and *twist* mutants, exhibits defects in ventral or "mesodermal" cells. The mesoderm (middle layer) includes internal structures such as muscle, heart, and fat. Although the mutant screen was designed to identify mutants with patterning defects in the exposed outer surface of embryo (i.e., the cuticle), mutants grossly disrupting formation of the internal mesoderm were also isolated, because in the absence of muscle, embryos lack rigidity and their cuticles assume a spiral shape (Fig. 3.3J).

The other two groups of D/V axis mutants isolated in the screen have defects in the external portion of the embryo called the ectoderm (outer layer), which gives rise to epidermis (skin, or cuticle in the embryo) and the central nervous system. The ectoderm is naturally subdivided into two parts: the neural ectoderm (or neuroectoderm) comprising the lateral region of the blastoderm embryo, which becomes the ventral ectoderm once the mesoderm invaginates during gastrulation, and the nonneural dorsal ectoderm. The neuroectoderm generates both neural and ventral epidermal structures, whereas the nonneural ectoderm produces only dorsal epidermis. Of the two groups of D/V mutants affecting development of the ectoderm, one disrupts formation of the lateral neuroectoderm and the other has defects in the dorsal nonneural ectoderm. The neuroectoderm group of mutants includes *rhomboid* (abbreviated *rho*), *short gastrulation* (*sog*), and *achaete-scute* (AS-C). In *rho* mutants, ventral epidermal structures such as denticles are greatly reduced (Fig. 3.3K), whereas in *sog* or AS-C mutants formation of the nervous system is compromised. Among the mutants affecting the dorsal nonneural ectoderm, the most severe is *decapentaplegic* (*dpp*), which was described in Chapters 1 and 2. In *dpp* mutants, the dorsal ectoderm assumes a lateral neuroectodermal identity and denticle belts encircle the embryo (Fig. 3.3L).

Mother Tells the Egg What Is Front Versus Back

The mother initiates patterning along both the A/P and D/V axes by supplying the embryo with asymmetrically distributed morphogens. As mentioned in Chapter 1, a morphogen is a molecule that is most concentrated in one region of the embryo, where it is produced, and becomes less concentrated in a graded fashion the farther it is from its source. Recall also that a defining property of a morphogen is that it activates expression of distinct subsets of genes at different concentrations. The morphogen provided by the mother to pattern the A/P axis is a protein called Bicoid. The concentration of Bicoid is highest at the anterior pole of the embryo and fades progressively in more posterior regions (Figs. 3.4 and 3.5; see also Plate 1A for actual data). This asymmetric distribution of Bicoid protein is a result of *bicoid* RNA being localized to the anterior cap of the embryo. *bicoid* RNA is confined to the anterior end of the embryo before fertilization during assembly of the unfertilized egg (or oocyte) by the mother. Bicoid protein is synthe-

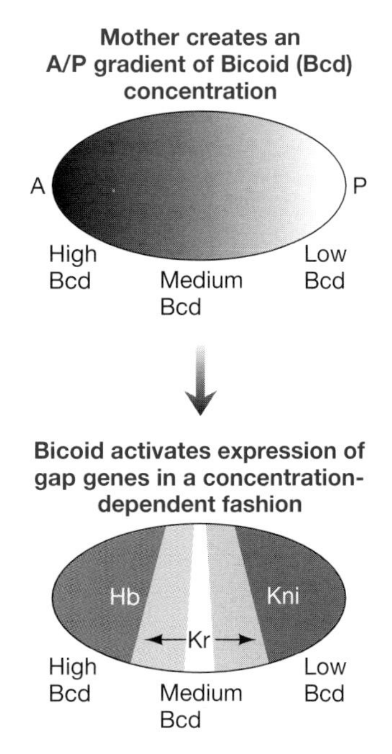

FIGURE 3.5. Bicoid concentration determines the pattern of gap gene expression.

sized at the anterior end of the embryo from the tethered *bicoid* RNA and then diffuses posteriorly. This polarized synthesis generates a steeply graded concentration of Bicoid protein, which tails off in the posterior third of the embryo.

Bicoid Is the A/P Morphogen

Bicoid is an example of a morphogen that is a transcription factor. Although most morphogens are secreted signals, transcription factors also can function as morphogens under circumstances where they are free to diffuse from one nucleus to another. The early *Drosophila* embryo is such a case, since cells are not fully enveloped by isolating membranes until gastrulation begins. Bicoid activates expression of several gap genes that are involved in establishing subdomains along the A/P axis. Different concentrations of Bicoid activate different gap genes. For example, high concentrations of Bicoid activate *hunchback* expression in the anterior region of the embryo, moderate Bicoid levels activate *Krüppel* in the middle portion of the embryo, and low Bicoid concentrations permit *knirps* expression in more posterior regions (Fig. 3.5; see also Plate 1B, C for actual data). Thus, Bicoid satisfies the three conditions for being classified as a morphogen: (1) It is synthesized in a confined region, (2) it diffuses from its site of synthesis and thereby becomes graded in concentration, and (3) it activates distinct subsets of genes at different concentrations.

As discussed in Chapter 1, transcription factors bind to specific sequences of DNA in the regulatory regions of genes and either increase or decrease transcription (i.e., expression) of these target genes.

The Difference between Maternal and Embryonic Mutants

Maternal mutants disrupt the formation of the egg in the mother prior to fertilization, whereas embryonic mutants interfere with the response of the embryo to maternally provided information. Genes disrupted in maternal mutants are referred to as maternal genes and those affected by embryonic mutants are referred to as embryonic genes.

Geneticists generally distinguish two types of mutants affecting embryonic development, referred to as maternal mutants versus embryonic mutants. *bicoid* is an example of a maternal gene because it is required only in the mother to initiate patterning along the A/P axis in the embryo. Because the father does not contribute to patterning the egg prior to fertilization, it does not matter whether he donates a functional or mutant version of *bicoid* to his offspring. Thus, if the mother is mutant for *bicoid* (i.e., *m/m*) and she mates with a normal male (i.e., +/+), all of her embryos will be mutant, even though they themselves carry one good copy and one bad copy of the *bicoid* gene (i.e., *m*/+).

In contrast to maternal genes, both parents contribute to embryonic genes that function in the embryo itself. Embryonic genes interpret information provided by the mother. To eliminate function of embryonic genes such as the gap genes *hunchback, Krüppel,* or *knirps,* it is necessary that the embryo inherits mutant copies of these genes from both parents (i.e., embryos must be *m/m* to be mutant). With the exception of maternal genes encoding morphogens, all the genes considered in this chapter disrupt the function of embryonic genes.

Making Stripes—Gap Genes and Pair-rule Genes

Gap genes encode transcription factors that function by controlling the expression of a second tier of eight genes—called pair-rule genes. Pair-rule genes are expressed in seven stripes, each of which is one segment wide (Fig. 3.4). Because there are 14 body segments in a fruit fly, the pair-rule genes are expressed in the primordia of every other segment. Some pair-rule genes, such as *even-skipped,* are expressed in the even-numbered segments, whereas others, such as *odd-paired,* are expressed in the odd-numbered segments. As mentioned previously, the expression pattern of a pair-rule gene typically matches well with the regions of the larvae exhibiting defects in mutants lacking function of that pair-rule gene. Gap genes and pair-rule genes then collaborate to activate expression of segment-polarity genes in segmentally repeated patterns as well as to restrict expression of homeotic genes to particular segments.

The mechanism by which gap genes generate the striped pattern of pair-rule gene expression is surprising at first. One might expect that an alternating pattern of gene expression would reflect some periodic property of the embryo that determined whether one or another type of pair-rule gene would be expressed in the odd-versus-even-numbered segments. (See box on Meinhardt's theory, below.)

Although Meinhardt's prediction that a periodic chemical wave initiated the segmentally repeated organization of the A/P axis was quite reasonable a priori, it turned out to be wrong. Rather, gap genes act in combination to determine the positions of each pair-rule stripe on a stripe-by-stripe basis. Different stripes of pair-rule gene expression are

■ The Chemical Wave Theory of Segmentation: A Good Wrong Idea ■

More than a decade before pair-rule genes had been discovered, theoreticians such as Hans Meinhardt predicted that there would be genes expressed in stripes. These predictions were based on work of the great British mathematician and World War II hero Alan Turing, who incidentally, cracked the Nazi Enigma Code used by U-boats to sink allied ships (a code so complex that it had been considered indecipherable). According to Meinhardt's mathematical formulations, striped patterns of gene expression would result from wave-like chemical diffusion. In these models, chemical standing waves, similar to vibrational modes of a violin string, had peaks and troughs corresponding to high versus low concentrations of a pattern-forming chemical. One difficulty with these models, which was evident at the time they were proposed, is that the pattern of stripes produced in an embryo depended very much on the exact shape of the embryo. Thus, the subtle variations in the shape of embryos that are actually observed would be predicted to result in significantly different patterns of gene expression.

controlled by separate regulatory elements (or switches). Each stripe is controlled by a distinct on–off type regulatory switch that is activated by a unique combination of transcription factors present in that particular region of the embryo. Thus, the periodic pattern of pair-rule stripes is created by the summation of several independent regulatory elements, each of which controls the expression of an individual pair-rule stripe (Fig. 3.6A). The general condition for turning on a particular regulatory switch is that all activators acting on that switch must be present and all relevant repressors must be absent. Because the collection of activators and repressors controlling each switch differs from one switch to the next, so do the rules for turning on various switches.

■ Distinct Regulatory Switches Control Expression of Stripes 2 and 3 of the *even-skipped* Pair-rule Gene ■

To illustrate how gap genes control pair-rule gene expression on a stripe-by-stripe basis, we can consider activation of the *even-skipped* (*eve*) gene in a pattern of seven stripes (Fig. 3.6B). Mike Levine and his colleagues showed that the *eve* regulatory region can be subdivided into several independent regulatory switches, each of which controls expression of the *eve* gene in a different stripe. The *eve* gene is expressed whenever any of these parallel-acting regulatory switches is flipped on. Levine analyzed the mechanism by which two separate *eve* switches control *eve* expression in two adjacent stripes (stripe 2 and stripe 3). The condition for activating the *eve* stripe-3 switch is simple: Two gap gene repressors, Hunchback and Knirps, must be absent (Fig. 3.6B; see also Plate 1C for actual data). An activator, which is present everywhere in the egg, can turn on *eve* stripe 3 in the narrow gap between the two broad domains of Hunchback and Knirps repression. The condition for activating the *eve* stripe-2 switch, however, is very different from that of *eve* stripe 3. Hunchback, which functions as a repressor of the *eve* stripe 3, is an activator of the *eve* stripe 2. Furthermore, Krüppel, which plays no role in regulating *eve* stripe-3 expression, represses expression of the *eve* stripe 2 and thereby determines the posterior edge of this stripe. The anterior edge of *eve* stripe 2 is set by yet another repressor known as *giant*, which is expressed in more anterior regions of the embryo (see Plate 1B for actual expression patterns of eve-stripe 2 and regulators). The parallel independent regulation of *eve* expression in stripes 2 and 3 illustrates a general principle for organizing the regulatory regions of genes expressed in multiple disconnected groups of cells. As a rule, complex gene expression patterns result from the summed activity of simple independently acting regulatory switches, any of which is sufficient to activate expression of the gene.

A. Independent regulatory elements activate *even-skipped* expression in different stripes

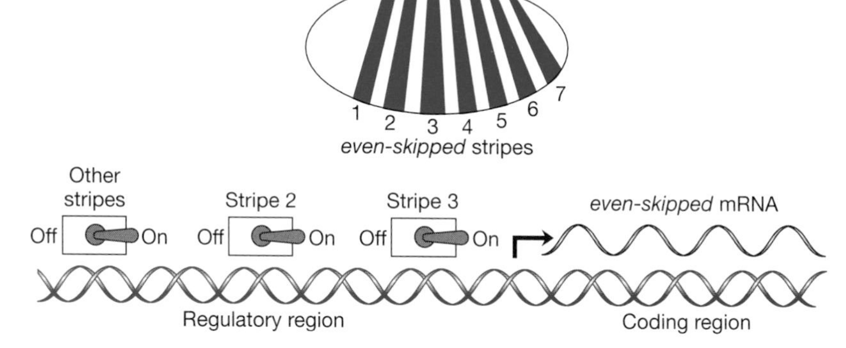

B. Condition for activating expression of *even-skipped* stripe 3

1. An activator that is expressed everywhere must be present.
2. Repressors **Hunchback** and **Knirps** must be absent.

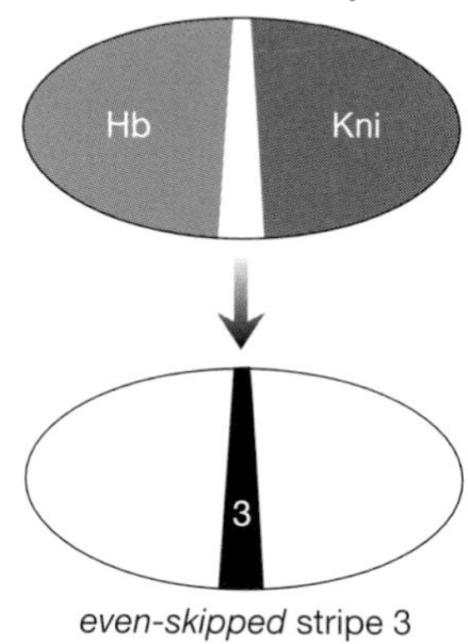

FIGURE 3.6. Gap gene control of even-skipped expression. (*A*) Independent regulatory elements activate *even-skipped* expression in different stripes. (*B*) Condition for activating expression of *even-skipped* stripe 3.

How Segment Identity Is Determined

In the previous two sections, we considered two classes of segmentation genes (gap genes and pair-rule genes), which together provide the necessary information to drive expression of homeotic genes in specific segments. Gap genes, which are expressed in a series of broad domains encompassing two or three contiguous segments, define broad subdomains along the A/P axis, whereas the periodically expressed pair-rule genes subdivide the embryo into alternating even-versus-odd-numbered segments. Pair-rule genes, in turn, collaborate to activate expression of segment-polarity genes in segmentally repeated stripes. A combination of gap genes and even- or odd-numbered pair-rule genes uniquely determine the position of homeotic gene expression in specific body segments (for an actual example, see expression of the *deformed* homeotic gene in Plate 1D). The control of *eve* stripe-3 expression in a central body segment provides an example of how this type of segment-specific regulation can be achieved.

As mentioned earlier, homeotic genes determine the identity of different segments and are expressed primarily in the segments whose identity they control. Homeotic genes were among the first patterning genes to be found to encode transcription factors. Bill McGinnis and Mike Levine first identified transcription factors in Walter Gehring's lab in Basel, Switzerland, and Matt Scott independently made the same discovery in his lab, then in Boulder, Colorado. Because all of these transcription factors contain a similar sequence of amino acids, which is required to bind to DNA, McGinnis dubbed this functionally important region of these proteins the homeobox. The term homeobox has stuck, and now all members in this particular family of transcription factors are referred to as homeobox proteins. Homeotic genes function by activating or repressing expression of a variety of target genes responsible for the differentiation of specialized segment-specific structures such as wings, legs, or antennae.

One important feature of homeotic genes is that those expressed in posterior regions of the embryo are dominant over those expressed in more anterior regions. For example, the *Antennapedia* gene is normally expressed in thoracic leg-bearing segments. If the *Antennapedia* gene is ectopically expressed (or misexpressed) in the head region of the embryo, the resulting flies have legs in place of antennae. This transformation occurs because the *Antennapedia* gene overrides the activity of the homeotic gene normally functioning in the head to specify antennal development. Because legs are the appendages produced in thoracic segments, antennae are converted into legs. Interestingly, in primitive insects, antennae look very similar to legs. Presumably, in these insects, homeotic genes functioning in the head are less effective in suppressing developmental programs active in the leg than are their fruit fly counterparts.

HOW BELLY-TO-BACK PATTERNING OCCURS

Mother Tells the Egg What Is Up from Down

In contrast to patterning along the A/P axis, which culminates in the formation of body segments, the tripartite subdivision of the D/V axis defines three basic tissue types (Fig. 3.7). The logic of progressive subdivision into smaller domains is quite similar along the A/P and D/V axes, even though these events are controlled by different morphogens. The morphogen provided by the mother to initiate D/V patterning is called Dorsal. The concentration of Dorsal is graded along the D/V axis with ventral cells experiencing peak levels of Dorsal, lateral cells containing lower levels of Dorsal, and dorsal-most cells being devoid of Dorsal (see Plate 1E for actual data).

Like the A/P morphogen Bicoid, Dorsal is a transcription factor that functions by binding to the regulatory regions of genes required for

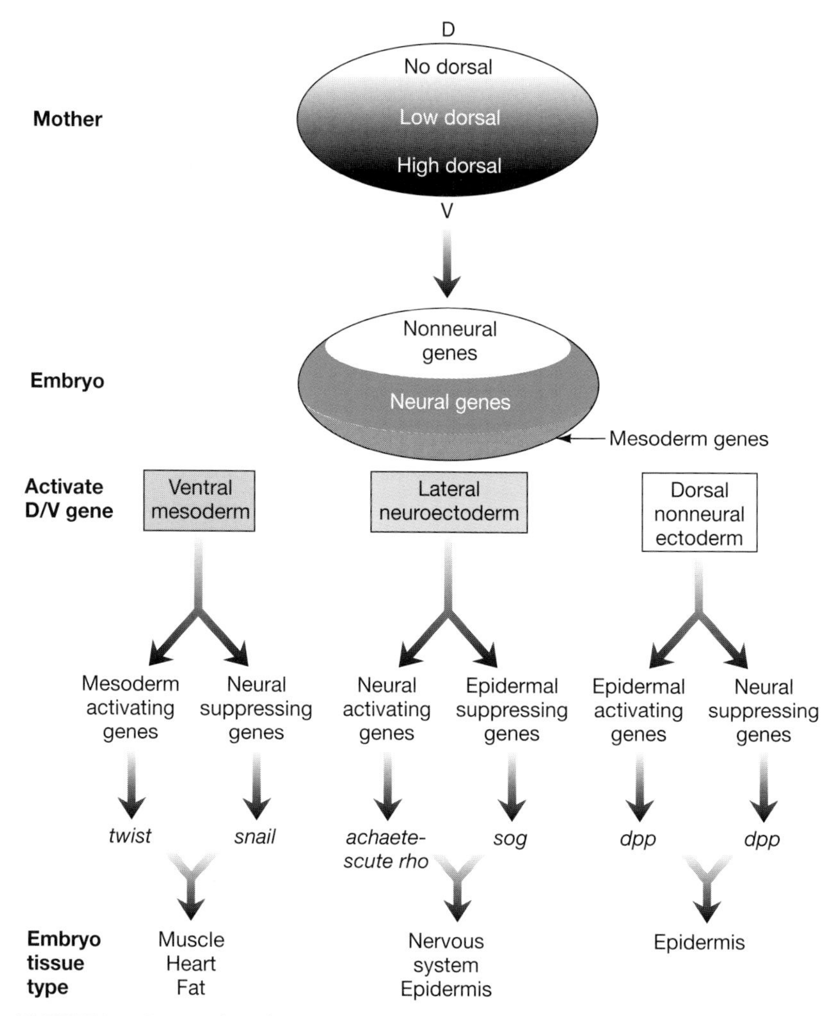

FIGURE 3.7. Hierarchy of gene action patterning the D/V axis.

establishing cell fates and turning them on or off. Through a series of elegant experiments, Levine and other investigators showed that different concentrations of Dorsal subdivide the D/V axis into three domains corresponding to primary tissue types. Ventral cells exposed to high levels of Dorsal form the mesoderm, which gives rise to muscle, heart, and fat; lateral cells having low levels of Dorsal generate a mix of nervous system and epidermis; and dorsal cells without any Dorsal contribute only to the epidermis (Fig. 3.7; see also Plate 1G for actual D/V gene expression patterns). Although it may seem that the *dorsal* gene has been misnamed based on the ventral distribution and activity of its protein product, recall that fly genes are named after what happens when the gene function is lost. In embryos lacking the function of the *dorsal* gene, all cells assume dorsal identities, since they lack Dorsal as do normal dorsal-most cells.

Genes That Act in the Ventral Mesoderm

Peak levels of Dorsal activate expression of genes required for formation of the mesoderm such as the *twist* and *snail* genes (Fig. 3.7; Plate 1G). The *snail* and *twist* genes themselves encode transcription factors (referred to as Snail and Twist, respectively). *twist* and *snail* are named for the appearance of mutant embryos lacking the function of these genes. These mutant embryos assume a spiral morphology due to the absence of muscle. Although the absence of *twist* or *snail* function leads to similar morphological defects, the mechanisms by which these two genes act are quite different. Twist provides a positive function by activating expression of mesoderm-specific genes. In contrast, Snail exerts a negative function by repressing expression of lateral neuroectodermal genes. These distinct genetic functions are revealed by examining early defects in *twist* or *snail* mutant embryos. In *twist* mutants, there is a failure to activate expression of genes in the mesoderm. In *snail* mutants, on the other hand, mesoderm genes are expressed as usual, but neuroectodermal genes are expressed inappropriately in ventral cells. Thus, the mesoderm is defined by a combination of positive and negative actions. Twist promotes mesoderm development, and Snail antagonizes the alternative course of neural development.

Genes That Act in the Lateral Neuroectoderm

Recall that the ectoderm (or outer embryonic layer) comprises the lateral and dorsal regions of the blastoderm embryo and that the lateral ectoderm (or neuroectoderm) gives rise to the central nervous system and ventral epidermis (the ventral region of the blastoderm embryo migrates into the interior of the embryo to give rise to the mesoderm). Expression of neuroectodermal genes, such as the *rho, AS-C,* and *sog* genes, is confined to lateral regions of the embryo through a combination of activation and repression (Figs. 3.7, 3.8, and Plate 1H). *rho* and *sog* are activated by either high or low levels of Dorsal. As discussed above, neuroectodermal genes are not expressed ventrally, where there are high levels of Dorsal, because Snail is present in ventral cells and represses expression of these genes.

The mechanistic basis for restricting *rho* expression to stripes of lateral cells is illustrated in Figure 3.8. The regulatory switch of the *rho* gene contains binding sites for both the Dorsal activator and the Snail repressor. The *rho* regulatory switch is turned on only when the Dorsal activator is present and the Snail repressor is absent. In cells on the dorsal side of the embryo, the *rho* switch is off because the Dorsal activator is absent. In ventral cells, the switch is off because both Dorsal and Snail are present and bound to the *rho* regulatory element. In lateral cells, the *rho* switch is activated since the condition for activity is met: Dorsal is present but Snail is absent.

Two different mechanisms are responsible for preventing neuroec-

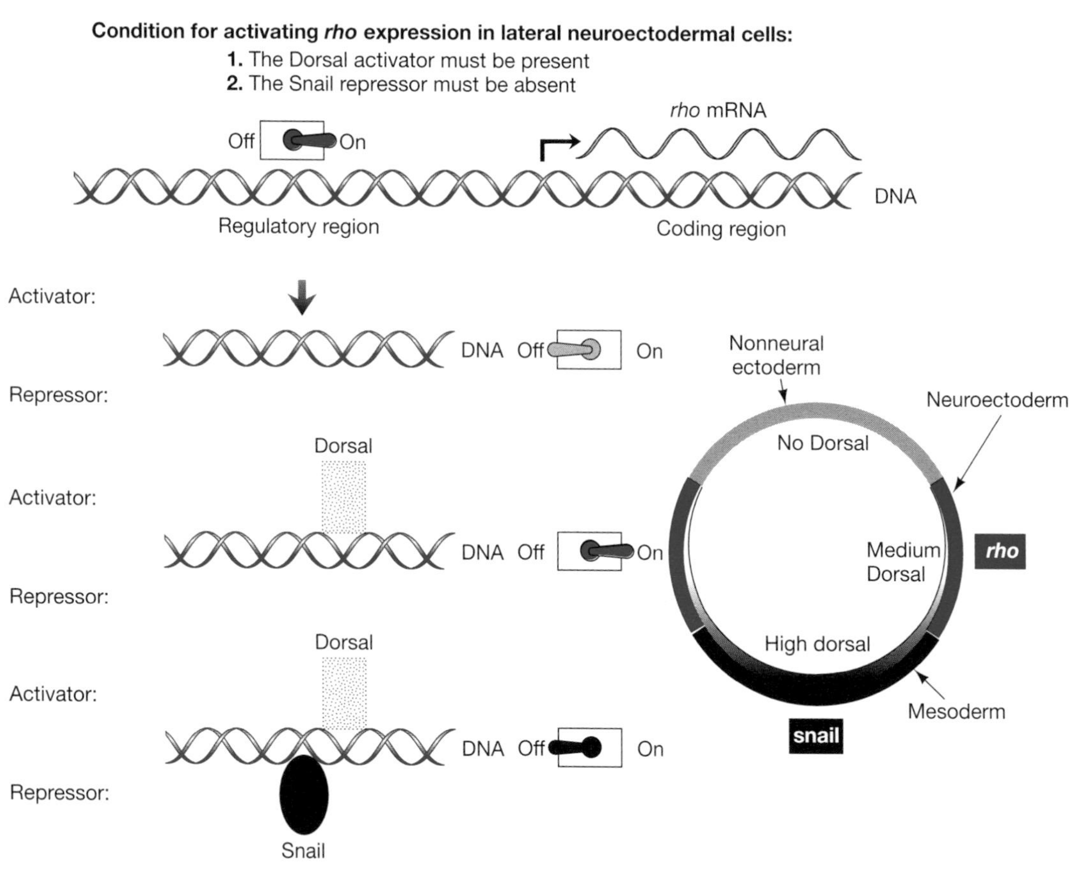

FIGURE 3.8. Making a lateral stripe.

todermal gene expression in dorsal cells. *rho* and *sog* fail to be expressed in dorsal cells because of the absence of the Dorsal activator (Fig. 3.8). AS-C expression, however, is actively excluded from the dorsal region (Figs. 3.9 and 3.10). Suppression of *AS-C* expression in dorsal cells is necessary because *AS-C* genes are activated by a transcription factor present throughout the egg. As discussed below, expression of *AS-C* genes is repressed in dorsal cells by Dpp signaling. Thus, restriction of *AS-C* expression to the lateral neuroectoderm is the result of negative regulation in dorsal and ventral regions of the embryo (i.e., Snail represses *AS-C* expression ventrally and Dpp signaling suppresses *AS-C* expression dorsally). The importance of localized repression in confining expression of genes to stripes along the D/V axis is reminiscent of similar mechanisms we discussed earlier for generating pair-rule stripes along the A/P axis.

Like mesodermal genes, neuroectodermal genes such as *rho*, *AS-C*, and *sog* function to define neural cell fates by a combination of positive

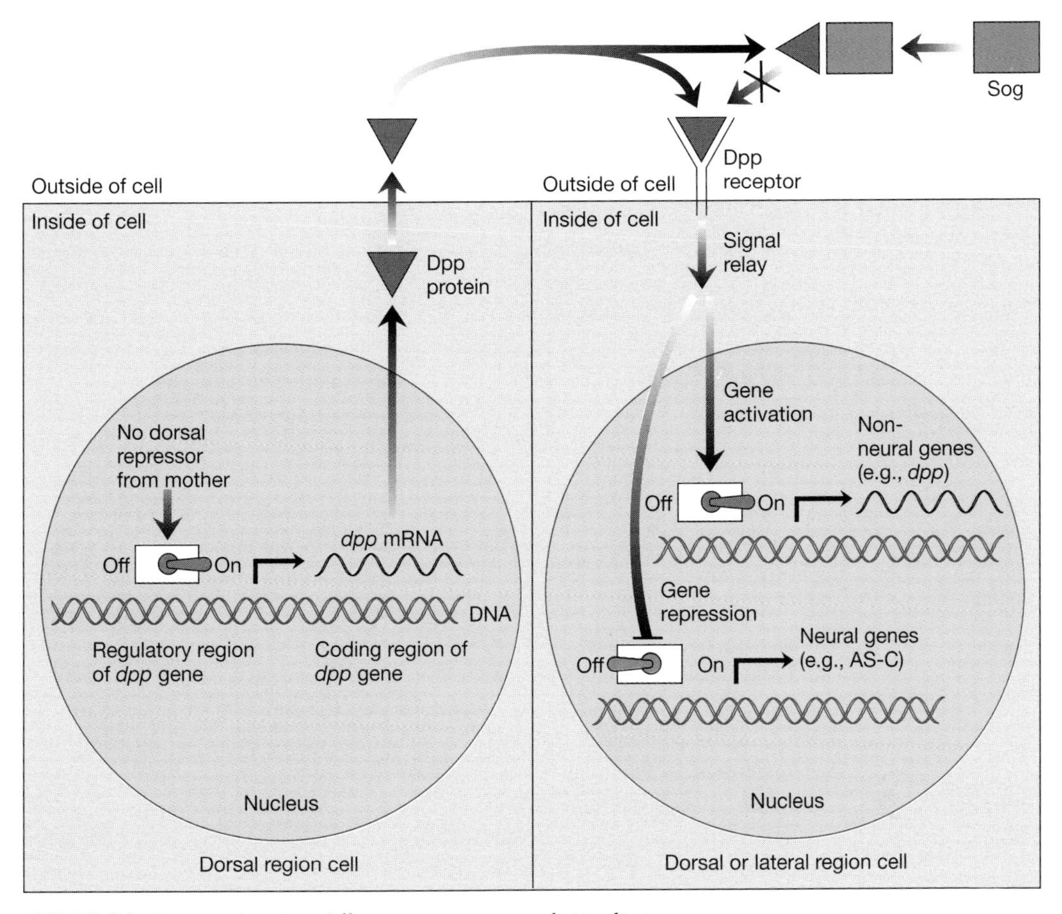

FIGURE 3.9. Dpp production, diffusion, reception, and signaling.

and negative activities. The *AS-C* genes encode a family of related transcription factors that promote neural development by activating expression of nervous system-specific genes. Mutants lacking *AS-C* function generate greatly reduced numbers of neurons. Thus, *AS-C* plays a role in lateral cells akin to that of *twist* in the mesoderm. *rho* also functions by promoting neuroectodermal cell fates. The *sog* gene, on the other hand, functions by blocking the alternative epidermal developmental program active in dorsal cells. This negative role of *sog* in the neuroectoderm is similar to that of *snail* in the mesoderm. The protein (Sog) encoded by the *sog* gene is a secreted diffusible molecule, which blocks the effect of Dpp, the key molecule required for the establishment of dorsal cell fates. One consequence of Dpp signaling is repression of lateral cell fates. By negating the inhibitory action of Dpp, Sog provides a permissive condition allowing lateral cells to follow the default neural developmental program. The interplay between Sog and Dpp in D/V patterning is explored in greater detail below.

Genes That Act in the Dorsal Nonneural Ectoderm

Expression of genes such as *dpp* is restricted to dorsal nonneural cells (see Plate 1G, H) because the Dorsal morphogen represses expression of these genes in ventral and lateral cells. An activator distributed throughout the embryo is responsible for turning on *dpp* expression in dorsal cells where Dorsal is absent. As mentioned earlier, transcription factors are not inherently activating or repressing. The only invariant property of a transcription factor is the specific DNA target sequence to which it binds. Whether a transcription factor activates or represses expression of a given target gene depends on what other transcription factors also bind to the regulatory region of that gene.

The *dpp* gene encodes a secreted signal, Dpp, which plays a pivotal role in specifying the nonneural identity of these cells (Fig. 3.10). In addition, Dpp diffuses from the dorsal region into the adjacent lateral region, where it also can influence neuroectodermal cell fates. As Dpp elicits different responses in cells depending on its concentration, it satisfies the criteria for being classified as a morphogen. Dpp protein secreted from cells binds to its receptor (the Dpp-Receptor) to initiate signaling (Fig. 3.9; see below). As the Dpp-Receptor is present on the surface of every cell in the embryo, all cells are potentially responsive to Dpp. When Dpp binds to its receptor, the receptor sends a signal to the nucleus to alter gene expression within that cell. The Dpp signal is propagated by altering the activity of preexisting transcription factors present in responding cells.

Dpp signaling promotes development of dorsal nonneural cell types both by activating expression of nonneural target genes and by repressing expression of neural genes (Figs. 3.9 and 3.10). One important gene activated by Dpp signaling is *dpp* itself. The ability of Dpp signaling to activate its own expression is termed autoactivation. The key group of neural genes repressed by Dpp signaling is the *AS-C* family of genes. In *dpp* mutants, *AS-C* gene expression is no longer limited to the lateral neural ectoderm but extends into dorsal cells. Because *AS-C* genes play an essential positive role in promoting neural development, misexpression of *AS-C* in dorsal cells of *dpp* mutants results in the formation of ectopic neural precursor cells. Thus, Dpp signaling specifies dorsal ectoderm by a combination of activating expression of dorsal genes and repressing expression of neural genes. These complementary actions of Dpp signaling are analogous to functions mediated by two distinct genes in the ventral mesoderm (i.e., by the Twist activator and the Snail repressor).

Neural Versus Nonneural Development

Although the mother initiates D/V patterning by creating a graded concentration of the Dorsal morphogen, it falls upon genes regulated by Dorsal to move development forward to the next step. Recall that neuroectodermal genes such as *rho* and *sog* are expressed in the lateral re-

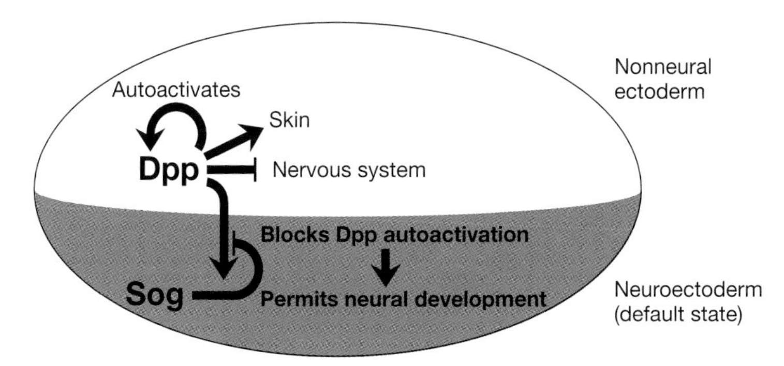

FIGURE 3.10. Subdivision of the ectoderm into neural versus nonneural regions.

gion of the embryo, where there are low levels of Dorsal, and that non-neural genes such as *dpp* are expressed in adjacent dorsal cells, where there is no Dorsal (Fig. 3.7). As shown below, a critical mechanism for maintaining the subdivision of ectoderm into neural versus nonneural domains is the antagonism of Dpp signaling by Sog in lateral cells.

Since the mother initially restricts *dpp* expression to dorsal cells through the repressive action of Dorsal, one could ask why there is a Dpp antagonist such as Sog in the neuroectoderm. After all, the problem already appears to have been solved by Dorsal. The difficulty is that Dpp protein produced in dorsal cells can diffuse into the neuroectoderm and activate its own expression (autoactivate) (Fig. 3.9). Recall that *dpp* mRNA expression is initially limited to dorsal cells as a consequence of Dorsal repressing *dpp* expression in ventral and lateral cells. This dorsal expression of *dpp* mRNA leads to Dpp protein synthesis being restricted to dorsal cells. Since the Dpp protein is secreted and can diffuse, however, some of the dorsally produced Dpp leaks down into the neuroectoderm. If this immigrant Dpp protein is permitted to reach and bind to Dpp-Receptors present on the surface of lateral cells, which are not expressing *dpp* mRNA at this moment, it will induce *dpp* mRNA expression in these cells as a consequence of autoactivation. The combination of Dpp diffusion and autoactivation creates a positive feedback cycle, which, if unopposed, will spread throughout the entire ectoderm, resulting in all ectodermal cells expressing *dpp* RNA and protein (see Figs. 3.9, 3.10, and Plate 1N). Thus, given its ability to diffuse and autoactivate, it is inherently difficult to maintain the dorsally restricted pattern of *dpp* RNA expression initiated by the mother.

One way to keep Dpp signaling from invading the lateral neuroectodermal region of the embryo is to oppose Dpp signaling in those cells. Sog plays such a protective role in the neuroectoderm (Fig. 3.10). Dpp produced in dorsal cells can diffuse into the lateral region of the embryo, but it cannot induce its own expression in those cells because Sog blocks that step. Sog is thought to block Dpp signaling by binding to Dpp and preventing it from activating the Dpp-Receptor (Fig. 3.9). In *sog* mutants, Dpp signaling can autoactivate in the neuroectoderm (see

Plate 1N). The resulting misexpression of Dpp in lateral cells represses expression of critical neural genes such as *AS-C* and, at the same time, induces expression of dorsal region genes. Thus, in the absence of Sog, Dpp signaling spreads into the neuroectoderm where it promotes expression of nonneural genes and suppresses expression of neural genes, thereby transforming neuroectodermal cells into nonneural dorsal ectoderm.

There are two major summary points to make regarding the function of Dpp and Sog in subdividing the ectoderm into neural versus nonneural components (Fig. 3.10). First, all ectodermal cells will develop as neuroectoderm unless they are actively prevented from doing so by Dpp signaling. In other words, the default state of ectoderm is neural, and nonneural development requires inhibition of the neural state. Second, Sog acts as a neuralizing agent through a double-negative mechanism by blocking the activity of a neural inhibitor (i.e., Dpp). As described in greater detail in Chapter 5, the mechanism by which Sog antagonizes Dpp to define neural versus nonneural cell fates appears to have remained essentially unaltered during the more than 500,000,000 years of evolution since the split between vertebrate and invertebrate lineages.

DEVELOPMENT PROGRESSES IN MANY SMALL STEPS RATHER THAN IN A FEW LARGE LEAPS

We have seen that the embryo responds to graded concentrations of Bicoid along the A/P axis and Dorsal along the D/V axis by activating expression of target genes in a few broad domains. Genes expressed in these broad domains then interact to generate finer patterns of gene expression. This gradual accumulation of patterning information in a series of simple steps is referred to as "progressive patterning." It may be surprising that continuously graded positional information provided by morphogens such as Bicoid and Dorsal defines only crude primary patterns of gene expression, since much finer-grained information appears to be carried in the initial graded distributions of these morphogens. Why not use this information directly to activate precise patterns of gene expression? A trivial explanation is that it is not possible to do so. Perhaps cells are unable to measure the small differences in morphogen concentration existing between neighboring cells. Such a practical limitation might prohibit the generation of finer patterns of gene expression than those observed. According to this hypothesis, it would not be possible for a researcher to design a gene-regulatory element driving differential gene expression on a cell-by-cell basis in response to graded morphogen concentrations. Alternatively, cells might be able to measure small differences in morphogen concentration, but might not do so for some reason. If this latter hypothesis were correct, it might be possible to design a regulatory element that responds to the small changes in morphogen concentration between neighboring cells.

Mike Levine designed a clever experiment to address this question. He constructed a hybrid regulatory element containing binding sites for both Bicoid and Dorsal. This was an artificial regulatory element, because no known genes have such regulatory elements. However, Levine found that this artificial regulatory element could respond to both Bicoid and Dorsal and could activate gene expression in a wedge-shaped domain (see Plate 1F). The wedge of gene expression was widest at the anterior end of the embryo and tapered continuously to a sharp point along the ventral midline of the ectoderm at about the midpoint of the embryo. This wedge shape can be attributed to the fact that Bicoid is most concentrated in the anterior end of the embryo and Dorsal is most concentrated in ventral cells. The revealing feature of this experiment, however, was that the wedge of gene expression declined continuously in width on a cell-by-cell basis. Thus, it is possible to make use of graded distributions of morphogens to distinguish cells on the finest possible scale. Why isn't this done in nature? It seems tedious and wasteful to produce pattern indirectly through a lengthy series of intermediate steps when the information is there to generate fine structure from the very start. One way to rationalize progressive patterning over one-step patterning is that it is more reliable to take many simple steps than to make one complex leap. In analogy to hiking, it is easier and safer to ascend a steep mountain by gradually walking up a trail making switchbacks than it is to climb straight up a treacherous cliff along the fall line.

Mike Levine (1955–)

Mike Levine was born in West Hollywood, California, which according to his own admission, explains a great deal about his well-appreciated outgoing style. He majored in genetics as an undergraduate at the University of California, Berkeley, and then went on to Yale for his graduate studies in the department of Molecular Biophysics and Biochemistry, where he did his first work on fruit flies in Allen Garren's lab studying hormone regulation of gene expression. He then joined Walter Gehring's laboratory in Basel, Switzerland, where he collaborated with a fellow postdoctoral fellow Bill McGinnis and a graduate student Ernst Hafen to identify and characterize the *Antennapedia* homeotic gene. Before starting up his own laboratory at Columbia University as an independent investigator, he returned briefly to Berkeley to work with Gerald Rubin on analyzing the expression of neural genes. After launching his research program at Columbia, Levine moved his laboratory to the University of California, San Diego, and then uprooted again to move back to his site of origin at the University of California, Berkeley.

Levine's first major contribution to understanding development, which he made in collaboration with McGinnis and Hafen while in Gehring's laboratory, was the co-discovery of the shared homeobox domain present in homeotic proteins such as Antennapedia and Bithorax. Levine and Hafen then worked out conditions to show that homeotic genes were expressed in stripes corresponding to the segments they specified. In addition, McGinnis and Levine showed that the homeobox portion of homeotic proteins was also present in other segmented animals such as vertebrates. Levine recalls this was something of a surprise. "The Lewis '78 *Nature* paper clearly predicted homology (amino acid sequence similarity) among homeotic genes in *Drosophila* (e.g., fruit flies). The more global conservation of Hox genes (e.g., in other species) was unanticipated. There was no sense of evolutionary conservation of basic mechanisms back in 83–84. Such conservation probably represents one of the most important insights in developmental biology during the past 15 years."

Levine hit the ground running as he set up his own laboratory at Columbia in New York. He set out to use fruit flies to study the function of DNA regulatory elements controlling expression of genes in multicellular organisms such as animals and plants, which differ in several key respects from regulatory elements present in more primitive organisms such as bacteria. These regulatory elements, often referred to as "enhancers," can activate expression of genes located several thousands of base pairs away from the beginning of the coding region. In contrast, bacterial regulatory elements act much more locally (i.e., 1–50 base pairs away from the coding region). He recalls that there were two strong prejudices about enhancers that came mainly from studies of gene expression in mammalian tissue culture cells. First, they are relatively simple (the prototypic enhancer is just 72 base pairs long). Second, enhancers interact solely with activators but not repressors. One outcome of Levine's work on analysis of fruit fly enhancers (e.g., the *eve-stripe 2, eve-stripe 3,* and *rhomboid* regulatory elements) is that repression has been consistently underestimated by those studying transcriptional control in mammalian cells. He notes "The view was (and still is to some extent) that the default state of a gene is generalized repression. Tissue-specific (or spatially localized) expression depends on the right combination of transcriptional activators. The demonstration that the *eve-stripe 2* enhancer, and shortly thereafter the *rho* lateral-stripe enhancer, contain both activator and repressor elements was somewhat unexpected. This observation was important for the notion of enhancers as integrators of complex regulatory information in development. As for the *Drosophila* field, the segmentation field was dominated by geneticists who were satisfied with a gene hierarchy, arrows, and progressively more refined patterns of gene expression. Most people who worked on the problem of segmentation stripes quit after looking at gene pattern A in mutant B, etc. to test the simple concept of a gene hierarchy."

Levine recalls two particularly exhilarating moments of discovery. The first highlight, which occurred when he was a postdoc, was the time he and his collaborators saw the localized expression patterns of new genes that had been identified via homeobox cross-hybridization. Later, as an independent investigator, he was particularly excited by finding that the Dorsal morphogen gradient was established via regulated transport of the Dorsal protein into the nuclei of cells, and then seeing a similar form of regulation of another fly protein related to Dorsal that mediates insect immunity. Regarding these latter experiments, Levine reminisces "these were the last real experiments I did with my own hands, and there is no substitute for doing it yourself. Most of our other experiments are rather stringent tests of formal hypotheses, and so do not offer the kind of serendipitous insights that underlie many unexpected discoveries."

Levine has kept his lab tightly focused on gene regulation in the early fruit fly embryo. He notes that "It is sometimes a tedious grind and while this style can provide deep insights, there is less potential for a truly unexpected discovery." He believes that it is healthy to strike a balance between focusing on a particular question and remaining open to the unexpected by "designing experiments that test specific hypotheses, but are not too rigidly designed to preclude serendipitous findings." Recently, in addition to continuing his productive studies of fruit fly regulatory elements, Levine has initiated an analysis of notochord specification in a primitive chordate. He was elected into the National Academy of Sciences in 1998 and serves on the review boards of several prestigious journals.

■ Summary of Fruit Fly Embryonic Development ■

Christiane Nüsslein-Volhard, Eric Wieschaus, and Gerd Jürgens conducted a comprehensive genetic screen for mutants affecting patterning along the A/P and D/V axes of the embryo. They recovered mutants in approximately 150 different genes, which fell into well-defined categories based on the nature of defects observed in mutant embryos. Analysis of groups of mutants affecting the patterning of the basic body axes revealed that development of the *Drosophila* embryo is controlled by a series of simple hierarchical events. The mother provides initial polarity in the egg by creating graded distributions of morphogens along the A/P and D/V axes. These maternal morphogens are transcription factors that activate gene expression in a concentration-dependent fashion. In reponse to different concentrations of morphogen, embryonic genes are activated in one of a few broad domains. Along the A/P axis, the Bicoid morphogen drives expression of gap genes in overlapping domains to initiate segmentation. Gap genes encode transcription factors and act directly upon a second tier of genes called pair-rule genes, which are expressed as a series of stripes in every other segment. The striped expression of pair-rule genes is controlled by the summed action of several independent regulatory elements, each responding to particular combinations of gap genes. The pair-rule genes in turn regulate expression of segment-polarity genes in segmentally repeated stripes. Pair-rule and gap genes also activate expression of a series of segment-identity genes known as homeotic genes in specific segments. Finally, segment-polarity and homeotic genes collaborate to generate a segmented embryo and adult fly adorned with distinct structures such as legs, wings, or antennae in appropriate segments.

Along the D/V axis, the Dorsal morphogen defines three adjacent domains of gene expression corresponding to the primordia for primary tissue types. Genes expressed in a given domain of the D/V axis (i.e., ventral mesoderm, lateral neuroectoderm, and dorsal nonneural ectoderm) function by activating expression of genes appropriate to that region of the embryo or by suppressing expression of genes directing alternative developmental programs in adjacent domains. On the positive side, Twist activates expression of mesoderm genes in ventral cells, *achaete-scute* activates expression of neural genes in lateral cells, and Dpp activates expression of nonneural genes including *dpp* itself in dorsal cells. On the flip side of the coin, Snail represses neuroectodermal gene expression in ventral cells, Sog prevents Dpp signaling from invading the lateral neuroectoderm, and Dpp suppresses neuroectodermal gene expression in dorsal cells. The idea is reminiscent of political strategy. To succeed, you both promote yourself and attack your competition.

4 Patterning Fly Appendages and Eyes

In the previous chapter, we discussed how the two primary body axes of the fruit fly embryo—anterior–posterior (A/P) and dorsal–ventral (D/V)—are subdivided into a series of discrete domains. In this chapter, we focus on small groups of embryonic cells, known as imaginal discs, that give rise to structures in the adult fly such as legs, wings, eyes, and antennae. Formation of fly appendages takes place during metamorphosis and differs from embryonic development in two important respects. First, patterning information in the embryo is elaborated de novo on the basis of crude asymmetries in the egg that were created by the mother, whereas appendage formation takes place in the context of an already polarized organism, the larva, which has precisely defined primary axes. Second, because the embryo is encased by an inelastic egg shell, embryonic development in fruit flies occurs in the absence of growth. Genesis of appendages from imaginal discs, on the other hand, involves more than a thousandfold increase in cell number and volume.

The primary topics featured in this chapter are the development of adult wings and eyes during metamorphosis. We first consider how an adult fly is assembled in a patchwork fashion from separate imaginal discs and then concentrate on how positional information is elaborated along the A/P axis of the wing. As with embryonic patterning described in Chapter 3, the A/P axis of the wing is subdivided into a series of discrete domains by a sequence of simple steps. A/P patterning of the wing provides a clear example of the logic by which morphogen signals generate patterns in growing tissues such as appendages. Finally, we consider specification of the eye primordium and the development of facets in the eye by a series of inductive events.

■ Cast of Characters ■

Terms

Compartments Domains of cells that do not intermix with each other (e.g., cells of the anterior and posterior compartments of the wing which are separated by the A/P boundary).

Furrow A morphologically visible crease in the imaginal disc in the eye that moves across the disc from posterior to anterior during development.

Genetic mosaics Animals that are heterozygous for a mutation but contain clusters of homozygous mutant cells referred to as clones.

Imaginal discs A small group of embryonic cells that gives rise to an adult structure of the fly such as a leg, wing, eye, or antenna.

Master gene A gene that acts as a single regulator to specify a certain cell fate and can redirect other cells to adopt that fate.

Metamorphosis The transformation of a larva into an adult fly.

Mutant clone analysis A genetic method in which the defects observed in genetic mosaics (individuals carrying small patches of homozygous mutant cells surrounded by normal cells) are analyzed.

Organizer A region of a developing organism that sends signals to neighboring cells to organize the formation of a morphological structure (e.g., the narrow stripe of cells running up the center of the wing disc that organizes the A/P axis of the wing).

Patterning The process by which cells acquire distinct identities during development.

Photoreceptors Light-sensitive neuronal cells in the eye that respond to light by producing an electrical impulse.

Genes

Antennapedia A gene encoding a homeobox transcription factor that defines the second thoracic segment (T2), which has wings and legs.

apterous (ap) A homeobox gene required for the outgrowth of appendages.

bicoid A gene encoding the maternal morphogen (Bicoid) that is provided by the mother to pattern the A/P axis.

bride-of-sevenless (boss) A gene encoding a ligand for the Sevenless receptor that is expressed in the R8 photoreceptor.

dachshund A gene required for initiating eye development.

decapentaplegic (dpp) A gene encoding a secreted signal required for patterning the A/P axis of adult appendages.

distalless A homeobox gene that plays an organizing role in defining the proximal distal axis of appendages.

engrailed A segment-polarity gene required for the formation of the posterior portion of each segment.

even-skipped A pair-rule gene encoding a transcription factor that is required for the formation of even-numbered segments.

eyeless A homeobox-encoding gene required for initiating eye development that is required to activate expression of eye-specific genes.

eyes absent A gene required for initiating eye development.

hedgehog A segment-polarity gene encoding a secreted signal required for the formation of posterior structures in the segment.

hunchback A gap gene encoding a transcription factor that is required for formation of the most anterior region of the embryo.

knirps A gap gene encoding a transcription factor that is required for formation of the second wing vein in adult flies.

Notch A gene encoding a receptor that is required for outgrowth of the wing and for cells along the wing margin to assume proper dorsal versus ventral identities.

optimotor blind (omb) A very sensitive target gene of Dpp signaling in the wing that is expressed in a broad domain centered over the A/P organizer and contains the *spalt* expression domain.

pax 6 The vertebrate counterpart of the fly *eyeless* gene that plays an essential role in initiating eye development in mice and humans.

sevenless A gene encoding a receptor protein for the Bride-of-Sevenless signal that is required for R7 photoreceptor development.

sine oculus A gene required for initiating eye development.

spalt (sal) A moderately sensitive target gene of Dpp signaling in the wing that is expressed in a domain centered over the anterior-posterior organizer and is contained within the wider *omb* expression domain.

vestigial (vg) A gene required for specifying wing cell fates.

GENESIS OF ADULT FLY APPENDAGES—WINGS

Adult Flies Are Assembled from Pieces like a Quilt

Imaginal discs consist of 10–20 cells organized as flat pancakes that are set aside during embryonic development (colored circles in Fig. 4.1A). Imaginal disc cells remain quiescent until the onset of metamorphosis, when they are induced by a hormone to undergo a series of cell divisions. Proliferation of imaginal disc cells creates a sack-like structure of 50,000 cells resembling a pita bread. Separate imaginal discs give rise to various adult fly structures such as legs (L), wings (W), eyes (E), and the abdominal body wall (A). During metamorphosis, imaginal discs are transformed from rudimentary sacks into defined parts of the adult. Ultimately, the fragments formed by the various imaginal discs link up like pieces in a complex puzzle to generate a seamless adult fly (Fig. 4.1B). This eerie piecemeal assembly of the adult fly from imaginal disc is similar to running a movie of breaking glass in reverse.

Metamorphosis entails the destruction of more than 90% of the larva followed by growth and fusion of imaginal discs. The adult fly is resurrected from the ashes of the larva by a mind-boggling redesign of the body plan. For example, virtually the entire collection of embryonic muscles is destroyed and replaced with a completely different set of muscles that attach to targets in the adult fly unrelated to any structures present in the embryo. Similarly, nearly all cells in the larval brain die and are replaced with neurons connecting to new adult muscles and to each other in novel patterns appropriate for operating a fly versus a larva. It is not an exaggeration to state that an adult fly differs

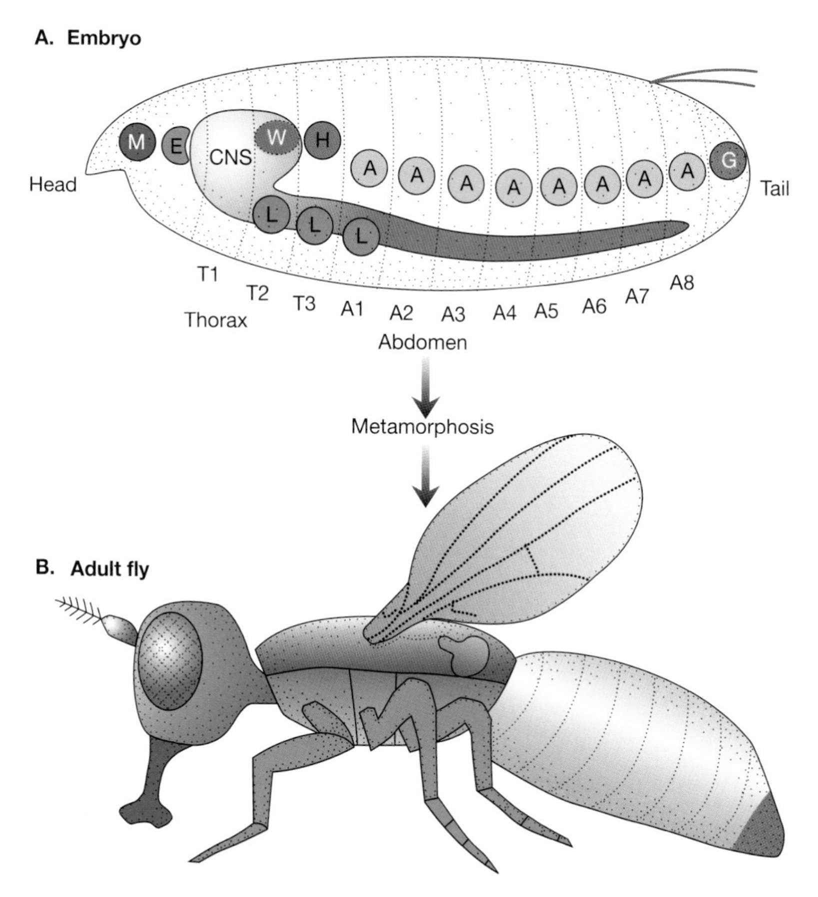

FIGURE 4.1. Quiltwork assembly of a fly.

more from its larval state than it does from adults of distantly related insect species. A metamorphosing insect puts even the most sophisticated children's transformer toy to shame.

A/P and D/V Information in the Embryo Determines the Identity of Imaginal Discs

Imaginal discs acquire their identities when they are first formed in the embryo (Fig. 4.1A). Because imaginal discs are generated after the primary body axes have been established, they form in response to preexisting A/P and D/V positional information. For example, the homeotic gene *Antennapedia* is responsible for specifying the second thoracic body segment. Like other structures derived from the second thoracic segment, formation of wing and leg imaginal discs depends on *Antennapedia* function. Recall that mutant flies misexpressing the *Antennapedia* gene in the head have legs in place of antennae. This monstrous structural substitution results from the antennal imaginal disc being partially transformed into a second thoracic leg disc.

D/V information also is important in specifying imaginal disc identities. For instance, with respect to the wing and leg discs in the second thoracic segment, the wing disc forms more dorsally than the leg disc. These differences in D/V positional information result in the activation of a gene called *vestigial (vg)* in the dorsally positioned wing primordium, but not in more ventral cells comprising the leg disc. As discussed below, *vg* plays a key role in specifying wing identity. Thus, A/P and D/V positional information established in the first half of embryonic development is used to define the identities of different imaginal discs that ultimately generate specialized structures in stereotyped positions of the adult fly.

A critical question regarding the formation of adult insects during metamorphosis is how the crude A/P and D/V positional information is used to generate different imaginal discs. In the case of the wing and eye discs, regulatory genes expressed in localized regions of the embryo determine the identities of these two very different adult primordia. For example, Sean Carroll at the University of Wisconsin has shown that the *vg* gene plays an essential role in establishing wing identity. Work in his laboratory demonstrated that loss of *vg* function results in the absence of wings, whereas misexpression of *vg* in other imaginal discs such as the eye disc redirects these cells to form wings. Similarly, as discussed further below, a gene called *eyeless* plays a central role in specification of eye cell fates. The action of genes involved in assigning imaginal disc identities is formally similar to that of homeotic genes in specifying distinct segmental identities.

Sean Carroll (1960–)

Sean Carroll was born in Toledo, Ohio. He attended college at Washington University in St. Louis and earned a Ph.D. in Immunology from Tufts University Medical School. He then joined Matt Scott's laboratory at the University of Colorado as a postdoctoral fellow, where he contributed to analysis of segmentation genes. Among other things, he examined the hierarchical relationship between A/P patterning genes in various A/P patterning mutants and provided some of the first evidence that pair-rule genes were regulated by gap genes and that they, in turn, regulated the expression of segment-polarity genes.

Carroll started his own research group at the University of Wisconsin, Madison, where he continued working on segmentation. He also initiated studies of the genes that are required for early neural development and made key contributions to this field before shifting his interest to patterning of adult appendages. He began studying wing development by dissecting the genetic control of dorsal–ventral patterning of the wing, and then turned to analyzing genes, such as *vestigial (vg)*, that define the identity of wings versus other appendages. His work on *vg* was inspired by the idea that understanding the development of adult body parts was key to understanding the morphological diversification of animals such as the arthropods and insects.

Carroll chose to focus on analysis of *vg* for three reasons: "First, it [i.e., *vg*] was expressed in all cells in the developing wing field (an exciting day when we first saw that pattern). Second, the discovery that it contained an enhancer [regulatory element] that responded to compartmental signals told us that it was a direct link between compartments and the organization of a whole structure. Third, *vg* could induce outgrowth of wing tissue in other places (pretty dramatic)."

As his interest in the genesis of appendages deepened, Carroll became increasingly intrigued by the evolution of developmental mechanisms and has been a major founding figure of this new field, which is now sometimes referred to as Evo-Devo in pop-science jargon. This latter interest spurred Carroll to begin a molecular analysis of butterfly wing development, where he has focused on understanding how eyespots form concentric rings of pigmentation in characteristic species-specific locations.

As is apparent from Carroll's varied interests, he describes his style as that of a generalist. "My interests span biological levels (molecular to organismal) and include deep mechanistic as well as historical questions. The strength of this approach is that there is cross-fertilization of ideas and the possibility of synthesis across disciplines."

Among his many seminal discoveries, Carroll singles out three as particularly memorable. "The first was when I saw green stripes of *ftz* [a pair-rule gene] protein through the microscope for the first time. Eighteen months of unsuccessful work preceded that night. I drank a lot of champagne. The second was the image of *vg* in the wing field. The third was the discovery of Distal-less expression in the developing butterfly eyespots. That was entirely unexpected. In all three cases, these patterns gave us a handle on the developmental processes we were pursuing—segmentation, appendage formation, and the evolution of novelties. In each case, the patterns (or success) were unexpected but as soon as we saw them, we knew we were in business."

Carroll's leading role in unraveling many mysteries of development have won him broad accolades, such as the prestigious National Science Foundation Presidential Young Investigator Award. In 1994, he was also selected as one of the top 50 leaders in America by *Time* magazine on the basis of his groundbreaking work in pattern formation and his founding of a biotech company producing antivenin. Carroll describes the key elements of discovery as "passion, perseverance, support from mentors and colleagues, LUCK, and a pinch of intuition about where the gems are buried and how to find 'em."

Appendages Are Subdivided into Anterior Versus Posterior Territories

Adult appendages have A/P and D/V axes, which are aligned with the corresponding axes of the fly. Each appendage spans one complete body segment. Appendages also have a proximal–distal (P/D) axis (proximal being closest and distal being farthest from the body). To illustrate the principles by which appendages are patterned, we focus on generation of the A/P axis in developing wing discs. The A/P axes of other imaginal discs are determined by similar mechanisms. Although the specific genes involved in establishing the D/V and P/D axes of appendages are different from those used to create the A/P pattern, the principles underlying these developmental events are much the same.

As mentioned above, an appendage, such as a wing, spans a single segment, which is subdivided into anterior and posterior domains (Fig. 4.2). As described in more detail below (Fig. 4.3) subdivision of ap-

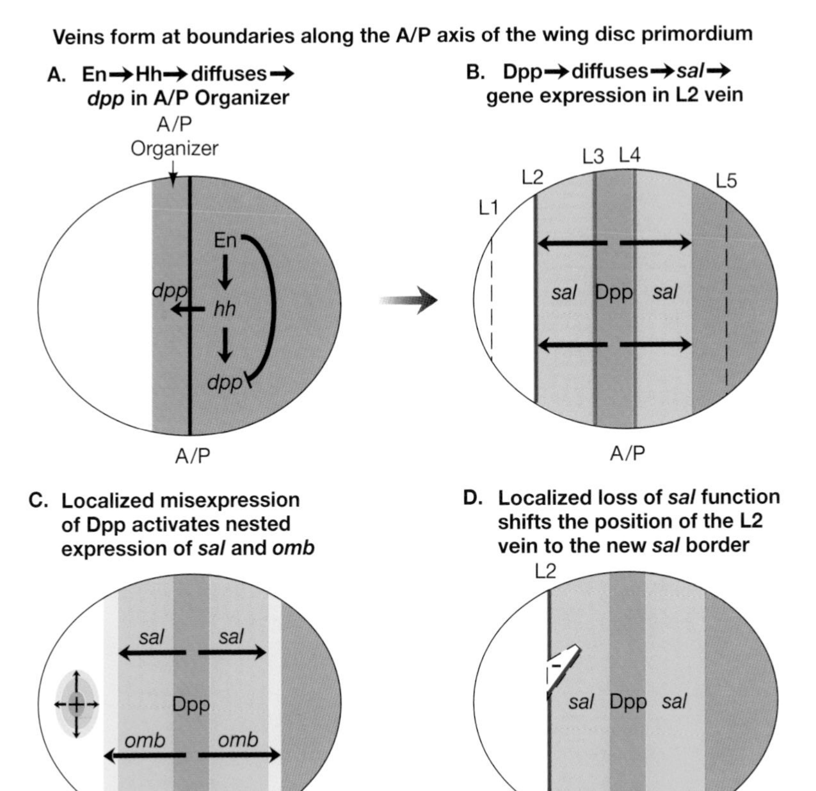

FIGURE 4.2. Wing vein development.

pendages into two territories can be traced back to formation of imaginal discs during embryonic development. Imaginal disc cells straddle the A/P boundary of the embryonic segments from which they derive. The position of the A/P boundary within imaginal discs is maintained with great precision in the face of the thousandfold increase in cell number that takes place during larval development.

As discussed in Chapter 3, the segment-polarity genes *engrailed (en)* and *hedgehog (hh)* are expressed in the embryo in stripes confined

■ Anterior Versus Posterior Cells ■

Cells lying in the anterior and posterior portions of a segment can be distinguished from one another by several criteria. The most obvious difference between anterior and posterior cells is that, like oil and water, they do not mix with one another. Because anterior and posterior cells cannot intermingle, they are said to lie within discrete compartments. There is a sharp line dividing imaginal discs into anterior versus posterior compartments which is referred to as the A/P boundary. The position of the A/P boundary in a given imaginal disc is invariant. For example, in the wing, there are four major longitudinal veins running the length of the wing that are numbered L2–L5, L2 being the anterior-most vein and L5 being the posterior-most vein. There also is a vein circumnavigating the anterior margin of the wing known as the L1 vein. The A/P boundary in the wing disc runs in a stereotyped position just anterior to L4 (Fig. 4.2B).

to the posterior portion of each segment. These two genes play essential roles in specifying the identity of posterior cells. In mutants lacking function of either of these two genes, the posterior portion of each segment develops as a mirror duplication of the anterior half of the segment (see Fig. 3.3F). *en* encodes a transcription factor (En) that activates expression of various target genes including the *hh* gene. *hh* encodes a secreted signal (Hh) that plays an essential role during embryonic development in maintaining the subdivision between the anterior and posterior halves of the segment. The *en* and *hh* genes continue to play critical organizing roles during the development of imaginal discs.

Juxtaposition of Anterior and Posterior Cells Induces an A/P Organizing Center

Expression of the *hh* gene in posterior compartment cells initiates A/P patterning of the wing imaginal disc during early larval development. The primary effect of Hh protein signaling is to activate expression of several target genes in a narrow stripe of cells running up the center of the wing disc along the A/P compartment boundary (Fig. 4.2A). This stripe of Hh-responding cells, which I refer to as the A/P organizer, lies in the anterior compartment and directly abuts the A/P boundary. The reason Hh only activates gene expression in this narrow central stripe of cells is that the response to Hh protein signaling is blocked in the posterior compartment by the En transcription factor. Hh is a diffusible signal and can leak a short distance into the anterior compartment. Because anterior compartment cells do not express *en,* they are able to respond to Hh signaling. Thus, only anterior compartment cells within close enough range of the A/P boundary can receive the Hh signal. Cells stimulated by Hh activate expression of various genes in the A/P organizer.

The narrow stripe of Hh-responding A/P organizer cells in the anterior compartment plays two essential roles in patterning the A/P axis of the wing. First, the anterior and posterior borders of A/P organizer define the positions of the two central wing veins L3 and L4 (Fig. 4.2B). The L3 vein forms in the anterior compartment at the anterior edge of the A/P organizer, and the L4 vein forms just within the posterior compartment along the A/P boundary. Second, one of the targets of Hh signaling in the A/P organizer is the now-familiar gene *dpp,* which plays a pivotal role in organizing lateral and peripheral regions of the wing.

A Signal Emanating from the A/P Organizer Determines A/P Positions

Although *dpp* RNA expression in the A/P organizer is bounded by the primordia of the L3 and L4 veins, the encoded Dpp protein is a long-range diffusible signal that can travel significantly farther from its site of production than Hh. As a result of its localized production and diffusion, Dpp protein levels are highest in the central region of the wing disc (i.e., where it is synthesized between the L3 and L4 veins), intermediate in cells lying between the primordia of the L2 and L3 veins and between the L4 and L5 veins, and lowest or absent in the most anterior and posterior extremes of the wing that are farthest from the Dpp source. Because Dpp activates expression of different target genes in the wing disc depending on its concentration, it functions as a morphogen in the wing as well as in the embryo.

To illustrate how Dpp acts as a morphogen in the wing, we can consider two known target genes of Dpp signaling called *spalt* (*sal*) and *optimotor blind* (*omb*). These genes are expressed in two broad nested domains centered over the A/P organizer. The *sal* expression domain is contained within the wider *omb* expression domain (Fig. 4.2C; see also Plate 3A for actual expression of *sal* relative to *dpp*). Activation of the *sal* and *omb* genes requires Dpp signaling because cells carrying crippled receptors for Dpp fail to express either gene. Gary Struhl at Columbia University invented a genetic method to test the idea that different doses of Dpp activate expression of *sal* versus *omb*. He devised a way to activate expression of the *dpp* gene randomly in small patches of cells in the wing disc through a mechanism that is independent of Hh protein signaling. Wings derived from such flies contain small islands of cells misexpressing the *dpp* gene in addition to the normal central stripe of *dpp* expression in the A/P organizer (Fig. 4.2C; see also Plate 3B, C for actual data). Struhl observed that concentric circles of *sal* and *omb* expression surrounded the small patches of cell misexpressing *dpp*. Most importantly, he found that *omb* was expressed in a larger circle than *sal* (Plate 3C, arrow). This result suggested that Dpp diffused out from the small patch of *dpp* expression where it was produced and activated expression of the *sal* and *omb* genes at different doses. Expression of the *sal* gene could only be induced by high Dpp concentrations in neighboring cells, which were close to the source of Dpp, whereas expression of *omb* could be induced by lower levels of Dpp at a greater distance from the A/P organizer.

Gary Struhl (1954–)

As a graduate student and then a postdoc, Struhl performed a series of ingenious experiments to examine the role of homeotic genes and other A/P patterning genes in the fly embryo. One of the many interesting observations that Struhl made during this very productive period was defining the default segmental state obtained when all homeotic genes are expressed throughout the developing embryo. He found that in this situation all segments developed as the most posterior abdominal segment. This result is the opposite of that caused by eliminating homeotic gene function, in which segments adopt anterior identities (e.g., every segment looks like an antennal segment in beetles lacking all homeotic genes). These and other experiments, such as those in which Struhl compared the reciprocal sets of defects in loss-of-function versus misexpression mutants of the *Antennapedia* gene, ultimately led him to realize that comparing the consequences of ectopic activity with loss of activity of such genes should be a powerful general strategy for demonstrating that the product of a patterning gene confers spatial information and could also provide insights into the nature of that information.

Gary Struhl was born in Brooklyn, New York. He obtained bachelor's and master's degrees from the Massachusetts Institute of Technology and moved to Cambridge, England, for his graduate studies, where he worked with Peter Lawrence at the University of Cambridge to study the regulatory relationship between homeotic genes. Following his graduate studies, Struhl worked briefly with Nüsslein-Volhard and Wieschaus in Tübingen on their screen for embryonic patterning mutants, which galvanized his interest in the segment-polarity genes and embryonic patterning as ways for getting at morphogens and gradients, concepts he was already well aware of from being a student of Lawrence. Struhl then moved to Harvard, where he continued his work independently in Tom Maniatis's lab on anterior–posterior patterning in flies. After completing these studies, he accepted a faculty position at Columbia University Medical School in New York, where he currently resides.

Beginning in the late 1980s, Struhl became frustrated by the fact that long-range organizing events—in which he was, and remains, most interested—were occurring under the unusual circumstance of the early fruit fly embryo in which cells were not fully enclosed by membranes. He realized that it would be important to understand how spatial information is generated and interpreted in cell populations and decided that the best approach, at least for him, would be to develop methods that would allow potential signaling molecules, or components of their receptor systems, to be activated or inactivated at will during imaginal disc development. This need, and his background in comparing loss-of-function and misexpression mutants, led Struhl and his postdoctoral collaborator Konrad Basler to develop a method called the Flp-out technique, in which genes are ectopically expressed in isolated patches of cells. (This method is the converse of mutant clone analysis pioneered by García-Bellido, in which patches of cells are produced that lack the function of a gene of interest.) This ingenious Flp-out method has proven very successful for studying the organizing roles of secreted signals like Dpp and Hh, which have been a major focus of Struhl's work over the past several years. It is worth noting that the grade school adage "success is 1% inspiration and 99% perspiration" often applies to scientific success as well. Struhl recalls that this was true regarding development of the Flp-out technique. "The Flp-out method, unfortunately, did not come easily—the Struhl and Basler paper of 1993 represents the culmination of around 4 years of work—but it was well worth the investment."

Although the existence of gradients and morphogens was not unanticipated when Struhl performed his Flp-out experiments, it was not clear what molecules were the morphogens, and more importantly, whether any molecules really behaved like hypothetical morphogens. Two general types of models had been proposed, and they can be traced back to Boveri and Spemann. In the first model (the one Struhl has demonstrated to act in the wing), a morphogen acts as a long-range signal to activate expression of different genes in a concentration-dependent fashion, whereas the second model proposed that morphogens act through a series of short-range inductive events that tend to peter out. Regarding this point, Struhl comments, "The Flp-out method was critical for doing this in *Drosophila* (fruit fly), as was the recognition that a way to discriminate between morphogen and sequential inductive models was to compare the effects of manipulating ligand production with signal transducing activity."

A predominant theme in Struhl's work is that he often invents a clever genetic method (e.g., the Flp-out technique) to answer an important question. Regarding this style of experimentation, Struhl remarks, "I get my kicks out of trying to understand things I cannot see or touch—and in particular by setting genetic traps for obtaining information about how things work in vivo. I can't explain why I find this so challenging and engrossing, but this has always been the case since I began research as an undergraduate. In terms of subject matter, I am indebted to Peter Lawrence for making me aware of the general problem of understanding how cells know where they are and how they use such spatial information to decide what to do.

This is a fundamental problem of animal development, and it is one which is particularly amenable to the sort of approach I like to take. It is also a broad problem—there are many kinds of spatial information and many different contexts in which it is generated and used. As a consequence, my work tends to cover a broad range of patterning problems which has the advantage of allowing unexpected connections to emerge, but the disadvantage that it is difficult to think creatively about more than a few problems at a time." Although Struhl has tackled a broad range of patterning problems, he believes that there are many viable approaches to science and that style depends much on individual temperament, approach, and scientific question. He notes "There are many paths through the forest."

Struhl is somewhat unusual among lab heads in that he works at the bench himself much of the time (Ed Lewis and Eric Wieschaus are notable examples of this hands-on approach). Struhl comments "The most fulfilling experiences I have had in science have been the moments of actual discovery. The experience is not quite the same when someone else, even under my guidance, or in collaboration, makes the discovery. This is one of my main motivations for continuing to work at the bench."

It is interesting to consider Struhl's penchant for inventing genetic traps in light of his family background. His father is an entrepreneur/inventor who prospered in business by thinking about unrequited needs that average people or retailers might have and then creating products to fill those needs. In pondering whether this Rugrats-type household had any effect on him or his brother Kevin (also a talented geneticist) as children, Struhl comments "...there is a loose analogy with identifying missing links in a scientific problem and designing strategies to resolve the problem. My parents certainly value 'sciences' in a general sense...." Perhaps the most intriguing Struhl family parallel is the remarkable similarity between the approaches of the two Struhl brothers. Struhl remarks, "What is actually more strange to me is that Kevin and I both generally develop in vivo, genetic strategies to address biological problems, which is not the tactic most people use. And what makes it strange is that Kevin and I arrived in our present areas by opposite routes, in his case via Mathematics, then Chemistry, then Molecular Biology, and in my case via Natural History then Biology. We both did go to MIT (two years apart), and we both did attend a few of the same interesting and challenging classes in phage and bacterial genetics, one in particular given by Ethan Signer. So perhaps that is where we both got interested in using genetic strategies for understanding biological problems." Or, could scientific style just be genetic?

Veins Form along A/P Boundaries in the Wing Primordium

Work performed in my own laboratory has suggested that wing veins form along boundaries between discrete territories along the A/P axis. For example, as mentioned above, the L3 and L4 veins run along the borders of the narrow A/P organizer. The best example of vein induction at a boundary is that of the L2 vein along the anterior border of the broad domain of *sal* expression (Fig. 4.2B; see also Plate 3D for actual data). We used a classic genetic method called mutant clone analysis, which was pioneered by one of the great geneticists of our time, Antonio García-Bellido, to investigate the hypothesis that the anterior border of the *sal* expression domain induced formation of the L2 vein in neighboring cells (Fig. 4.2D; see also Plate 3F for actual data). The mutant clone method is essentially the converse of the strategy used by Struhl to assess the consequences of misexpressing *dpp* in isolated groups of cells. The idea is to generate a small patch of mutant cells lacking the function of a gene that is surrounded by normal cells. Mutant clone analysis is necessary in situations where elimination of the function of the gene in all cells of the organism would kill the animal before it developed into an adult. For example, the *sal* gene is required during embryonic development for patterning along the A/P axis. As a consequence of this early requirement for *sal,* mutants entirely lacking function of the *sal* gene die during embryonic development and therefore cannot give rise to adults. Mutant clone analysis permits one to generate heterozygous animals composed predominantly of normal cells along with a few clusters of homozygous mutant cells. These animals are called mosaics to reflect the fact that their cells are not all genetically identical. Because most of the cells comprising a mosaic animal are normal (i.e., heterozygous for the mutation), the fly can survive to adulthood. One can then assess the consequence of eliminating function of the gene of interest during adult development by analyzing the defects generated in and around the small patches of homozygous mutant cells.

When we used the mutant clone method to generate small islands of *sal* mutant cells in otherwise normal wings, we often created new

Antonio García-Bellido (1936–)

It is difficult to catalog the many contributions that Antonio García-Bellido has made to the field of development during his prolific and brilliant career. One of his major contributions to the field, which led to the idea that genes existed which functioned to control development, was his role in developing the mosaic technique (referred to here as mutant clone analysis). He refined and elevated this powerful technique into an art form, which he then systematically exploited to study genes he termed selector genes (genes such as *engrailed* and *Bithorax* that function as high-order genetic switches to define cell identities). He also employed mosaic analysis to pioneer the fields of neural patterning, appendage growth control, and wing vein development.

García-Bellido was born in Madrid. He did his undergraduate studies at the Complutense University of Madrid and then worked briefly with Wigglesworth at Cambridge on larval molting in flies before returning to Madrid for his Ph.D. work, which he pursued independently on a gene now believed to play a role in adhesion between cells during development. García-Bellido did his postdoctoral studies with Ernst Hadorn in Zurich and then worked with Allan Sturtevant (one of the founders of the fly field) and one of his eminent students, Ed Lewis, at CalTech.

Many of the trailblazers who have contributed to the modern molecular analysis of development were profoundly inspired by García-Bellido's masterful genetic dissection of development, as well as his ability to deduce large amounts of information from his ingenious indirect genetic studies. To take just one example, in what is perhaps the most important paper written about the genetics of wing vein development, García-Bellido was able in 1977 to define the order in which key genes act by determining the latest stage at which eliminating the function of a gene led to developmental defects in the vein pattern. These deductions, based in part on intuitive leaps, have proven to be remarkably accurate now that it has become possible to examine the expression and developmental function of these genes in detail. Paradoxically, García-Bellido designed critical experiments which provide some of the first genetic evidence that cells engage in inductive interactions, to show the contrary (i.e., that vein defects would be limited to cells which were mutant). He recalls, "This led to the discovery of inductive interactions in clones."

In discussing his mosaic technique, García-Bellido vividly remembers the moment when he realized the potential of this method. "From the experiments on pattern reconstruction of dissociated cells arises the question of how many cell divisions *take place* from mature imaginal discs to actual cuticular differentiation. I irradiated *mwh*/+ larvae (*mwh* is a mutation that causes wing cells to make extra hairs) and collected them every eight hours at pupariation. The wings of the resulting adults carried hundreds of clones of 1–2 cells when irradiated at pupariation, but the number of clones appeared reduced by 1/2× and their size increased by 2× every eight hours. Obviously a clonal description of development was possible. Moreover, the association of cell markers and morphogenetic mutants allowed genetic mosaic analysis *to be performed* at an unprecedented level of resolution."

García-Bellido's scientific approach differs from that of most other workers in that he is more of a problem-seeking person than a problem-solving person. He finds that it is most challenging to look for new questions to be answered and relies on imagination to connect separate findings and pose questions. In his mind, the most valuable elements of discovery are rigor and obsession!

García-Bellido is currently Professor of Research of the Spanish Research Council. He continues to be very productive and maintains his provocative and imaginative style in pursuing his current interest to link the processes of morphogenesis and growth control.

boundaries between *sal*-expressing (sal^+) and *sal* nonexpressing (sal^-) cells (Fig. 4.2D, Plate 3F). In these cases, we found that the L2 vein followed the borders of sal^- patches of mutant cells. The displaced L2 vein always formed just within the patch of cells mutant for *sal*. This configuration of mutant cells mimicked the normal situation in which the L2 vein comprises cells not expressing *sal* that are adjacent to cells that do.

Induction of the L2 vein in cells lying just anterior to the *sal* expression domain is similar in several respects to induction of the A/P organizer in anterior compartment cells bordering *en*-expressing cells of the posterior compartment. In the case of the A/P organizer, *en* activates expression of the secreted signal Hh and simultaneously prevents *en*-expressing cells from responding to Hh (i.e., by activating expression of genes such as *dpp* in the A/P organizer; Fig. 4.2A). In the case of induction of the L2 vein, a simple explanation for results of the mutant clone analysis (Fig. 4.2D) is that *sal* activates expression of a short-range signal, while suppressing the response to that signal. The only cells that can respond to this hypothetical L2 inducing signal (i.e., cells that do not express *sal* but are within range of the signal) are those immediately anterior to the *sal* expression domain (see Fig. 4.3F). Thus, the primary patterning event (i.e., formation of the A/P organizer) and subsequent secondary events (e.g., induction of the L2 vein) rely on similar mechanisms to elaborate progressively finer positional information. The iterative use of a simple line-drawing mechanism in the wing is reminiscent of pattern formation in the embryo through a sequence of simple subdividing events (revisit Chapter 3).

Linking Embryonic and Adult Fly Development

It is a reflection of the maturity of the fruit fly development field that the frontiers of embryonic development and adult development have met. To illustrate this continuum of embryonic and adult development, we can consider the series of events leading from the fertilized egg to the formation of the L2 vein in the adult wing (Fig. 4.3).

Recall that where Bicoid is present at intermediate levels (Fig. 4.3A), there is a narrow stripe of cells lying between the *hunchback* and *knirps* expression domains that does not express either gap gene (Fig. 3.5). The third stripe of *eve* expression (*eve-stripe 3*) falls precisely in the small gap between the expression domains of the *hunchback* and *knirps* gap genes (see Fig. 3.6). The anterior and posterior limits of *eve-stripe-3* expression are determined, respectively, by the repressive action of *hunchback* and *knirps* (Fig. 4.3B).

The main thing to remember from this brief recapitulation of embryonic patterning is that the anterior border of *eve-stripe-3* expression is determined by the posterior extent of the *hunchback* expression domain, which in turn is set by high levels of the maternal morphogen Bicoid. The anterior border of *eve-stripe 3* is important for wing development because it corresponds to the position of the A/P border in the second thoracic segment (i.e., the segment forming wings

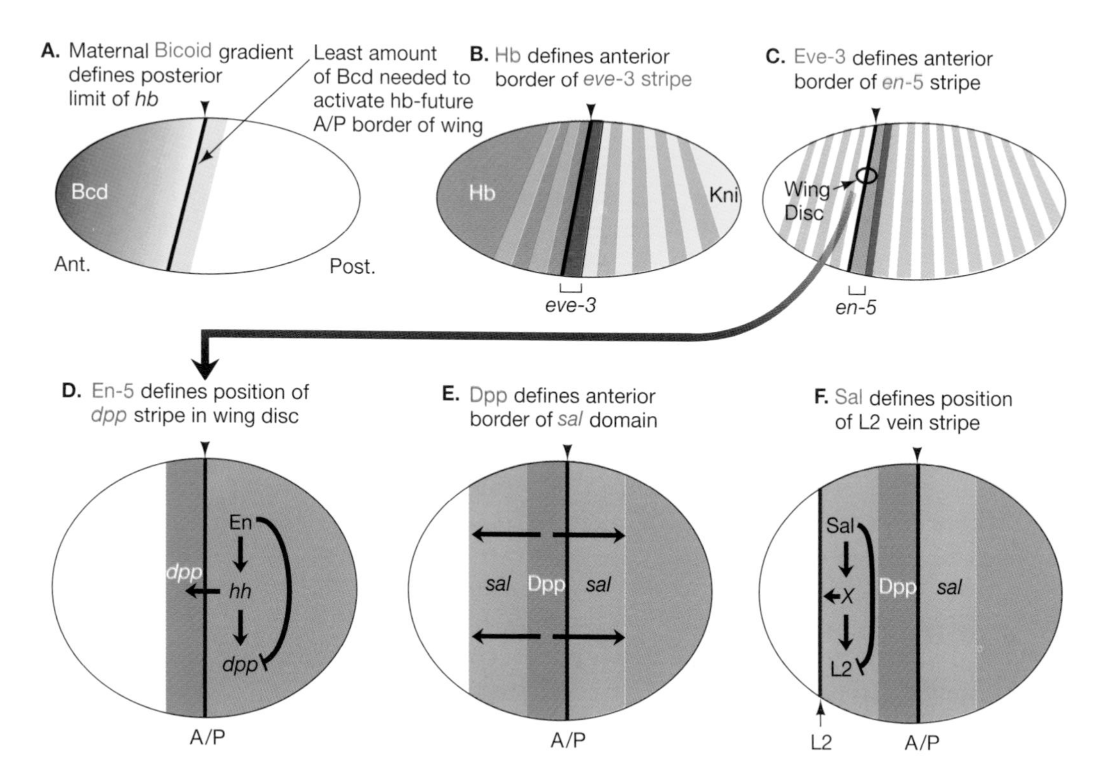

FIGURE 4.3. From Bicoid in the embryo to the L2 vein in the adult.

■ Bicoid Recap ■

Recall from Chapter 3 that the mother creates a graded distribution of the Bicoid morphogen in the egg. Bicoid is most concentrated at the anterior pole of the embryo and diminishes in concentration in progressively more posterior regions of the embryo (see Fig. 3.4). Bicoid is a transcription factor that activates expression of different gap genes at particular concentrations (Fig. 3.5). High levels of Bicoid activate expression of the *hunchback* gene in a broad anterior domain of the embryo, whereas low levels of Bicoid lead to expression of the *knirps* gene in posterior regions.

in adult flies; Fig. 4.3B). *eve* collaborates with other pair-rule genes to activate expression of the segment-polarity gene *en* in the posterior half of each segment. As discussed above, *en* defines the posterior compartment of each segment. In the second thoracic segment (corresponding to the fifth stripe of *en* expression or "*en-5*"), the anterior edge of *en* expression (i.e., the A/P boundary) is determined by the position of the anterior edge of *eve-stripe 3* (Fig. 4.3C). Thus, the position of the A/P compartment boundary in the wing can be traced back to the anterior edge of *eve-stripe 3,* which is defined by the posterior border of *hunchback* expression, which is set by a particular level of Bicoid, which is supplied in a graded concentration at the anterior end of the embryo by the mother.

As described earlier in this chapter, formation of the second longitudinal wing vein is induced at the anterior border of the *sal* expression domain (Fig. 4.2B). The width of the *sal* expression domain is determined by a particular level of Dpp emanating from the A/P organizer. Because the A/P organizer forms just anterior to the A/P compartment boundary, we now can link formation of the L2 vein to the earliest steps in A/P patterning in the embryo (Fig. 4.3). This domino effect may sound a bit like a rendition of a "Partridge in a Pear Tree," but here goes: Mom creates a graded distribution of Bicoid (Fig. 4.3A), which activates *hunchback* in the anterior portion of the embryo (Fig. 4.3B), which sets the anterior edge of *eve-stripe-3* expression (Fig. 4.3B), which determines the location of *en* expression along the A/P compartment boundary in the second thoracic segment bearing the wing disc (Fig. 4.3C), which triggers formation of the A/P organizer in the wing disc (Fig. 4.3D), which secretes Dpp (Fig. 4.3E), which activates *sal* expression in a broad central domain in the wing disc (Fig. 4.3E), which induces formation of the L2 vein at its anterior border (Fig. 4.3F). Pretty cool, eh?

Juxtaposition of Dorsal and Ventral Cells Induces a D/V Organizer

Dorsal–ventral patterning of appendages in flies provides another example of formation of an organizer at the interface of two domains of cells. In the wing, dorsal and ventral cells, like anterior and posterior cells, are separated into non-intermixing compartments. The line along which the dorsal and ventral compartments abut will form the future edge of the wing (or wing margin). Experiments performed in Steve Cohen's laboratory demonstrated that apposition of dorsal and ventral cells initiated formation of the margin at their interface. At the time, Cohen was studying a homeobox-containing gene called *apterous (ap)* which was known to be required for wing formation because mutants lacking the function of this gene have no wings. A reason for thinking that *ap* might exert its effect by defining some critical aspect of dorsal cell fate is that it is expressed only in dorsal cells. Cohen demonstrated that *ap* is required for development of dorsal cells by generating patches of cells lacking *ap* function on the dorsal surface (Fig. 4.4) using the mutant clone method described above. Cohen observed that when he generated such ap^- mutant patches of cells on the dorsal surface of the wing, an island of marginal tissue surrounded the group of mutant cells. The most telling feature of these islands of marginal tissue was the nature of the structures produced by cells at the interface between normal and mutant cells. ap^- mutant cells facing normal cells formed structures such as curly bristles, which are normally formed by

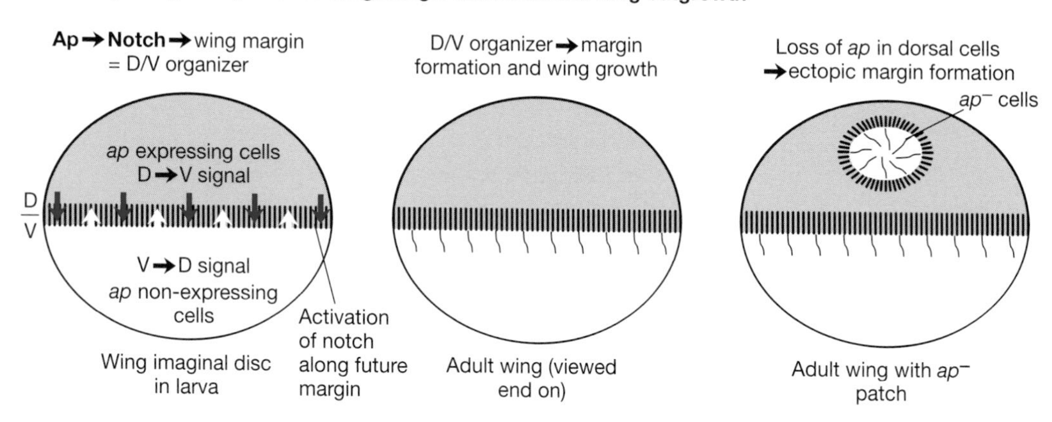

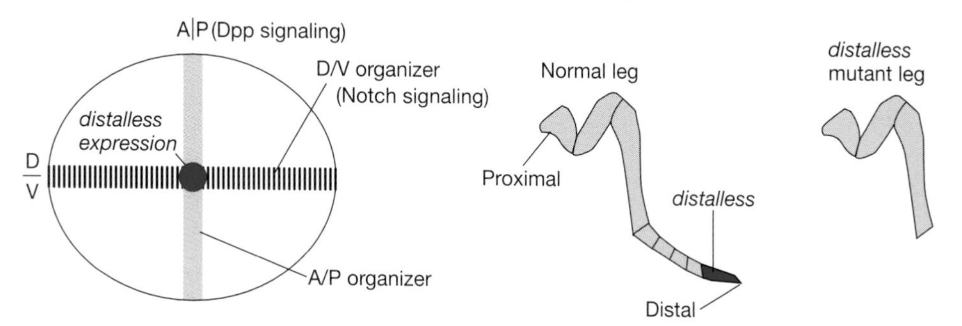

FIGURE 4.4. D/V patterning in fly appendages. *(A)* D/V signaling is required for wing margin formation and wing outgrowth. *(B)* Intersection of the D/V and A/P organizers induces leg outgrowth.

ventral cells at the wing margin, whereas normal cells facing *ap*$^-$ cells formed structures such as thick stubby bristles, which are normally produced by dorsal cells at the margin. Thus, the artificial apposition of cells having dorsal identities with those having ventral identities (i.e., the *ap*$^-$ cells) was sufficient to entrain a series of developmental events required for formation of a complex structure such as the wing margin.

Formation of a D/V Organizer Is Required for Appendage Outgrowth

As mentioned above, flies lacking the function of the *ap*$^-$ gene entirely lack wings. Because *ap* is expressed only on the dorsal surface of the wing and is thought to act by inducing the formation of a margin at the interface of the dorsal and ventral compartments, researchers such as Steve Cohen and Sean Carroll wondered whether the margin might play an important role in regulating the growth of appendages. Work from these and several other labs has confirmed this prediction and has provided a fairly detailed understanding of what takes place at the margin to promote appendage outgrowth. The essence of this model of wing margin formation is that a receptor known as Notch is activated by two different ligands tethered to the surfaces of dorsal and ventral cells of the wing, respectively (Fig. 4.4A). Activation of Notch by these ligands is required for marginal cells to assume their proper dorsal versus ventral identities and is also required to promote outgrowth of the wing. Mutants entirely defective for this pathway, such as *ap*$^-$ mutants, lack all wing structures because the wing disc cells fail to proliferate. Partial loss-of-function mutants manifest various defects such as notches along the wing margin (hence the name of the Notch receptor), which

■ Double Role of Dpp Signaling ■

It turns out that Dpp signaling along the A/P compartment boundary also plays a critical role in regulating wing growth as well as in patterning the A/P axis of the wing. In fact, the intersection of the A/P and D/V axes defines the center of appendage outgrowth, a phenomenon which has been particularly well studied in the leg. In the leg, Cohen's group has shown that cells lying at the intersection of the A/P and D/V organizers express another homeobox gene called *distalless,* which plays an organizing role in defining the proximal–distal axis of appendages (Fig. 4.4B). For example, legs lacking the function of *distalless* lack the distal portions of the leg. As shown in Chapter 6, specification of the distal tip of appendages by the *distalless* gene is a general feature of appendage development in both invertebrates and vertebrates. It is perhaps not too surprising that organizing centers of developing appendages not only control cell fates, but also regulate cell growth. After all, these two processes—growth and patterning—must be coordinated to generate an appendage of the correct size and shape. Antonio García-Bellido has also played an important role in analyzing the relationship between growth and patterning during development. His studies using mutant clone analysis have revealed the remarkable degree to which the proliferation of cells in one part of the wing depends on the growth potential of cells in adjacent territories.

reflects the fact that cells along the margin are most dependent on the full operation of this pathway (see Plate 3G for actual Notch mutant wing). We show in Chapter 8 that a remarkably similar process seems to be operating during development of leaves in plants to define marginal structures of the leaf and to promote its outgrowth. Thus, the D/V organizer is involved not only in patterning the margin and nearby structures, but also in regulating growth of the wing.

HOW TO MAKE FLY EYES

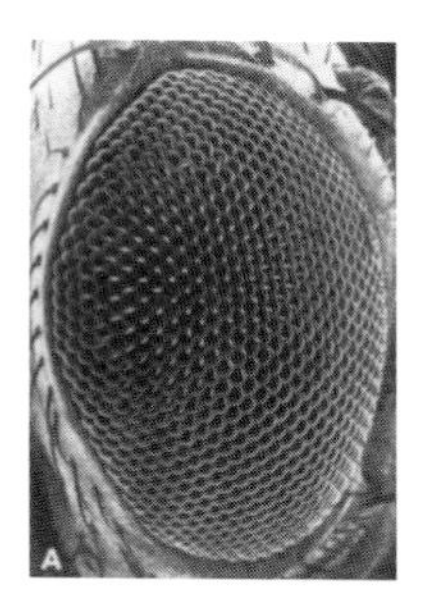

Fly eye

The eye, like other adult structures in the fly, derives from an imaginal disc. Flies have typical insect compound eyes composed of an array of more than 800 identical facets. Each of these facets contains its own tiny lens, an array of different light-sensitive neurons called photoreceptors, and several other cells such as pigment cells that serve structural roles. Facets of the compound eye send out bundles of neural processes that connect to the primary optic center in the brain in an orderly pattern. Thus, nerves emanating from two adjacent facets contact neighboring regions of the brain, and nerves originating from facets on opposite sides of the eye connect to opposing regions of the brain. As the mapping of facets to the brain is orderly and reflects the relative positions of facets in the eye, a given facet senses a specific region of the fly's visual world. The vertebrate eye contacts the brain in a similarly organized fashion. Thus, contrary to the portrait of insects typically rendered in science fiction movies, flies are likely to see the world much as we do and not through a multitude of separate images combined as in a kaleidoscope.

Cells comprising the eye imaginal disc are set aside during embryonic development and proliferate during larval stages to generate a sack-like structure similar to other imaginal discs (Fig. 4.5A, B). During late larval development, small clusters of cells in the eye disc communicate with one another to determine which type of light-sensitive photoreceptor cell they will become (Fig. 4.5C, D). Each of these interacting clusters of photoreceptor cells will become a single facet in the adult eye. The clusters are assembled in a defined sequence, which is organized in a spatial and temporal pattern. Cells in the posterior region of the eye disc are the first to initiate formation of evenly spaced photoreceptor clusters, and then more anterior cells begin to undergo development. This temporal delay in the development of posterior and anterior cells means that clusters in the posterior region of the eye represent a later phase of the developmental process than those in more anterior regions (Fig. 4.5C). Thus, a single eye primordium contains the full spectrum of stages of facet development. The outcome of the successive developmental events in cluster formation is the creation of a compound eye comprising an array of identical facets, which is discussed in more detail below.

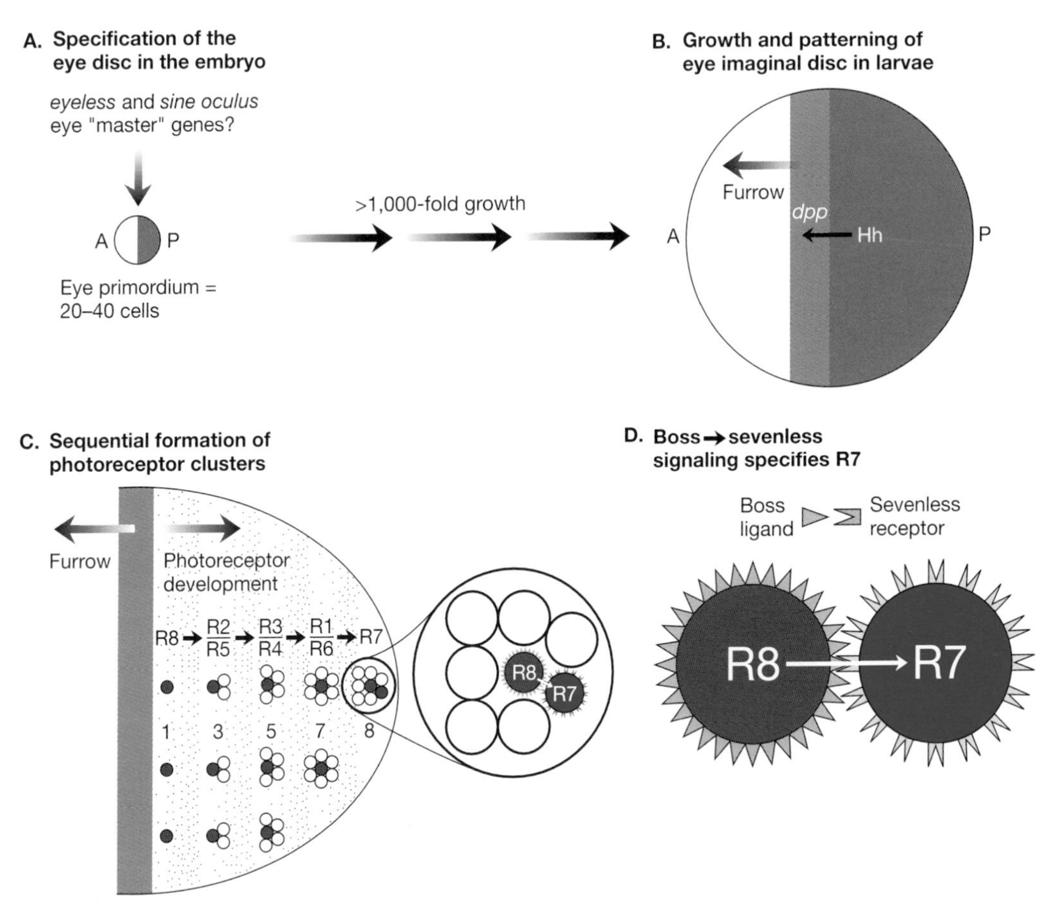

FIGURE 4.5. Eye development in flies.

Genes Required for Making Eyes

Several genes are essential for making eyes in flies. These genes go by eye-opening names such as *eyeless, sine oculus, eyes absent,* and *dachshund.* Loss-of-function mutants in any of these genes have reduced eyes or no eyes at all. Significant attention has been focused (so to speak) on the *eyeless* gene since there is a corresponding vertebrate gene called *pax-6,* which also plays an essential role in initiating eye development (see Chapter 6). The *eyeless* gene is expressed in an anterior region of the embryo including the region from which the eye disc derives. *eyeless* encodes a transcription factor distantly related to homeotic genes and is required to activate expression of eye-specific genes such as those encoding light-sensing molecules in photoreceptor cells. *eyeless* also is required for activating or maintaining expression of other genes involved in early eye formation. Thus, the *eyeless* gene plays a positive role in eye development akin to that played by the *twist* gene in mesoderm formation during embryonic development.

Making Eyes in Wings

The most striking evidence that the *eyeless* gene plays an important role in initiating eye development derives from misexpression experiments. Recall that in general, one expects opposite effects to result from reducing the function of a gene versus overexpressing or misexpressing that gene. Misexpression of the *eyeless* gene was performed in Walter Gehring's laboratory in Basel, Switzerland. The Gehring group made the dramatic discovery that misexpression of the *eyeless* gene in several imaginal discs other than the eye disc transformed them into eye primordia, which then resulted in well-organized ectopic eyes in the adult. For example, when *eyeless* was misexpressed in groups of cells in the wing imaginal disc, adult flies emerged with little eyes popping out of the surface of their wings (see Plate 3H for example of a wing derived from a fly expressing the human counterpart of *eyeless*, *pax*, in the developing wing primordium). These freaky wing eyes were fully formed and even sent nerves out to contact the brain! Imagine how weird it would be if you had little eyes on your finger tips that saw everything you touched!

More recently, it has been shown that misexpression of another eye gene called *dachshund* can also induce formation of ectopic eyes. These tissue transformation experiments indicate that a genetic system determines the formation of eyes versus other structures. It should be noted, however, that misexpression of either the *eyeless* gene or the *dachshund* gene alone cannot transform all parts of the fly into eyes. Thus, additional factors must contribute to defining eyes.

Is There a Master Gene for Eye Development?

Because loss-of-function mutations in the *eyeless* gene result in lack of eyes and misexpression of *eyeless* in other imaginal discs can generate ectopic eyes, this gene has been proposed to act as a "master" gene for eye development. Although it is clear that *eyeless* plays a central role in eye development, many investigators in the field, including myself, believe that the term "master gene" is somewhat misleading and overstates the actual role of the *eyeless* gene. There are two major problems with the idea that this gene can single-handedly determine the identity of eyes. First, *eyeless* does not function in a void. For a gene such as *eyeless* to act, other genes such as *sine oculus, eyes absent,* and *dachshund* also must function. Loss-of-function mutations in any of these genes lead to elimination of eyes. In addition, misexpression of at least one of these other genes (*dachshund*) can generate ectopic eyes. The functions of the *eyeless* gene and *dachshund* genes are very intertwined because misexpression of either gene activates expression of the other. Furthermore, induction of ectopic eyes by misexpression of *eyeless* requires the function of the *dachshund* gene. Thus, if one appoints *eyeless* the master eye gene, several other genes must also be admitted into the hall of governors. One typically does not think of a master as sharing power to such a great degree.

The second problem regarding the master gene hypothesis is that misexpression of *eyeless* and *dachshund* can only transform certain imaginal discs, such as wings or legs, into eyes. Other discs are indifferent to the presence or absence of *eyeless*. Thus, the parliament of eye-specifying genes apparently wields much less power than a king because only certain provinces respond to orders issued from the governing body. Although the image of an all-powerful monolithic ruler may derive from our familiarity with primitive political regimes, the reality of eye development seems best approximated by a collaborating group of genes that interact to stabilize decisions made by committee. Among its distinguished colleagues, *eyeless* is undoubtedly a significant voice, but probably not a master. In analogy to electronics, the *eyeless* gene could be likened to an essential circuit component in a multifunctional device such as a stereo receiver, which is required for the function of the CD player, rather than to a controlling toggle switch that determines whether the receiver will activate the CD player, tape deck, or radio.

Signals Initiate Photoreceptor Development

Photoreceptor development consists of an orderly series of events spanning several hours. As mentioned above, cells in posterior positions in the eye imaginal disc begin to organize themselves into clusters of photoreceptors earlier than cells in more anterior positions of the disc. Once the process of photoreceptor cell development is initiated, the eye disc can be subdivided into three dynamic regions: an anterior-most domain in which cells are yet unpatterned, a middle region where photoreceptors are assembled into clusters, and a posterior-most domain consisting of fully formed mature facets (Fig. 4.5C). Throughout eye development there is a dividing line in the disc separating unpatterned cells from cells beginning to organize themselves into photoreceptor clusters. This line is marked by a morphologically visible crease in the imaginal disc called the furrow (Fig. 4.5B, C). Early during photoreceptor development, the furrow is located in the posterior extreme of the disc. Over the course of a day, the furrow moves progressively in an anterior direction until it reaches the most anterior edge of the eye disc. At any given point during its excursion across the eye, the furrow can be viewed as a developmental wave moving forward into unpatterned cells. If one uses the position of the furrow as a reference point, cells lying just behind the furrow will be initiating photoreceptor development and cells in more posterior positions will be at progressively later stages of cluster assembly, since they will have initiated development earlier (Fig. 4.5C).

Given that the furrow marks the beginning of photoreceptor development, it is important to understand how it moves from posterior to anterior in the eye disc. Because cells in the eye disc are stationary, the anterior propagation of the furrow into unpatterned cells must be mediated by secreted signals that are liberated by cells behind or within the furrow. Perhaps it will come as little surprise that two key genes involved in progression of the furrow along the A/P axis are *hh* and *dpp* (Fig. 4.5B). *hh* is expressed in all cells posterior to the furrow, and the *dpp* gene is expressed in cells within the furrow itself. This arrangement of gene expression in the eye disc is very similar to that observed in other imaginal discs that have fixed anterior and posterior compartments. For example, in the wing imaginal disc, *hh* is expressed in the posterior compartment and *dpp* is expressed in a narrow stationary strip of abutting cells just within the anterior compartment comprising the A/P organizer (Fig. 4.2A).

In several respects, the furrow in the eye disc can be viewed as a propagating A/P organizer. To begin with, the relationship between Hh and Dpp protein signaling in the eye disc is essentially the same as that in the wing disc. The Hh protein leaks into virgin unpatterned disc cells and activates expression of Dpp in cells comprising the furrow. These furrow cells are close enough to the *hh*-expressing cells to receive the Hh signal. As in the case of the wing disc, *hh*-expressing cells are refractory to the Hh signal and therefore do not express Dpp themselves in response to Hh. A unique feature of the relationship between Dpp and Hh in the eye disc is that Dpp signaling can activate *hh* expression to create a positive feedback loop. Posterior-to-anterior progression of the furrow results from the combined action of Hh diffusing forward to activate *dpp* expression and Dpp diffusing forward to activate *hh* expression. In contrast to the furrow, the A/P organizer in the wing disc

■ Transforming a Stationary A/P Organizer into a Moving Furrow ■

Gary Struhl and his collaborator Conrad Basler did a very clever experiment to show how easy it is to transform a stationary A/P organizer into a propagating furrow-like developmental wave. They hooked the regulatory region of the *dpp* gene responsible for *dpp* expression in the A/P organizer of the wing disc to the coding region of the *hh* gene. This hybrid *dpp/hh* gene was then inserted into flies. Since the *dpp* regulatory region is activated by Hh signaling, the *dpp/hh* hybrid gene is likewise activated in cells receiving the Hh signal. These cells then send the Hh signal to the next group of anterior cells, which respond by turning on *dpp,* and so on. In wing discs carrying the *dpp/hh* hybrid gene, Struhl and Basler witnessed an anterior progression of the A/P organizer. In essence, the *dpp/hh* hybrid gene short-circuited the need for Dpp in completing a positive feedback loop by creating a direct Hh→*hh* autoactivation loop. This artificial experiment in the wing illustrates how a single evolutionary invention resulting in Dpp being able to activate *hh* expression in the eye disc may have been sufficient to convert a stable A/P organizer into a moving furrow.

does not move forward because Dpp signaling cannot activate *hh* expression in the wing.

Photoreceptor Cell Types Are Induced Sequentially in the Eye

Because there is a moving wave of development in the eye disc, photoreceptor clusters located at different distances from the furrow are at different stages of development (Fig. 4.5C). As mentioned earlier, clusters immediately behind the moving furrow are undergoing the very first stage of patterning, cells a bit farther back are in intermediate stages of development, and cells far behind the furrow are in the terminal stages of development. Thus, by looking at cluster formation at different distances from the furrow, one gets a snapshot of the full sequence of developmental events in a single eye imaginal disc.

Each mature facet of the eye consists of a collection of eight different photoreceptor cells, which are referred to as R1–R8. The first of these photoreceptors to form is R8. R8 can be distinguished from the other photoreceptor cells by virtue of its position within clusters and by its pattern of gene expression. The R8 cells are first visible just posterior to the furrow as a single row of cells. A short distance back behind the furrow, the R8 cell is part of a three-cell cluster that also contains R2 and R5. A bit farther yet from the furrow, five-cell clusters form consisting of R8, R2, R5, R3, and R4. The next cluster group in this developmental series contains seven photoreceptors (R8, R2, R5, R3, R4, R1, and R6). Finally, far from the furrow, fully formed clusters of all eight photoreceptors can be found.

The sequential addition of photoreceptors to growing clusters is the result of a series of inductive events. The first row of photoreceptor cells (R8 cells) is specified by its direct proximity to the furrow. R8 cells in this first row are spaced evenly from one another by virtue of a mutual inhibitory form of cell–cell signaling (discussed in Chapter 1). All cells in the row behind the furrow would like to become R8 cells. These initially equivalent cells engage in mutual inhibitory signaling in an attempt to suppress each other from becoming R8 cells. Ultimately, some cells win this competition and become R8 cells, which then force their neighbors in the same row to remain developmentally uncommitted. A stable pattern of evenly spaced R8 cells emerges from such a mutual inhibitory mechanism. In such an array, no two R8 cells are close enough to inhibit each other from becoming R8 cells. In addition, neighboring R8 cells collaborate to suppress their intervening neighbors from becoming R8 cells. Mutants lacking one of several genes required for this mutual inhibitory mechanism form a solid row of R8 cells behind the furrow because all cells will assume the R8 identity unless told to do otherwise by their neighbors.

■ The Sevenless Story ■

One of the most interesting mutations disrupting eye development is the *sevenless* mutant. *sevenless* was discovered by Bill Harris, a talented graduate student in the laboratory of Seymour Benzer, who is one of the founders of the field of neural genetics. As its name implies, the *sevenless* mutation results in the specific loss of the R7 photoreceptor cell. The remaining photoreceptors are formed in the usual sequence, however, and facets of adult eyes appear normal except that they lack R7 photoreceptors. Because the R7 photoreceptor is responsible for flies being able to sense ultraviolet light, these mutants can see visible light, but are blind to ultraviolet light frequencies. Harris used the method of mutant clone analysis described above to show that the only cell in the developing eye disc that required the function of the *sevenless* gene was the R7 cell itself. This result suggested that the *sevenless* protein product functioned within the R7 cell to receive an inductive signal. Consistent with the Harris hypothesis, the *sevenless* gene was subsequently shown by workers in Seymour Benzer's and Gerald Rubin's lab to encode a receptor type protein which is expressed in R7 as well as in several other photoreceptors.

Because the Sevenless protein seemed likely to be involved in receiving an inductive signal from a neighboring cell(s), a search was undertaken for the hypothetical signal. The search was successful when Larry Zipursky at UCLA identified a second mutant which, like *sevenless,* specifically lacked the R7 cell in the eye. Zipursky humorously named this new mutant *bride-of-sevenless* or *boss* for short. Mutant clone analysis of *boss* revealed that this gene differed from *sevenless* in that it was only required in the R8 photoreceptor. On the basis of this result, Zipursky considered the possibility that the *boss* gene might encode a signal produced by the R8 cell which was received by the Sevenless receptor in a neighboring cell. When Zipursky cloned the *boss* gene, he confirmed this genetic prediction by showing that the *boss* gene was expressed only in the R8 cell and that Boss protein could bind to and activate the Sevenless receptor in adjacent cells, causing them to differentiate as R7 cells.

Once an evenly spaced row of R8 cells is established, the R8 cells signal to their immediate posterior neighbors to become R2 and R5. The result of this signaling event is the formation of a three-cell cluster consisting of R8, R2, and R5, which lies just posterior to the row of single R8 cells. If one disrupts the R8 → R2, R5 inductive signaling event, multiple rows of single R8 cells form posterior to the furrow. In a second inductive event, cells in the three-cell cluster induce two neighboring undifferentiated cells to become the R3 and R4 photoreceptors. These five-cell clusters then recruit R1 and R6 to generate a seven-cell cluster missing only the R7 cell. To complete the formation of the photoreceptor cluster, the R8 cell induces an undifferentiated cell to become R7. This last stage in photoreceptor development is the most heavily studied and best-understood step in the series of inductive events required for facet formation.

Genetic analysis of R7 development by many labs (see box above) has provided the best-understood example of a signaling pathway in which nearly all components are known, cloned, and characterized. Like the Wieschaus, Nüsslein-Volhard, and Jürgens genetic dissection of embryonic patterning, analysis of the Sevenless signaling pathway in the eye is one of the shining facets of *Drosophila* genetics.

■ Summary ■

The adult fly is formed during metamorphosis by the amalgamation of several independently forming primordia known as imaginal discs, which give rise to structures such as legs, eyes, and wings. Imaginal discs arise during early embryonic development and remain quiescent in the form of small sacks of cells until larval development, when they grow and become patterned to generate distinct structures. Genes such as *vg* or *eyeless,* which are expressed in localized regions of the embryo, determine the identity of discs (e.g., the wing versus eye, respectively) in much the same way that homeotic genes determine segment identity.

Patterning of imaginal discs takes place along three interrelated axes referred to as the anterior–posterior (A/P) axis, the dorsal–ventral (D/V) axis, and the proximal–distal (P/D) axis. The A/P axis of imaginal discs is initially subdivided into anterior versus posterior parts—or compartments—by the action of the *en* gene, which is expressed in posterior cells and defines posterior cell identities. *en* initiates the next phase of A/P patterning by simultaneously activating expression of the *hh* gene, which encodes a short-range diffusible signal (Hh) while suppressing response to the Hh signal. The result of this dual activity of *en* is activation of genes responsive to Hh signaling, such as *dpp,* in a narrow stripe of cells known as the A/P organizer lying within the anterior compartment of the wing. These A/P organizer cells are close to, but not within, the posterior compartment where Hh is produced. The *dpp* gene, which is expressed in the A/P organizer, encodes a long-range diffusible morphogen (Dpp) that activates expression of different genes in a concentration-dependent fashion along the entire A/P axis of the wing. One example of how the A/P position of a specific adult structure can be tied to a particular level of Dpp activity is the second wing vein. This vein forms just anterior to the border of a broad central domain of expression of the *sal* gene, which is activated by intermediate levels of Dpp.

As with the A/P axis, juxtaposition of dorsal and ventral cells of the wing disc induces formation of an organizer along the interface between these cells. In this case, the *ap* gene, which is expressed only in dorsal cells, defines the identity of dorsal cells and leads to asymmetric activation of the Notch receptor in cells on the two sides of the D/V boundary. In addition to specifying the identities of cells along the A/P and D/V axes, the A/P and D/V organizers play a role in appendage outgrowth along the proximal–distal axis. The *distalless* gene, which is activated at the intersection of the perpendicular A/P and D/V organizers, organizes outgrowth of appendages such as legs along the P/D axis.

The fly eye also forms from an imaginal disc, which is specified by the concerted action of a group of genes including *eyeless* and *dachshund.* The eye consists of a repeating array of photoreceptor cells that are organized into facets. Each facet arises from a cluster of cells in the eye imaginal disc, which self-assemble in response to a series of inductive signaling events. Photoreceptor development is initiated at the furrow, which is a morphologically visible crease in the eye disc. The furrow sweeps across the eye disc in a posterior-to-anterior direction as a result of a positive feedback loop between Hh signaling posterior to the furrow and Dpp signaling within the furrow. As a consequence of the progressive motion of the furrow, each eye disc contains clusters at different stages of development. Clusters nearest the furrow are in the earliest stages of development, and those in the most posterior portion of the disc are most developed. The final and best-studied step in photoreceptor specification is induction of the R7 photoreceptor cell by the R8 cell. The critical event in this induction is activation of the Sevenless receptor in R7 cells by the membrane-tethered Boss signal that is expressed in R8 cells.

Development of vertebrate appendages and eyes is considered in Chapter 6. An important theme of Chapter 6 is that many of the mechanisms discussed in this current chapter for patterning the primary axes of appendages and specifying eyes are shared between vertebrates and invertebrates.

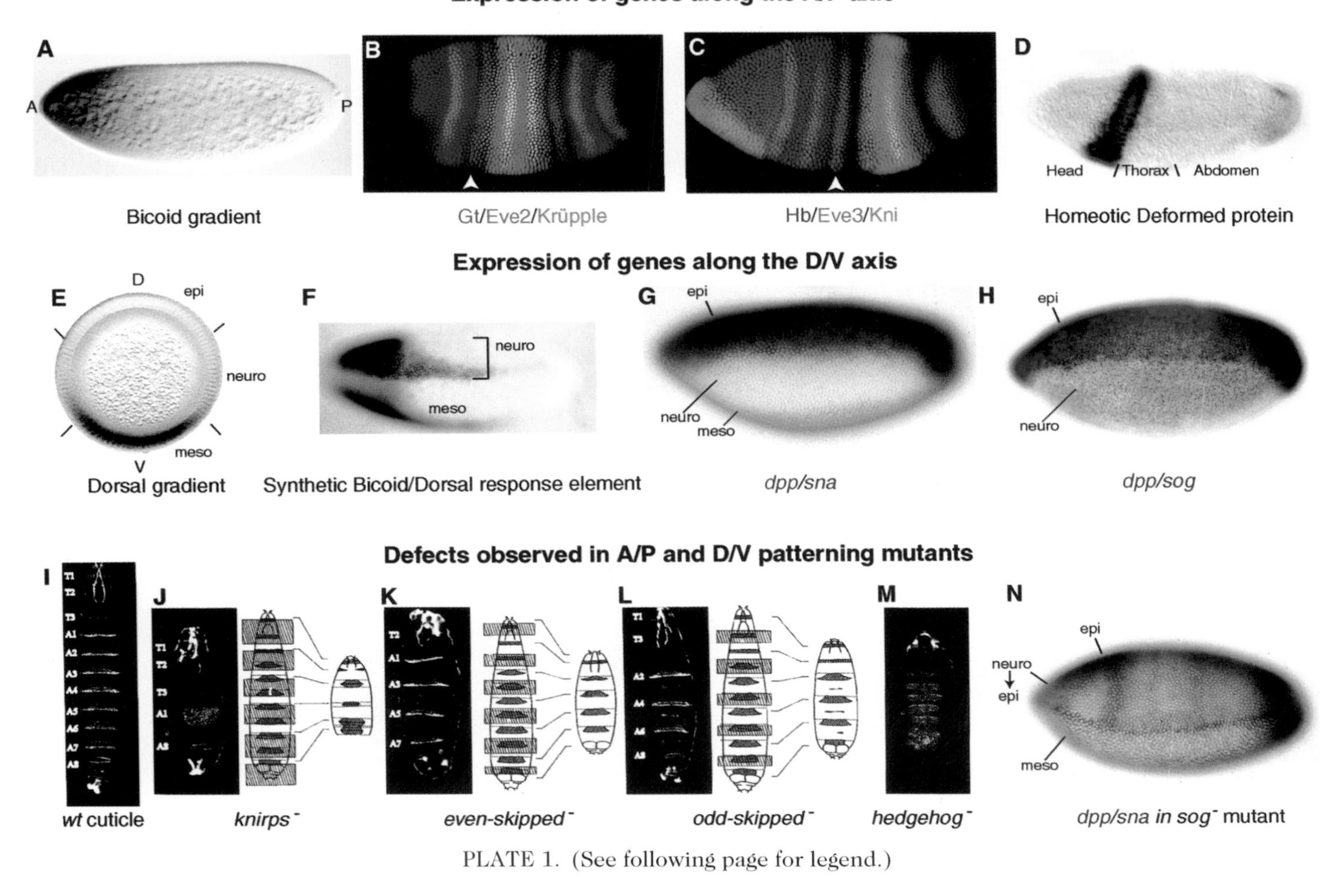

PLATE 1. (See following page for legend.)

PLATE 1. (*A*) Bicoid protein is most concentrated anteriorly (A) in the head and fades progressively in the posterior (P) direction. Dorsal to the top in panels *A–H* and *N*. (*B*) A triple-labeled blastoderm-stage fly embryo showing the relative expression patterns of the Gt (*purple*), Eve (*red*), and Krüpple (*green*) proteins. The arrowhead points to the Eve-2 stripe between the Gt and Kr expression domains. (*C*) A triple-labeled blastoderm-stage fly embryo showing the relative expression patterns of Hb (*blue*), Eve (*red*), and Kni (*green*). The arrowhead points to the Eve-3 stripe between the Hb and Kni expression domains. (*D*) Expression of the Deformed homeotic protein in a blastoderm embryo, which specifies the identity of a posterior head segment. (*E*) A cross-section through a blastoderm embryo. Dorsal protein levels are high in ventralmost cells (meso = mesoderm), low in lateral cells (neuro = neuroectoderm), and absent in dorsal cells (epi = epidermis). (*F*) A synthetic fusion gene consisting of a regulatory region with both Bicoid and Dorsal DNA-binding sites joined to the bacterial *lacZ* coding region is expressed in a wedge shape that tapers off in both the posterior and ventral directions on a cell-by-cell basis. (*G*) Expression of *dpp* (*blue*) and *sna* (*brown*) RNA in a blastoderm-stage embryo. Dorsal *dpp*-expressing cells give rise to epidermis (epi), lateral unstained cells give rise to neural and epidermal cells (neuro), and ventral *sna*-expressing cells (meso) give rise to mesoderm (e.g., muscle and heart). (*H*). Expression of *dpp* (*blue*) and *sog* (*brown*) RNA in a blastoderm-stage embryo. The *dpp* and *sog* domains abut. Ventral mesodermal cells lie out of the plane of focus. (*I*) A wild-type (e.g., normal) larval cuticle (i.e., skin) viewed from a ventral perspective. The periodic pattern of ventral denticles (hairs) appears white against the dark background. T1–T3 = thoracic segments and A1–A8 = abdominal segments. The head segments have folded back inside the larva to form internal structures by this stage. Anterior is to the top in panels *I–M*. (*J*) A *knirps*$^-$ mutant cuticle (*right*) and interpretive diagram (*left*). Note the missing abdominal but not thoracic segments. (*K*) The cuticle of an *even-skipped*$^-$ mutant (*right*) and interpretive diagram (*left*). Note the absence of even- but not odd-numbered segments. (*L*) An *odd-skipped*$^-$ mutant cuticle (*right*) and diagram (*left*). Note the absence of odd- but not even-numbered segments. (*M*) A *hedgehog*$^-$ mutant cuticle (*right*) and diagram (*left*). Anterior portions of each segment, which normally do not make denticles, do so in this mutant because these cells have been transformed into posterior-like cells. (*N*) Expression patterns of *dpp* (*blue*) and *sna* (*brown*) RNA in a *sog*$^-$ mutant. In this embryo, there is no intervening white neuroectodermal domain between *dpp*- and *sna*-expressing cells as there is in wild-type (compare with panel *G*).

PLATE 2. (*A*) β-Catenin protein expression (*red*) in an 8-cell-stage frog embryo is highest in nuclei of dorsal cells (*arrow*). The embryo is oriented with dorsal (D) to the top, ventral (V) to the bottom, animal pole (An) to the left, and vegetal pole (Vg) to the right. (*B*) *chordin* expression (*blue*) in a gastrulating frog embryo viewed from a vegetal perspective. *chordin*-expressing dorsal mesodermal cells have folded underneath the neuroectoderm by this stage. (*C*) *bmp4* expression (*blue*) in a gastrulating frog embryo viewed from a vegetal perspective is largely complementary to that of *chordin* (panel *B*). (*D*) A wild-type tadpole viewed from the side with anterior (head) to the left. (*E*) A tadpole derived from an embryo in which *sog* RNA was injected into ventral cells. This tadpole has an additional neural axis (*arrow*), which is very similar to that observed in the classic transplantation experiments performed by Hilde Mangold. (*F*) A wild-type heart-stage mustard embryo oriented with the apical (A) pole to the top and basal (B) to the bottom. Cotyledons (c), embryonic leaves, and the suspensor (s), which connects the embryo to the yolk are indicated. (*G*) A *gurke* mutant embryo lacks the primordia for the cotyledons. (*H*) A *monopteros* mutant embryo lacks the basal primordium that gives rise to roots. (*I*) A section of a wild-type heart-stage embryo labeled to reveal the pattern of *shootmeristemless* (*stm*) expression (*purple*), which is confined to the apical shoot meristem that generates all above-ground structures of the plant. (*J*) A wild-type seedling viewed from an apical perspective. The arrow points to a stem-like structure produced by a normal apical shoot meristem. (*K*) A *stm* mutant seedling viewed from an apical perspective. The arrow points to the lack of the stem-like structure produced by a normal apical shoot meristem (see panel *J*). (*L*) A cross-section of a wild-type heart-stage embryo labeled to reveal the pattern of *cup-shaped cotyledon 2*(*cuc2*) expression (*blue*). The plane of this section is indicated in the inset diagram. (*M*) An intact wild-type heart-stage embryo with the lobed cotyledons at the top. (*N*) A wild-type seedling viewed from the side with apical to the top and basal to the bottom. (*O*) A *cuc1; cuc2* double mutant embryo in which the prominent lobes of the cotyledons are replaced by a circularly symmetric lip (compare with panel *M*). (*P*) A *cuc1; cuc2* double mutant seedling in which the cotyledons fuse to form a single continuous cup-shaped leaf (compare with panel *N*).

Frog Embryonic Development

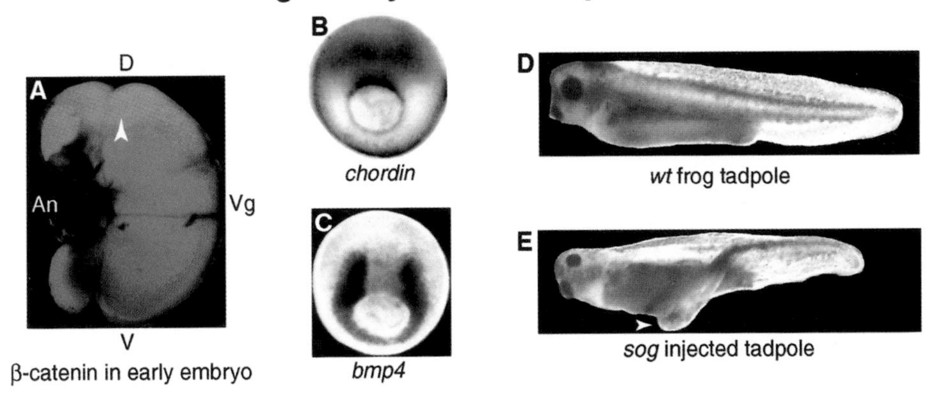

Mustard Embryonic Development

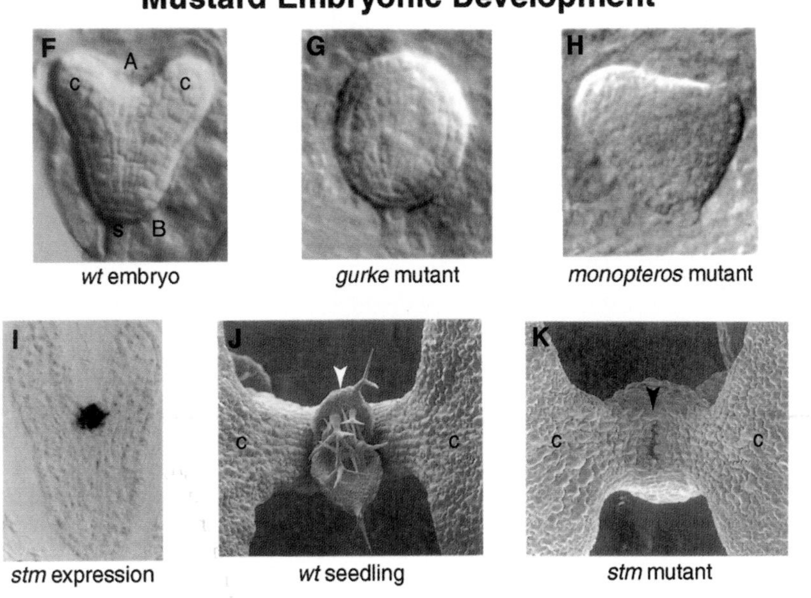

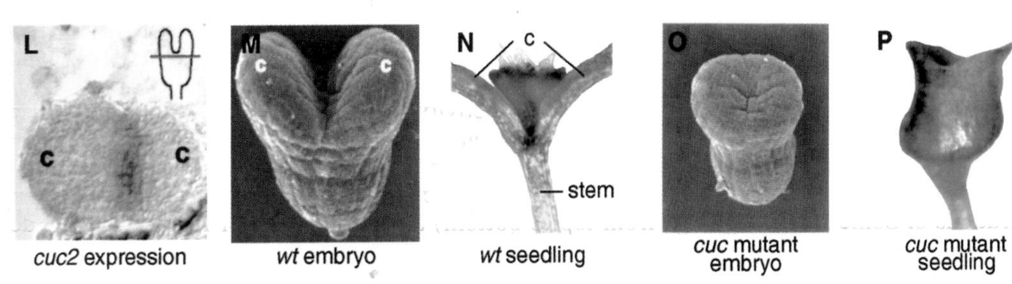

PLATE 2

PLATE 3. (*A*) Wild-type wing imaginal disc from a late-stage larva double-labeled to reveal the pattern of *dpp* (*blue*) and *spalt* (*sal*, *brown*) expression. The disc is oriented with anterior (A) to the top and posterior (P) to the bottom. The compartment border is indicated as A/P. (*B*) Wing imaginal disc misexpressing the *dpp* gene (*arrow*). This disc is double-labeled and oriented as in *A*. The patch of *dpp*-misexpressing cells is contained within a larger circle of *sal*-expressing cells, consistent with the Dpp product diffusing from its site of production into adjacent cells where it can activate *sal* expression. (*C*) High-magnification view of a wing disc containing flip-out clones misexpressing *dpp* in isolated patches of cells (e.g., *arrows*). (*Left panel)* Pattern of Sal protein expression (*green*); (*right panel)* pattern of Omb protein expression (*red*); (*middle panel*) combined expression of Sal and Omb (cells expressing both products are yellow and lie within a red Omb-expressing halo). The key point of this experiment is that the patches of Sal-expressing cells lie within larger Omb expressing domains, consistent with the view that lower levels of Dpp (not seen directly in this panel) are required to activate *sal* than *omb* expression. (*D*) Wild-type wing imaginal disc double-labeled to reveal the pattern of vein primordia (*blue*) and *sal* expression (*brown*). Note that the primordium for the second vein (L2) forms just anterior to the *sal* expression domain. Primordia of the second through fifth veins are indicated L2–L5. (*E*) Wild-type wing. The second through fifth veins are indicated L2–L5. The wing is oriented with anterior to the top and posterior to the bottom. (*F*) High-magnification view of a wing containing a sal^- clone between the L2 and L3 veins. An ectopic L2 vein (L2′) forms within the clone of cells lacking *sal* (outlined in red). The clone of cells that is mutant (e.g., sal^-/sal^-) is indicated by the minus sign, and neighboring wild-type or $sal^{-/+}$ cells are indicated by a plus sign. (*G*) Notch mutant wing that is elongated and lacks structures normally located near the edge (or margin) of the wing. (*H*) Wing derived from a fly misexpressing the human *pax6* gene (= fly *eyeless* gene) in the developing wing primordium. A fully formed ectopic eye is popping out of the wing. (*I*) High-magnification view of a fly head showing the antenna. (*J*) Fly leg with the distal segments (*bracket*) and claw indicated. (*K*) Head derived from a fly that misexpressed the mouse *hox6* gene during development. The antenna has been partially transformed into a leg, as revealed by the presence of distal leg segments (*bracket*) and a claw. (*L*) Chick wing bud labeled to reveal the pattern of Sonic hedgehog expression (*blue*). Anterior (A) is to the top and posterior (P) is to the bottom. Sonic Hedgehog is expressed in the zone of polarizing activity (ZPA). (*M*) Wild-type chick wing (*upper*) and a wing from a chick that received a ZPA graft in the anterior portion of the wing bud (*lower*). The chick that received the ZPA graft has a wing with mirror duplicated digits (*bracket*). (*N*) Series of wings from chicks that were exposed to progressively higher concentrations of Sonic Hedgehog (e.g., from top to bottom) applied to anterior regions of their developing wing buds. Note the dosage-dependent effect on the extent of digit duplication. Low doses of Hedgehog result in duplication only of anterior digits whereas high doses generate a complete A/P pattern duplication.

Fly appendage development

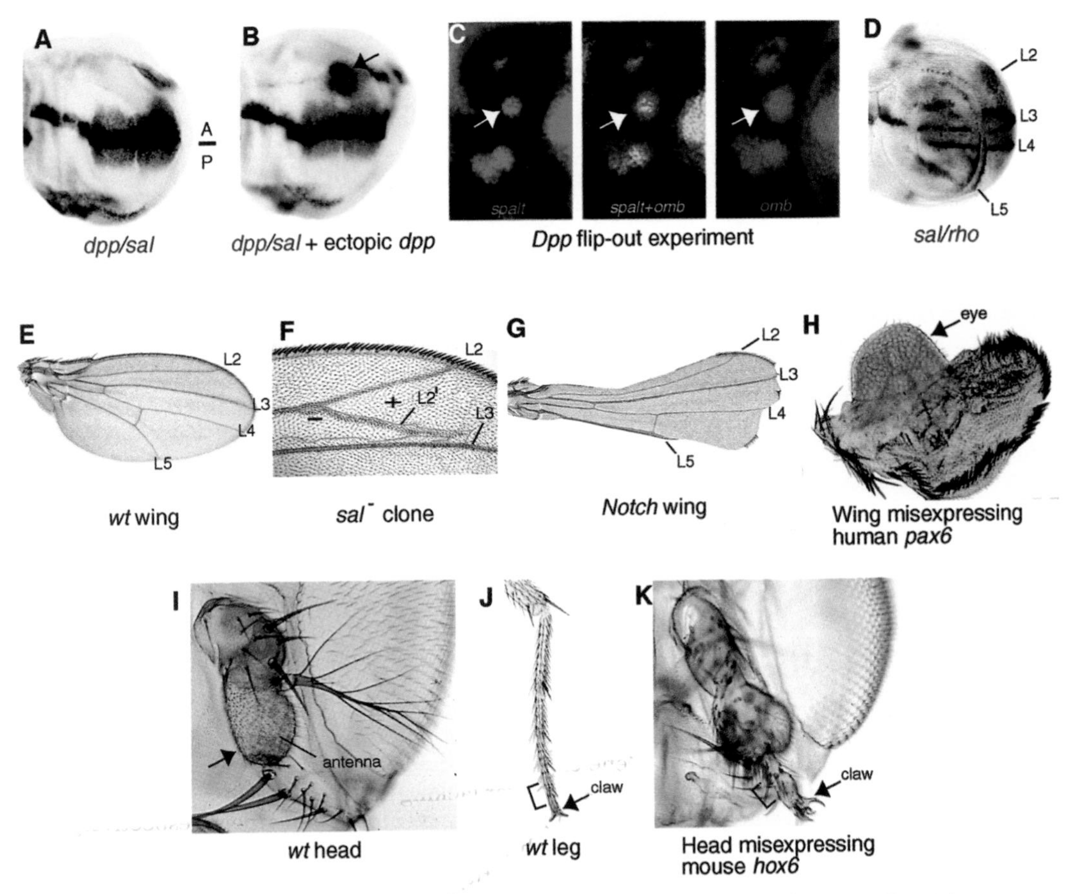

Vertebrate appendage development

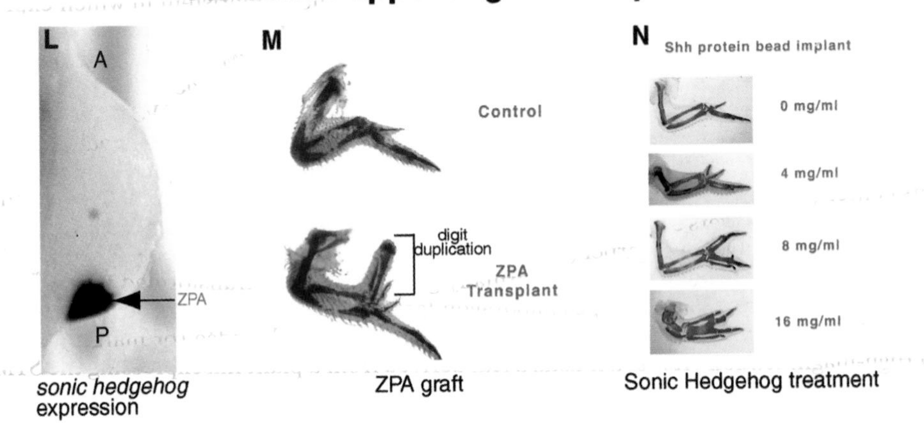

PLATE 3

PLATE 4. *(A) Left:* Diagram of the organization of floral organs viewed as a cross section from above. Floral organs are organized in four concentric whorls, which are from outside to in: sepals (se), petals (pe), stamens (st), and carpels (ca). *Right:* Diagram of the ABC model for homeotic gene function during floral development. According to this model, the behavior of three different classes of floral patterning mutants (e.g., *A*, *B*, and *C* mutants) is explained by proposing that there are *A*, *B*, and *C* function genes that act in a pairwise fashion to specify the four floral organ identities: sepals (*A* function alone), petals (*A* plus *B* functions), stamens (*B* plus *C* functions), and carpels (*C* function alone). *(B)* Wild-type mustard flower. Sepals (se), petals (pe), stamens (st), and carpels (ca) are indicated. *(C)* Cross section of a floral primordium in which expression of the *AP1* gene (an *A*-function gene) is revealed as red staining in the outer whorl primordia generating sepals (se) and petals (pe). *(D)* Diagram showing *A*-gene expression superimposed on the image of a real floral primordium. *(E)* An *A*-class mutant mustard flower lacking function of the *AP2* gene. The outer whorl sepal is partially transformed into a carpel (ca). *(F)* An *AP1;CAL* double-mutant mustard flower that lacks the ability to suppress indeterminate stem development in secondary shoot meristems. The result of this mutant condition is a flower resembling the cauliflower dinner vegetable. *(G)* Cross section of a floral primordium in which expression of the *AP3* gene (a *B*-function gene) is revealed as red staining in the intermediate whorl primordia generating petals (pe) and stamens (st). *(H)* Diagram showing *B*-gene expression superimposed on the image of a real floral primordium. *(I)* Mutant *B*-class mustard flower lacking function of the *AP3* gene. The intermediate whorls develop as sepals (se) and carpels (ca) instead of petals and stamens, respectively. *(J)* Double *B;C* mutant mustard flower in which all whorls develop as sepals (the organ specified by *A*-function alone). *(K)* Cross section of a floral primordium in which expression of the *AG* gene (a *C*-function gene) is revealed as red staining in the inner whorl primordia generating stamens (st) and carpels (ca). *(L)* Diagram showing *C*-gene expression superimposed on the image of a real floral primordium. *(M)* Mutant *C*-class mustard flower lacking function of the *AG* gene. The inner whorls develop as petals (pe) and sepals (se) instead of stamens and carpels, respectively. *(N)* A triple *A;B;C* mutant mustard flower in which all whorls develop as leaves, consistent with Goethe's hypothesis that the leaf is the default identity of a secondary meristem. *(O)* Cross section of a floral primordium in a *C*-class mutant (*AG*) in which *AP1* expression (red stain) expands into the inner whorls (cf. C). *(P)* Cross section of a floral meristem in which expression of the *clavata1* (*CLV1*) gene is restricted to the central apex of a floral meristem. Note that *CLV1* expression is excluded from the outer layer (L1, *white arrowhead*) of the meristem. *(Q)* Cross section of a floral meristem in which expression of the *clavata3* (*CLV3*) gene (purple staining) is restricted to the central apex of a floral meristem. Note that *CLV3* differs from *CLV1* in that it is expressed in the outer (L1) layer of the meristem (*black arrowhead*) as well as in deeper layers (cf. *P*). *(R)* Wild-type cluster of developing flowers surrounding a small meristem in its center (not visible at this magnification). *(S)* *CLV1* mutant with a greatly increased number of developing flowers surrounding a greatly enlarged meristem (*arrowhead*). *(T)* Wild-type tomato with sepals (se) and carpel (ca) showing. *(U)* Tomato derived from a plant misexpressing *AG* in the sepal primordium. The normally green sepals are transformed into a fleshy red carpel-like structure. *(V)* Wild-type snapdragon leaf. *(W)* *phantastica* mutant snapdragon leaf that is elongated and lacks structures normally located near the edge (or margin) of the leaf. *(X)* High-magnification view of a mustard leaf derived from a plant misexpressing the *STM* gene in the developing leaf primordium. An ectopic apical shoot meristem with normal-looking floral meristems (fm) developing along its flank is popping out of the surface of the leaf.

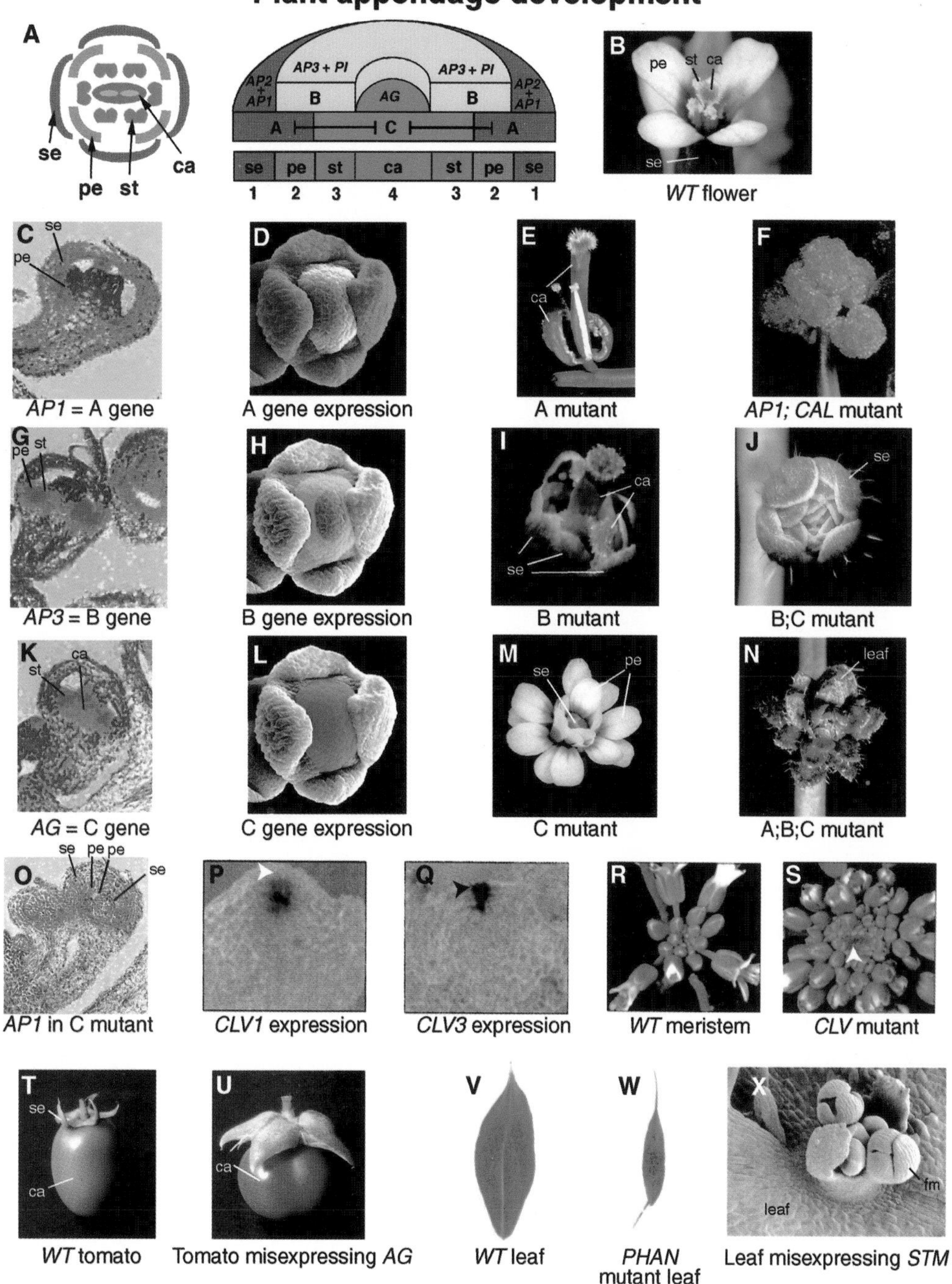
Plant appendage development
A
se
pe
st
ca
AP3 + PI
AP2 + AP1
B
AG
A
C
se pe st ca st pe se
1 2 3 4 3 2 1
B
WT flower
C
AP1 = A gene
D
A gene expression
E
A mutant
F
AP1; CAL mutant
G
AP3 = B gene
H
B gene expression
I
B mutant
J
B;C mutant
K
AG = C gene
L
C gene expression
M
C mutant
N
leaf
A;B;C mutant
O
AP1 in C mutant
P
CLV1 expression
Q
CLV3 expression
R
WT meristem
S
CLV mutant
T
WT tomato
U
Tomato misexpressing AG
V
WT leaf
W
PHAN mutant leaf
X
fm
Leaf misexpressing STM

PLATE 4

5 Establishing the Primary Axes of Vertebrate Embryos

A remarkable finding of the last several years is that many genetic pathways have been preserved in great detail during the course of evolution and guide equivalent aspects of vertebrate and invertebrate development. The resistance of these pathways to evolutionary change is surprisingly deep given the significant apparent differences between embryonic and final adult body plans of vertebrates and invertebrates.

As we turn to considering vertebrate development, an important theme is that many basic patterning mechanisms are uncommon to vertebrate and invertebrate embryos. In the previous two chapters, we discussed mechanisms by which the fruit fly embryo and an adult appendage are subdivided into primary units of organization. In this chapter, we examine patterning of the primary body axes in vertebrate embryos.

■ **Cast of Characters** ■

Terms

Animal hemisphere The darkly pigmented half of the oocyte from which the ectoderm and mesoderm derive.

Anterior–posterior (A/P) axis The anterior (head)–posterior (tail) axis of the frog ectoderm forms parallel to the animal-vegetal axis of the early egg.

β-Catenin A transcription factor required for dorsal–ventral patterning of the frog embryo that is concentrated in the nuclei of dorsal-most cells of the early fertilized embryo.

Blastoderm embryo An early embryo consisting of a hollow ball of cells prior to gastrulation.

Dorsal–ventral (D/V) axis The dorsal (back)–ventral (belly) axis of the frog embryo.

Ectoderm The outer embryonic germ layer of cells that gives rise to skin and nervous system (derived from the animal hemisphere of a frog embryo).

Endoderm The inner embryonic germ layer of cells of the embryo that gives rise to gut (derived from the vegetal hemisphere of a frog embryo).

Gastrulation The organized movement of cells during midembryonic development that creates a laminated embryo with distinct tissue layers.

Invertebrate An animal without a backbone such as insects or worms.

Marginal zone The equatorial zone of the frog embryo that forms within the animal hemisphere and gives rise to mesoderm.

Maternal information Information provided by the mother that has a role in establishing the position of the primary body axes of the embryo.

Mesoderm The embryonic germ layer that gives rise to muscle, heart, and fat (the marginal zone of a frog embryo).

Neural inducing substance A secreted signal liberated by the Spemann organizer that promotes neural over epidermal development.

Notochord A rod-like structure that serves as a rigid support.

Oocyte An unfertilized egg.

Patterning The process by which cells acquire distinct identities during development.

Primary body axes The perpendicular (anterior–posterior and dorsal–ventral) axes of animals established during early embryonic development.

Spemann organizer The organizing region of a frog embryo that induces neighboring cells to form the central nervous system.

Vegetal hemisphere The nonpigmented half of the oocyte from which the endoderm derives.

Vertebrate An animal with a backbone such as a human, mouse, or frog.

Genes

Antennapedia A fly gene encoding a homeobox transcription factor that defines the second thoracic segment (T2) of flies which has wings and legs.

BMP4 The frog version of *Dpp*, expressed in the nonneural ventral ectodermal region of the frog.

Chordin The counterpart of the *Sog* gene produced by cells in the neutralizing dorsal Spermann organizer.

deformed A fly homeotic gene that functions to specify a region of the fly head.

Hox genes Segment-identity genes of vertebrate embryos.

Hox4 The vertebrate *Hox* gene counterpart of the fly homeotic gene, *deformed,* which is expressed in the mouse head.

Hox6 The vertebrate *Hox* gene counterpart of the fly homeotic gene, *Antennapedia,* which is expressed in the upper trunk region of the mouse embryo.

VegT A gene encoding a transcription factor that is required for endoderm development of vegetal cells and for induction of mesoderm in adjacent cells of the animal hemisphere.

ESTABLISHING BASIC TISSUE TYPES IN VERTEBRATES

In contrast to the fruit fly, where the mechanisms employed by the mother to polarize the egg and initiate patterning of the primary body axes are being rapidly uncovered, the early steps in patterning of the vertebrate body axes are less well defined. It is not clear that great generalities can be made about vertebrate embryonic development since distinct maternal mechanisms or external conditions appear to guide early axis patterning in different vertebrate species. In frogs, which are discussed below in more detail, maternal information does play a role in establishing the position of the primary body axes of the embryo. In some fish embryos, however, gravity seems to be the primary early polarizing agent. Furthermore, in other vertebrates, including mammals, there is no evidence for axis patterning of any kind at the blastoderm stage because it is possible to dissociate cells from two different mouse embryos at this stage, mix them together and have these harshly treated balls of cells go on to develop into normal mice. The limitations in our current state of knowledge of maternally controlled vertebrate embryonic development derive primarily from the impracticality of conducting large-scale genetic screens for maternally acting mutations in vertebrate systems. In contrast, such genetic screens have been possible in fruit flies due to their short generation time, and have served as cornerstones for analyzing maternal mechanisms in flies. As shown below, significantly more is known about how various vertebrate embryos interpret maternally provided positional information than about how the egg is initially polarized by the mother or other external factors.

The vertebrate egg that is best understood with respect to early axis formation is the frog oocyte (i.e., unfertilized egg). The frog oocyte is a visibly polarized structure consisting of two differently pigmented hemispheres. The darkly pigmented half of the egg is called the animal hemisphere and the nonpigmented half is referred to as the vegetal hemisphere. Although the difference in pigmentation does not appear to play any role itself in defining the tissues deriving from these two halves, it does reflect a fundamental asymmetry in the egg that will ultimately define the anterior–posterior (A/P) axis of the embryo. Despite what the names seem to imply, the animal and vegetal portions of the embryo do not give rise to a frog versus a cucumber, but rather to different tissue layers within the frog embryo (Fig. 5.1). Most of the animal hemisphere becomes ectoderm (i.e., the outer layer) giving rise to skin and nervous system, whereas the vegetal hemisphere primarily becomes endoderm (i.e., inner layer), which generates the gut. The mesoderm (i.e., middle layer) derives from the equatorial or "marginal" zone of the animal hemisphere and gives rise to a structure called the notochord, which is a rod-like structure that serves as a rigid support (like a backbone), as well as heart, muscle, and blood. Thus, the three layers of the embryo formed during gastrulation correspond to basic tissue types. These primary tissue layers derive from a series

■ VegT, Mesoderm, and Ectoderm ■

Stratification of the frog embryo into the three basic tissue layers depends on the activity of a transcription factor called VegT, which functions both to define endodermal cell fates in the vegetal hemisphere and to induce mesodermal fates in adjacent animal hemisphere cells (Fig. 5.1). VegT is expressed in the vegetal cells, where it serves a dual function to promote endoderm development and to suppress mesoderm formation. In addition to controlling the expression of genes relevant to the development of vegetal endodermal cells, VegT activates expression of secreted signals, which are related in structure to the fly morphogen Dpp. These secreted signals activated by VegT function to promote mesoderm development. As a result of its combined activities within vegetal cells, VegT specifies endoderm by activating expression of genes required for endoderm development while preventing these cells from responding to the mesoderm-inducing signals also produced in these cells.

A critical consequence of VegT activation of mesoderm-inducing signals is induction of mesodermal cell fates in adjacent animal hemisphere cells. These animal hemisphere cells are close enough to the VegT-expressing vegetal hemisphere cells to receive the mesoderm-inducing signals activated by VegT. The future marginal cells do not express VegT, which also functions to prevent a response to these mesoderm-inducing factors. In many respects, the dual activity of VegT in defining endoderm locally and inducing mesoderm in adjacent cells parallels that of *engrailed* during fly wing development, in specifying posterior compartment cell fates and inducing formation of central organizer in adjacent anterior compartment cells (see Chapter 4).

of adjacent domains lying along the animal–vegetal axis of the earlier blastoderm-stage embryo.

ESTABLISHING THE PRIMARY AXES IN VERTEBRATES

In frogs, unlike flies, formation of the embryonic A/P axis is coupled to establishment of the dorsal–ventral (D/V) axis. Factors that dorsalize the egg also tend to specify anterior positions, whereas factors functioning to ventralize the egg promote more posterior cell fates. The future A/P axis of the frog forms along the animal–vegetal axis of the early egg, with the animal pole of the egg becoming the future anterior extreme of the ectoderm. The relationship of the A/P axis to other tissue layers such as the mesoderm and endoderm is more complex, however, because it depends on cell migrations during gastrulation. The major cell movements involve cells in the equatorial marginal zone folding underneath the ectoderm to form a two-layer embryo. Cells folding in first (i.e., those closest to the vegetal endoderm) migrate the farthest and come to underlie anterior parts of the ectoderm, whereas marginal cells migrating later (originally farthest from the boundary between the animal and vegetal hemispheres) move in only a short distance to take up residence beneath the posterior portion of the ectoderm. Signals passing between the mesoderm and ectoderm transfer A/P positional information from the mesoderm to the overlying ectoderm. After the future mesoderm enters the embryo, the vegetally derived endoderm

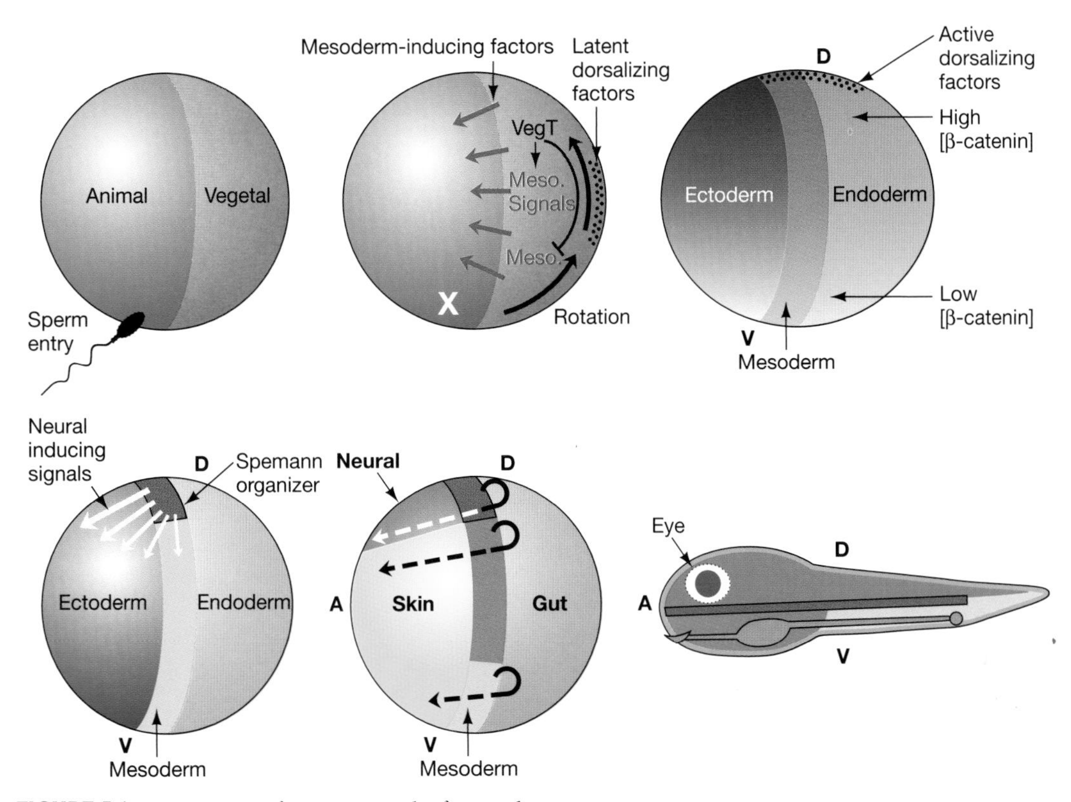

FIGURE 5.1. Primary axis formation in the frog embryo.

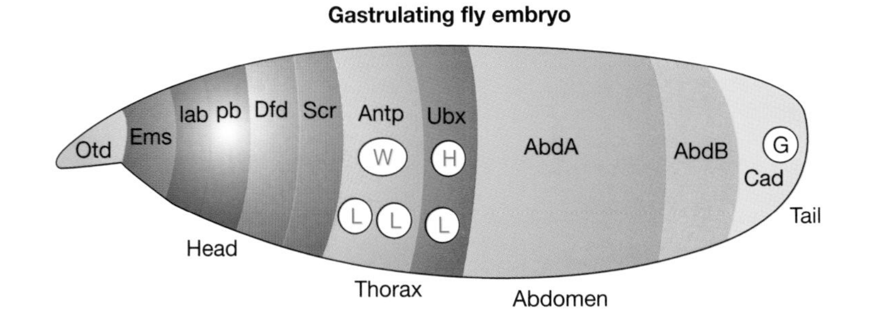

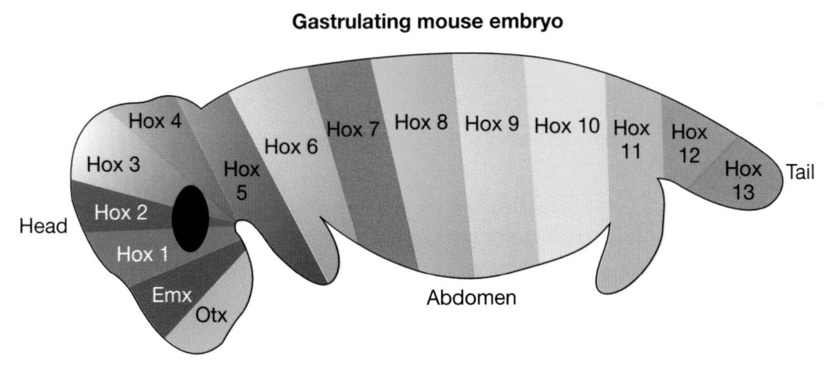

FIGURE 5.2. Equivalent expression of homeotic/Hox proteins in vertebrates and invertebrates.

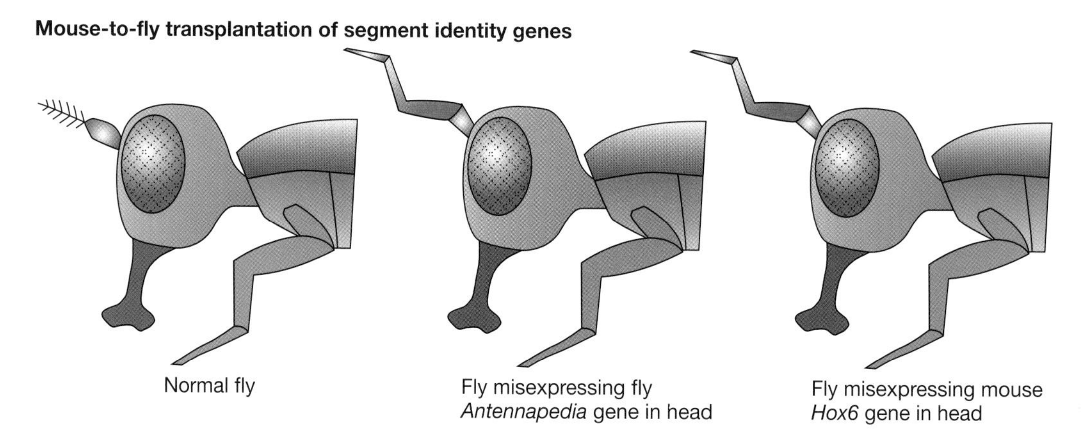

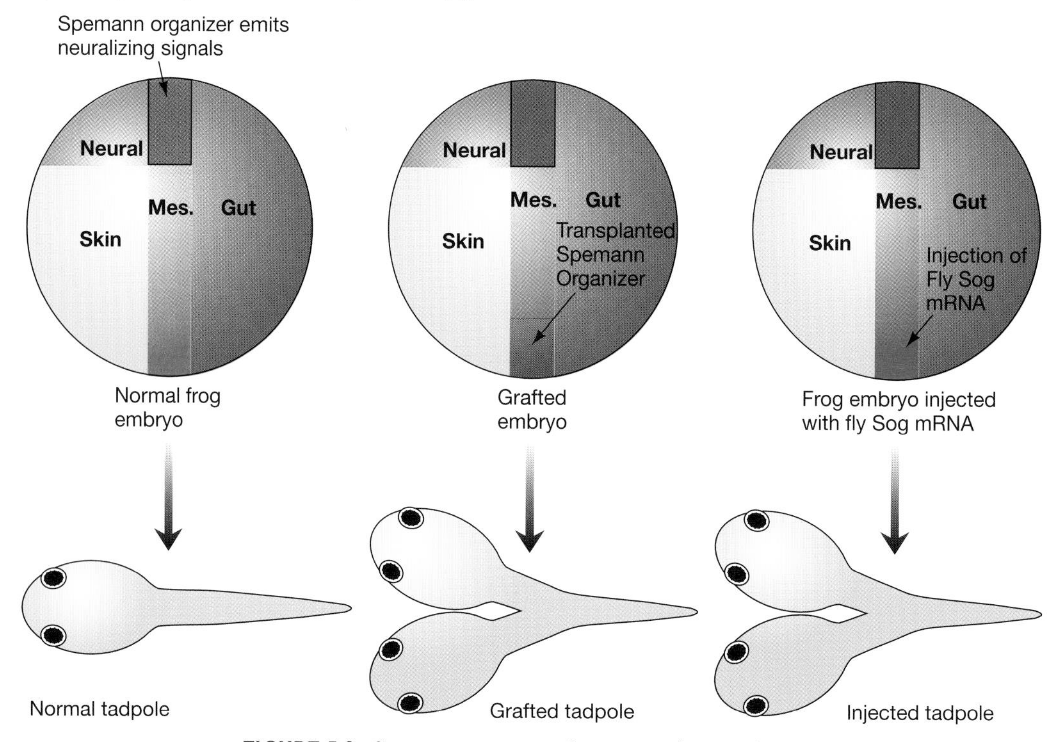

FIGURE 5.3. Cross-species transplantation of equivalent genes.

moves into the interior of the embryo to form the gut. As in the case of the mesoderm, endodermal cells achieve their ultimate A/P positions by migration.

Axis formation is initiated in the frog when the egg is fertilized in the animal hemisphere near the equator (i.e., the boundary between the animal and vegetal hemispheres). Why the sperm only fertilizes the egg in the animal hemisphere is unknown, but presumably it results from the same primary asymmetry that generates the difference in pigmentation. This subdivision into animal versus vegetal halves is reflected in the distribution of key patterning molecules, several of which are confined to the vegetal versus animal hemisphere of the unfertilized egg (e.g., VegT). The point of sperm entry defines the future ventral pole of the embryo. The sequence of events by which sperm entry polarizes the egg is summarized in Figure 5.1. Shortly after the sperm fuses with and injects its nucleus into the egg, an outer layer of the egg lying just beneath the cell surface in the vegetal region of the embryo rotates away from the sperm entry point. An important result of this rotation is that a dorsalizing substance concentrated in the vegetal pole of the egg is moved to a point approximately opposite to that of the sperm entry point in the vegetal hemisphere. This motion not only displaces the dorsalizing substance to the future dorsal pole of the embryo, but also activates it. If one experimentally prevents the rotation of the outer layer of the egg, the dorsalizing substance remains in a latent inert form and the resulting embryo lacks dorsal structures.

Once the dorsalizing factor has been activated and transported to a position opposite to that of the sperm entry point, the picture begins to resemble that of D/V axis formation in the fly embryo. The first detectable indication of D/V polarity is the formation of a graded concentration of a transcription factor known as β-catenin along the future D/V axis. The highest levels of β-catenin are found at the future dorsal pole, and lower levels are found at progressively greater distances from this point (Fig. 5.1, see also Plate 2A for actual data). This graded concentration of β-catenin is reminiscent of the graded distribution of the Dorsal transcription factor, which determines D/V positions in the fly embryo (see Chapter 3). Although β-catenin and Dorsal are structurally unrelated transcription factors, they function analogously in that they both activate distinct subsets of patterning genes at different doses. Specific mesodermal cell fates along the D/V axis are determined by the joint action of β-catenin, which determines dorsal–ventral position, and induction of mesodermal fates in marginal zone cells adjacent to the vegetal hemisphere by VegT (see box above). High levels of β-catenin activate expression of genes in the dorsal region of the marginal zone comprising Spemann organizer, whereas lower levels of β-catenin lead to expression of genes involved in specifying ventral structures of the embryo. Thus, the position of the Spemann organizer is determined jointly by high levels of β-catenin and proximity to VegT-expressing vegetal cells.

It is important to note that the Spemann organizer gives rise to notochord, a structure that is mesodermal rather than neural in origin. This fact is particularly relevant with respect to the seminal experiments of Hilde Mangold, which demonstrated that the Spemann organizer acts as a source of diffusible neural inducing substances, but does not itself contribute to the nervous system (see Chapter 1). Thus, formation of the nervous system relies on two sequential inductive events: induction of the mesoderm by vegetal cells, and neural induction by the dorsal mesodermal cells comprising the Spemann organizer. The mechanism by which the Spemann organizer induces dorsal structures, such as the nervous system, is discussed below in greater detail.

We first consider patterning along the A/P axis in vertebrates, however, as this analysis provided the first clear evidence that the function of key developmental genes has remained largely unaltered throughout enormous periods of evolutionary time.

A/P PATTERNING BY SEGMENT-IDENTITY GENES IN MICE: DÉJÀ VU

In Chapter 3 we discussed how segment-identity genes (or homeotic genes in the fly) translate early positional information provided by segmentation genes into unique labels for each segment of the embryo. Vertebrate embryos also have segment-identity genes, known as *Hox* genes, that are highly related in structure to fly homeotic genes. A given vertebrate *Hox* gene can be matched up with its fruit fly homeotic gene counterpart on the basis of two criteria. The first measure of relatedness between particular Hox and homeotic proteins is the degree of similarity in amino acid sequence of the two proteins. Recall that segment-identity genes encode transcription factors. Each homeotic gene can be distinguished on the basis of a unique sequence of amino acids. In species as far apart on the evolutionary scale as flies and humans, it is possible to tell which vertebrate *Hox* gene corresponds to which fly homeotic gene using similarity in the sequences of amino acids among these various proteins. For example, among a group of 13 mouse genes, the *Hox6* gene is clearly most related to the fly *Antennapedia* gene, which is involved in formation of the wing-bearing thoracic segment T2. Similarly, each of the vertebrate *Hox* genes can be assigned to a specific fly homeotic gene based on relatedness of amino acid sequence.

Another defining characteristic of homeotic genes in flies is that they are primarily expressed as RNA in the segments they define. The localized expression of segment identity in genes in both vertebrates and invertebrates serves as an independent criterion for assigning a correspondence between vertebrate *Hox* genes and fly homeotic genes. Equivalent fly and vertebrate segment-identity genes are expressed in the same relative positions in vertebrate and fly embryos (Fig. 5.2). For example, the fly *Antennapedia* gene is expressed in the T2 segment of the fly embryo, and the corresponding mouse *Hox6* gene is expressed in the upper trunk region of the developing mouse embryo. Similarly, vertebrate *Hox* genes corresponding to fly homeotic genes expressed in more anterior or posterior positions are expressed, respectively, in head and tail regions of vertebrate embryos. Segment-identity genes have been found in all segmented animals examined including flies, worms, and humans. In each of these animals, structurally equivalent genes are expressed in the same relative positions in the embryo.

There are two major points to extract from the fact that vertebrate and fly segment-identity genes are uniquely related to each other on the basis of amino acid sequences and expression patterns. First, the most recent common ancestor of flies, worms, and humans had a set of segment-identity genes similar to those common to all present-day segmented animals. This means that this progenitor not only possessed a head and tail, but also had a series of distinguishable segments between its two ends. Second, segment-identity genes must perform important and highly constrained functions because critical amino acid sequences have remained virtually unchanged during the more than 500 million years of evolution separating flies, worms, and humans. This striking degree of amino acid sequence similarity between vertebrate *Hox* and invertebrate homeotic genes is not typical of other genes in such distantly related species.

Cross-species Function of Segment-identity Genes

■ The McGinnis Experiment ■

In 1990, Bill McGinnis, a colleague at UCSD, who was then at Yale University, did a very important experiment. He had been doing some elegant work showing that the DNA-binding preferences of different fly homeotic proteins reside in an invariant portion of the protein called the homeobox (i.e., the sequence of amino acids in this DNA-binding region of the protein has remained relatively unaltered during the course of evolution). McGinnis reasoned that if the homeobox was the critical region of the homeotic protein, a corresponding vertebrate *Hox* gene that had a very similar homeobox might be able to substitute for the fly homeotic gene. At the time this interspecies experiment was conceived, this was a wild idea from the fringes of science fiction. Undaunted by the seemingly low odds for success, McGinnis and his graduate student Jarema Maleki went ahead and inserted the mouse *Hox6* gene, which is equivalent to the fly *Antennapedia* gene, into a fly. They then asked what would happen if they misexpressed this mouse *Hox6* gene in the head of the fly. Remember that the fly *Antennapedia* gene defines the leg- and wing-bearing thoracic segment and can suppress the action of homeotic genes functioning in more anterior regions of the embryo (see Chapter 3). The *Antennapedia* gene is named for a naturally arising freakish mutant that misexpresses this gene in the head of the fly, the result being a transformation of the antenna (an appendage of the head) into a leg (i.e., a thoracic appendage). Remarkably, when the mouse *Hox6* gene was misexpressed in the head of a fly, McGinnis saw the same thing happen: Antennae were replaced with legs (Fig. 5.3; see also Plate 3I–K for an actual example of antenna-to-leg transformation by mouse *Hox6*).

The McGinnis result was stunning for two reasons. First, it meant that two genes that had been separated by more the 500,000,000 years of evolution still performed exactly the same function. Second, although the DNA-binding homeobox portions of these two related segment-identity genes are very similar, the rest of the proteins are not. Thus, a very small part of homeotic proteins is responsible for organizing major features of an organism's body plan.

Since the publication of the original McGinnis experiment, several groups have gone on to show that this is a very general result. Bill McGinnis and his wife, Nadine, provided the first evidence that this was likely to be the case by misexpressing another mouse *Hox* gene in flies. This second *Hox* gene (*Hox4*) is most similar to a fly homeotic gene called *deformed,* which functions to specify a region of the fly head. When Bill and Nadine McGinnis misexpressed the *Hox4* gene in the fly head, they observed defects in the head region that were nearly indistinguishable from those resulting from misexpression of the fly's own *deformed* gene. On the other hand, the antennae were not transformed into legs as occurred when the *Antennapedia* gene was misexpressed. Other investigators have repeated this kind of experiment in flies using nearly all of the different vertebrate *Hox* genes. In one case, it even has been technically possible to replace the fly's own homeotic gene with its vertebrate counterpart. This experiment is tantamount to cross-species gene transplantation! Outside of the fact that such flies have an unnerving desire to listen to Mozart and write science fiction novels, they are remarkably normal healthy creatures. The stringent criterion of viability in such cross-species gene-replacement experiments has also been achieved in roundworms, indicating that these genes have retained their positional labeling functions across vast expanses of evolutionary time in diverse species.

One question that often arises is, Why should a *Hox* gene in mice that specifies a trunk region in the mouse embryo generate fly legs and not mouse limbs when misexpressed in the head of a fly? The answer is that flies cannot make a mouse leg because they do not contain the genetic information necessary for making structures associated with mouse limbs such as bones or fur. In another analogy to electronics, the McGinnis experiment is like taking the starter from a Volkswagen and getting it to work in a Mercedes. It would be surprising if this kind of tinkering worked, but if it did, you would end up starting and then driving a Mercedes not a Volkswagen. Similarly, the mouse *Hox* gene can substitute for the fly *Antennapedia* gene in switching on a cascade of fly genes that function to create a leg. Assembling a mouse limb, however, would require activation of a different set of genes, present

Bill McGinnis (1952–)

E.B. Lewis had proposed that the cluster of homeotic genes might be due to duplication and divergence of a primordial homeotic gene. Bill McGinnis and Mike Levine found a piece of coding sequence of the *Antennapedia* gene that was common to all of the fruit fly homeotic genes. McGinnis coined the term "homeobox" for this shared piece of DNA, and this has become its enduring name. He went on, with Levine and Ernst Hafen, to make several landmark discoveries about homeotic genes, including that all segmented animals have highly similar segment-identity genes, and that these genes are expressed in single stripes corresponding to the segments of the embryo whose identity they define.

Bill McGinnis was born in Warrensburg, Missouri. He left the heartland to study molecular biology as an undergraduate at San Jose State University and then went on to do his graduate work in molecular biology at UC Berkeley in 1982, where he worked with Steven Beckendorf to study mutations disrupting the function of the regulatory region of a gene normally active in the fly salivary gland. He continued to pursue his interest in molecular mechanisms controlling gene expression and development in Basel, Switzerland, as a postdoctoral fellow in Walter Gehring's laboratory, the same lab that trained Christiane Nüsslein-Volhard, Eric Wieschaus, and many others who have become leaders in the field of developmental biology. Perhaps most importantly for McGinnis, Mike Levine and Ernst Hafen joined the Gehring group around the same time McGinnis did. At that time in the early 1980s, it was known that homeotic genes in fruit flies had a very important role in morphological development, but the biochemical functions encoded by the genes were still a mystery and speculation was rampant. The important next step was to clone and characterize the genes, which McGinnis set as his goal.

McGinnis vividly remembers the moment when he discovered the homeobox in other species and understood the profound evolutionary implications of this discovery. "The more astonishing and important finding was that homeobox sequences (as part of homeotic genes) were present in many other animal genomes, including humans. At the time, the paradigm was that since the shapes of embryos in different animal lineages were so dissimilar, the underlying genes that controlled those shapes would be completely different. In contrast, the discovery that homeotic genes were conserved in animals indicated that the molecules used by mammals and flies to pattern embryonic morphology were virtually the same. As mentioned above, the conventional wisdom taught in classrooms was that different animal shapes were controlled by very different genes and control proteins. I wasn't immune to this at the time of the homeobox conservation experiments, but I was trained as a molecular biologist. Thus, I also knew that some molecular aspects of gene structure and regulation were similar in all living organisms, and it was natural to speculate that perhaps the molecules controlling developmental pattern might also be similar, despite the lack of outward evidence for this idea. The most satisfying moment in science is when you get a surprising result, and the realization hits that you are first human ever on the planet to know about this breakthrough into a new way of thinking. By good fortune, this occurred for me when I developed the X-ray film showing that humans and other animals had homeotic genes that were similar to the fly homeotic genes. Just as satisfying is to show such a result to others, and I remember running around sticking the dripping wet X-ray film into the face of everyone in sight, excitedly telling them that the implication was that humans were constructed on the same molecular plan as flies. What a thrill!"

After completing his postdoctoral studies in Switzerland, McGinnis took a position at Yale University, where he continued to study the mechanism by which homeotic genes define the identity of specific segments. Prompted by his finding that the region of homeotic proteins responsible for this function was the DNA-binding homeobox, he went on to do the famous McGinnis experiment in which he, Nadine McGinnis, and his graduate student Jarema Malicki placed mouse and human homeotic genes in fruit flies and showed that mammalian homeotic genes could perform some of the same patterning functions as their fly counterparts. The finding that animal developmental patterning is controlled by the same set of molecules has been confirmed over and over with a wide range of developmental patterning genes in animals. McGinnis points out "these discoveries made in model genetic systems (flies and worms, for example) have been enormously interesting to all biologists, since the genes discovered in these simple animals can be used to identify mouse and human genes involved in development, behavior and human disease."

McGinnis's approach to science is to search for unifying principles and molecules because "it is incredibly interesting to find hidden similarities among obvious and confusing differences, much more so than cataloging the differences." McGinnis prefers to attack problems in depth instead of skipping around from one problem to the next. He also sees the importance of not being afraid to do some risky experiments. He notes "My colleagues and I also get a great thrill from trying experiments that seem too crazy to work, and occasionally making a truly novel discovery. This approach is seen as bold when the experiment works, and naive or stupid when the experiment doesn't work. It is also more interesting to work in fields where you can break new ground, instead of churning up a field that has been plowed many times before. Sometimes this isn't initially rewarding, but if you can get money to do such experiments, it's satisfying, since you know that you are discovering molecules and principles that would not be found within a couple of days by a few or a hundred other labs." McGinnis believes that curiosity is the most important influence on scientific discovery, commenting, "Being able to appreciate a mystery, and then put all your creativity and will into digging into the solution. It is only when you absolutely must know what lies beyond the next hill that you are capable of making discoveries. Along with this, one must learn to be very critical of evidence, incessantly looking for holes and misinterpretations in experiments, especially your own. Being able to stay excited about the possible solution to a mystery, and at the same time being able to relentlessly criticize and doubt the solution that you have temporarily embraced, is the paradoxical attitude that leads to genuine progress in science."

McGinnis has since moved to the University of California, San Diego, where he continues to pursue his passion for understanding how segment-identity genes function as genetic switches to entrain specific developmental programs in different segments.

only in mice. Thus, segment-identity genes function as high-level switches, but do not themselves contribute directly to forming a particular morphological structure.

Different Mechanisms Activate Segment-identity Genes in Flies and Vertebrates

As mentioned above, the McGinnis experiment provided compelling evidence for segment-identity genes functioning through virtually identical mechanisms in flies and vertebrates. Unless this were true, it is very hard to imagine how a mouse gene could substitute for a fly gene. Given that the high-level genetic switches controlling segment identity have been preserved in detail during the course of evolution, one might expect that earlier mechanisms involved in establishing positional information along the A/P axis would also be similar in different organisms. This reasonable expectation seems not to be the case, however. For example, the morphogen involved in establishing the primary axes in frogs, β-catenin, is unrelated in structure or function to either of the primary maternal patterning morphogens in flies (i.e., Bicoid along the A/P axis or Dorsal along the D/V axis). Flies do have a β-catenin gene, which functions later during development, but the fly gene β-catenin does not seem to play any role in early axis formation in flies. In addition, as mentioned above, other vertebrates such as some fish and mice do not even use maternal information to initiate formation of primary body axes. Furthermore, there is currently no evidence for the existence of vertebrate genes corresponding to the fly gap and pair-rule genes functioning to link crude maternal information to expression of *Hox* genes in specific segments.

The earliest time that a common A/P patterning process seems to be operating in both flies and vertebrates is during mid-gastrulation when the embryos are beginning to exhibit overt morphological signs

Matthew Scott (1953–)

The co-discovery of the homeobox by Matt Scott in Thom Kaufman's laboratory and by Bill McGinnis and Mike Levine in Walter Gehring's laboratory and the subsequent work by these pioneers and others provided a mechanistic basis for understanding the function of homeotic genes. The remarkable segmental transformations observed in homeotic mutants could now be explained by imagining that the expression of many genes could be coordinated by a single regulatory factor.

Matthew (Matt) Scott was born in Boston. He did both his undergraduate and graduate studies at MIT, graduating in 1980. During his graduate course work, he became interested in homeotic genes and decided to work on this problem in the laboratory of Mary Lou Pardue. Because his fascination with homeotic genes kept growing, Scott went on to pursue postdoctoral work with Thomas Kaufman and Barry Polisky, with whom he received critical training in both genetics and molecular biology. When Scott joined the lab, Kaufman was in the process of discovering the nonhomeotic A/P patterning genes that lie near the *Antennapedia* (*Antp* for short) gene. These genes included a pair-rule gene, known as *fushi tarazu* (which is Japanese for "segment-deficient"), a gene required for dorsal cell fates, and *bicoid,* which encodes the maternal A/P morphogen. Scott recalls that Kaufman had clearly recognized that the *Antp* gene cluster was the anterior version of the *Bithorax* gene. About that time, Christiane Nüsslein-Volhard and Eric Wieschaus published their first seminal paper on their comprehensive genetic screen for patterning mutations. It soon became clear to Kaufman and his colleagues that *fushi tarazu* (or *ftz* for short) was a pair-rule gene. Scott remembers that "the location of *ftz* next to *Antp* seemed peculiar, as there was no reason to think that such a gene involved in a different patterning function should be located near a homeotic gene. The view was that segmentation was quite separate from segment identity."

Scott then took a faculty position at the University of Colorado at Boulder, where he continued his work on homeotic genes. No regulatory gene sequences were known then, so there was discussion of whether these genes even encoded proteins (for example, one hypothesis was that homeotic genes encoded RNA that directly controlled gene expression). Scott's team, consisting of Amy Weiner, Bob Laymon, and himself, began the then arduous process of cloning the *ftz* gene. They first located mutations in the neighboring *Antp* gene, and then, greatly aided by Kaufman's collection of mutants, Weiner located *ftz.* The isolation of these two different patterning genes led to the unexpected finding that they had a domain in common (later to be dubbed the homeobox) because a region of the DNA from the *ftz* gene could hybridize (e.g., form a double helix with) a part of the *Antp* gene (see Chapter 2 for a refresher on DNA hybridization). Scott didn't trust this crude form of measuring DNA similarity because the result could have been due to contamination, so he checked it with many other hybridization experiments and narrowed down the cross-hybridizing pieces to a small region of DNA. Scott then compared the DNA sequences of the regions that seemed to be in common between these two genes and recalls, "In the first sequencing I'd ever done, I determined parts of the sequences we now know as the two homeoboxes. Late one night I decoded the sequences using the Cold Spring Harbor symposium volume in which the code (i.e., genetic code) was originally published. That was the only place I could find a copy of the code at that time of night. I translated all six frames for each sequence and color-coded basic and acidic residues. Right away I could see some similarities between the sequences of one reading frame for each protein. That was the most exciting moment I've had in science. It said three important things at once: that I had isolated coding DNA, that the regulatory genes encode proteins, and that the proteins of a homeotic gene and a segmentation gene (e.g., pair-rule gene) are related. I was amazed and thrilled." Scott then went on to find that the homeotic gene *Ultrabithorax* also encoded a protein which contained a homeobox domain.

In retrospect, Scott muses about his identification of the homeobox, "the discovery (like many in science) was due to paying attention to oddities. If the blots (i.e., hybridization experiments) had been dirty, I would not have seen the weak hybridization between the two genes. Even with my clean data, I could have ignored the weak bands. The discovery was not expected, despite all the parallels among the homeobox genes and Ed Lewis' wonderful work on the *Bithorax* genes, because the initial discovery was about *Antp* and *ftz* and not about multiple homeotic genes. However after the *Ultrabithorax* result (i.e., that it also had a homeobox), it all made sense in terms of Ed's ideas that the genes played similar roles in different places. Ed has been a great inspiration over the years, not only for his science but for his strength of character and kindness."

Scott's laboratory proceeded to perform many critical experiments to prove that the homeobox indeed functioned as a DNA-binding transcription factor. An important connection in this problem was made when Robert Sauer, from MIT, visited Boulder and talked about a certain family of bacterial transcription factors called helix-turn-helix proteins. Allen Laughton and Scott tried to align the homeobox sequence with that of helix-turn-helix proteins, and Laughton found a match that looked pretty good. They also found that invariant portions of the homeobox sequence lined up with the sequence with yeast mating-type proteins (yeast cells can exist in one of two mating states that are controlled by homeobox-like mating-type proteins). Laughton's finding led Scott's group to publish the proposal that the homeobox proteins are sequence-specific DNA-binding proteins with a helix-turn-helix structure, and that the yeast proteins had that structure, too. This hypothesis has since proven to be true. Scott comments, "This was very exciting to us because it tied together the best understood eukaryotic molecular genetics—mating type—with the classic work on bacterial and viral (gene control) and with homeosis (i.e., homeotic transformations). We had a good hypothesis for how homeotic genes work."

Scott believes that choosing a scientific problem which is clearly important and experimentally accessible is the first element of discovery: "The problem must have mystery, but it must be accessible to practical experiments. Integrating new techniques into solution of a problem is also important as it creates opportunities to resume analysis of studying long-standing problems that had run into a temporary dead end." A second critical element of discovery is clean abundant data that gives rise to clear questions about mysterious processes. Scott notes that science is also vastly more fruitful and fun when there is lively discussion and debate among people with different backgrounds and biases. A final essential component of discovery is that experiments should be done using a variety of approaches such as genetics, molecular biology, or biochemistry to answer any particular question. Scott credits his own success in part to others and to the special time in which he performed his experiments: "I feel incredibly fortunate to be a biologist at this time in history, and even more so to have worked with so many wonderful teachers and students."

In 1990, Scott moved his laboratory to Stanford University where he maintains a very active research program. His laboratory currently pursues many interests, including signaling, membrane trafficking, cancer, brain development, and proteins functioning as motors within cells, in addition to continuing his studies of homeotic genes. He points out that "All of these topics are linked because the pathways and processes work together in shaping embryos. The strength of this broad approach is that the work continues to be fascinating and nearly every day brings surprises."

of segmentation. Interestingly, this stage of development has been recognized for more than 100 years as a period during which embryos from diverse vertebrate species most resemble each other. For example, it was noted that during this period of maximal morphological similarity, which has been referred to as the phylotypic stage, mammalian embryos transiently form gill slits resembling our primordial fish ancestor. Ernst Haeckel's dictum that ontogeny (i.e., development of an individual organism) recapitulates phylogeny (the evolution of different organisms) originated with this similarity between gastrulating embryos of different species.

The fact that segments are labeled by an invariant mechanism in a wide variety of animals, while the activation of segment-identity genes in particular segments seems to be accomplished by diverse mechanisms, presents an apparent paradox. Unfortunately, we do not have a satisfying answer to this perplexing question at the moment. One can paraphrase nature by saying that there must be something special about labeling segments according to their position along the A/P axis that is very difficult to change, but that there are no great constraints on how the labels are placed where they belong. Undoubtedly, one of the most interesting and challenging questions remaining to be answered in the field of A/P patterning is why the emerging generic body plan observed during the phylotypic stage has remained so unaltered during the course of evolution.

ESTABLISHING THE D/V AXIS IN VERTEBRATES

As mentioned above, the earliest steps in patterning the A/P and D/V axes in vertebrate and invertebrate embryos appear to be based on quite different mechanisms. In the case of the A/P axis, these diverse early developmental strategies seem to converge on activation of a common set of segment-identity genes. Likewise, with respect to the D/V axis, distinct early maternal patterning systems converge on a common embryonic patterning mechanism.

Because formation of the A/P and D/V axes is linked in vertebrates, determining the position of the dorsal pole of the embryo in the vegetal hemisphere (i.e., at a site opposite to that of sperm entry) initiates both A/P and D/V patterning. The definition of the dorsal pole in the vegetal region of the embryo by high levels of nuclear β-catenin triggers expression of several key genes in adjacent cells within the animal hemisphere. This specialized dorsal region of the animal hemisphere is the renowned Spemann organizer, introduced in Chapter 1. In addition to activating expression of genes in the Spemann organizer required for establishing dorsal cell fates, information provided by the primary dorsalizing center in the vegetal region of the embryo also suppresses expression of genes within the Spemann organizer that are involved in specifying opposing ventral fates. The consequence of these early maternal patterning events is that the vertebrate embryo expresses different sets of genes in different positions along the D/V axis. In the following sections, we consider the function of several of these key embryonic D/V patterning genes. As in the case of the A/P axis, several of these vertebrate patterning genes turn out to be counterparts of fly genes discussed in Chapter 3. Related vertebrate and fly D/V patterning genes again have been found to control highly similar developmental programs in both vertebrate and invertebrate embryos and can substitute for each other functionally (see below).

Secreted Neural-inducing Factors: Déjà Vu

The classic organizer transplantation experiments in frogs performed by Hilde Mangold (see Chapter 1) demonstrated two important facts. First, there is a unique region of the frog embryo, the Spemann organizer, which can trigger formation of a secondary nervous system when transplanted to ectopic sites in another embryo. Recall that the Spemann organizer comprises the dorsal portion of the marginal zone in the animal hemisphere (see Fig. 1.11) and normally gives rise to a specialized dorsal mesodermal structure known as the notochord—a rigid structure that underlies the spinal cord and likely represents an ancestral form of a backbone. The second important fact extracted from Mangold's experiments is that the ectopic nervous system in embryos receiving organizer grafts was made up of host cells rather than donor cells. In other words, the organizer induced host cells to make a second nervous system, but did not itself contribute to it. This important observation suggested that the organizer was the source of a diffusible signal that could alter the identities of surrounding cells. In the presence of this "neuralizing" signal, cells that otherwise would become epidermal were reprogrammed to become neural.

Following the seminal organizer transplant experiments, a variety of subsequent work reinforced the idea that the Spemann organizer was the source of neuralizing factors that could convert epidermal cells into nervous system. For example, it was found that if one made a sandwich consisting of a piece of dissected organizer tissue with a piece of naive ectoderm, the induced ectoderm would generate neural tissue. A similar piece of ectoderm would only form epidermis if sandwiched with itself or with any other part of the embryo. These experiments also provided evidence for the effect of the organizer being mediated by diffusible substances, because if one placed a tight mesh filter between the two layers of the sandwich, the organizer tissue could still neuralize the ectoderm even though the filter prevented the two cell types from directly touching one another. The filters used in these experiments allowed small molecules to pass from one cell layer to the other, however. The generally accepted explanation for the outcome of these experiments was that the default state of ectodermal cells in the animal hemisphere was epidermal (i.e., in the absence of a positive neuralizing signal, cells would become epidermal). If such cells were exposed to "neural-inducing" substances produced by organizer cells, however, these cells could be redirected to generate neural cell types. Although the results of the Mangold/Spemann experiments and subsequent sandwich experiments were consistent with positive-acting, neural-inducing substances being liberated by the organizer, there was another explanation for the results, which was largely dismissed at the time. This alternative interpretation was that the organizer blocked the action of neural-suppressing substances present in the ectoderm. According to this second scenario, the default state of ectodermal cells in the absence of communication with their neighbors was neural. As shown below, this latter more convoluted possibility has turned out to be correct.

Dissociated Ectoderm Cells Become Neurons

The first hint that the default state of ectoderm was neural came from experiments in which pieces of ectoderm from the animal hemisphere of frog embryos were dissociated into a suspension of single cells (Fig. 5.4, left column). This manipulation effectively cuts all communication between cells. After a period of floating in culture media, these cells were compacted into a blob by spinning them in a centrifuge (see sidebar).

When cells are placed in a test tube and spun in a centrifuge the artificial gravity created by the circular motion pushes the cells to the bottom of the tube where they pile up on top of one another.

These experimental manipulations obviously represent drastic departures from normal embryonic development. Thus, when this experiment yielded the unexpected result that such reaggregated ectodermal cells became neurons rather than the anticipated skin, the generally accepted interpretation was that the experiment gave the wrong result and that the cells were behaving abnormally due to their harsh treatment.

Neural Inducers Are Really Anti-neural Inhibitors

In Chapter 3, we discussed how the *sog* gene, which is expressed in the neural ectoderm of the early fly embryo, functions to block the neural suppressive activity of the *dpp* gene. Cells in the nonneural region of the embryo secrete the Dpp signal onto each other, thus inhibiting one another from following the default developmental path of ectodermal cells which is to become neural. In the neural region of the embryo, however, this mutual neural inhibitory signaling is blocked by Sog, which allows these cells to develop as neurons.

One of the first indications that the standard interpretation of the classic frog experiments might be backward was that vertebrate counterparts of the fly *dpp* and *sog* genes were found. Intriguingly, it was observed that these genes were expressed (i.e., transcribed as mRNA) in the same cell types as flies. Thus, BMP4, the frog version of Dpp, was expressed in the nonneural region of the frog, and the counterpart of Sog called Chordin was produced by cells in the neuralizing Spemann organizer (Fig. 5.5; see also Plate 2B, C for actual expression of *bmp4* and *chordin* in frog embryos). The fact that these vertebrate genes were expressed in the same tissue types as their fly counterparts suggested that the mechanisms by which these genes acted might also be the same in flies and frogs, in which case the default state of the ectoderm should be neural, not epidermal. As an interesting aside, the neural ectoderm forms dorsally in vertebrates (i.e., the spinal cord runs along the dorsal

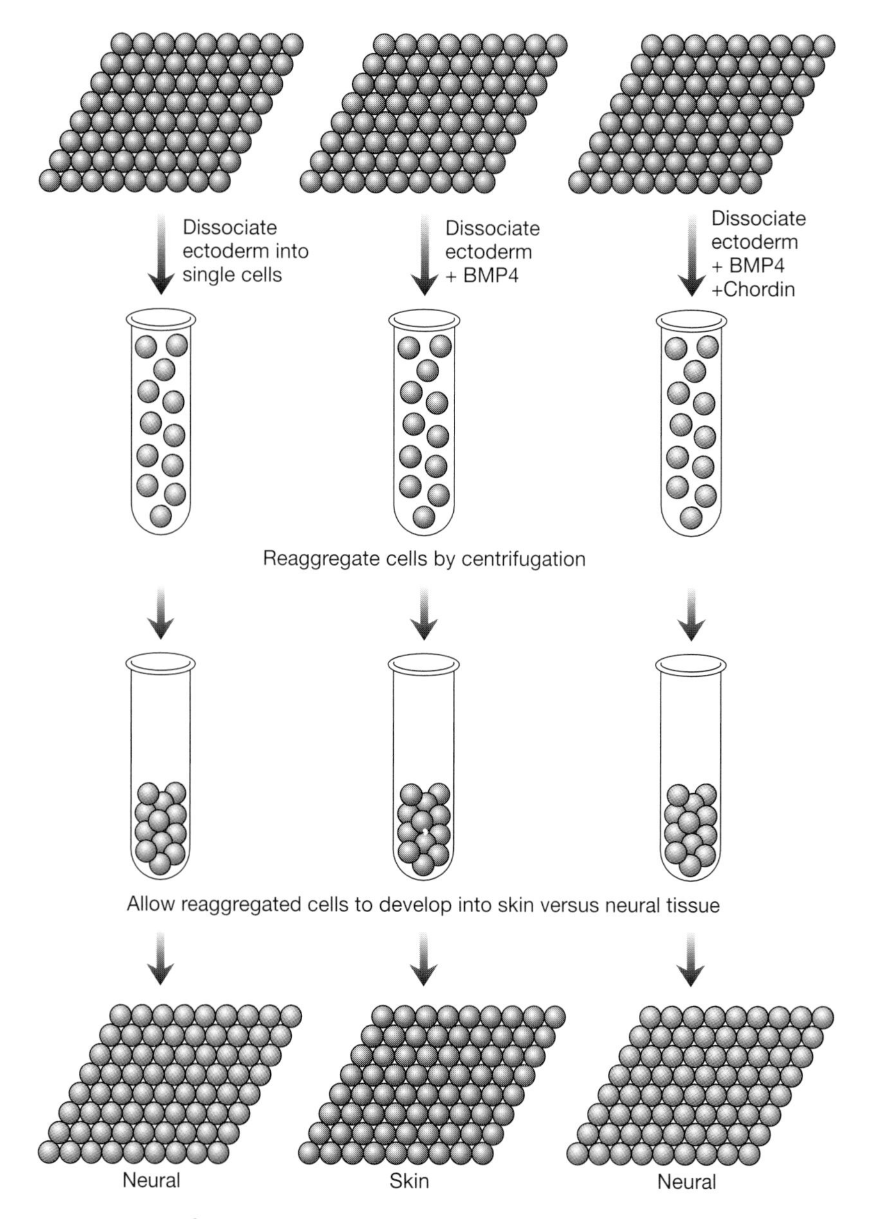

FIGURE 5.4. The ectoderm dissociation experiment.

midline), whereas in flies and other invertebrates, the nervous system forms ventrally. It is likely that during the course of vertebrate evolution the position of the neuroectoderm was flipped with respect to invertebrates, an idea I refer to as the "invertedbrate" hypothesis (Fig. 5.5). The position of the neural ectoderm notwithstanding, the main point to bear in mind is that cells in both vertebrates and invertebrates that will give rise to epidermis express BMP4 or its cousin Dpp, respectively, whereas Chordin and Sog are present in the neural ectoderm.

Cross-species Function of D/V Genes

■ Chordin Versus BMP4 ■

A series of very elegant experiments conducted in the laboratory of Eddy de Robertis at UCLA, where the *chordin* gene was isolated, suggested that Chordin functioned by a double-negative mechanism to overcome the antineuralizing activity of BMP4. His lab also showed that Chordin binds to BMP4 and prevents it from activating its receptor. Thus, it became apparent that Chordin, like Sog, promotes neural development indirectly by blocking the neural suppressive activity of BMP4. An experiment that confirmed this hypothesis and brought very satisfying closure to the vertebrate neural inducer field was performed in Hemmati-Brivanlou's laboratory at Rockefeller University. This group repeated the classic ectoderm dissociation experiment discussed above with the modification of adding identified molecules to the cell culture medium (Fig. 5.4). Hemmati-Brivanlou reasoned that in the original experiments, dissociated ectoderm cells were unable to inhibit one another from assuming the default neural state because the BMP4 they were producing was being greatly diluted by the large volumes of media in which they were incubated. As a consequence of this dilution factor, cells received insufficient levels of BMP4 to be prevented from becoming neurons. To test this idea, he added soluble purified BMP4 to the media into which the cells were dissociated. When these BMP4-treated cells were then reaggregated by centrifugation, they became epidermal rather than neural. He took the experiment yet one step further by showing that if Chordin was added to the media along with BMP4, the reaggregated cells once again became neural. These experiments reversed more than 50 years of thinking and established the current view that the default state of ectoderm in vertebrates as well as invertebrates is neural and that the derived epidermal cell state is dependent on active signaling (i.e., by BMP4/Dpp).

The fact that corresponding genes in flies and frogs perform the same functions in D/V patterning led Eddy De Robertis and his collaborators, as well as my own group and our collaborator David Kimelman at the University of Washington, to test the possibility that Dpp and Sog could pattern the D/V axis of frog embryos. We believed that these experiments were unlikely to succeed a priori, given the enormous evolutionary chasm between vertebrates and invertebrates, but we were motivated to try them anyway, partially because of the stunning success of the previously described McGinnis experiment which demonstrated that mouse *Hox* genes can function as their fly counterparts in flies. In addition, Richard Padgett, while working with Bill Gelbart (the discoverer of the *dpp* gene), had shown that a human version of the Dpp gene could substitute for *dpp* in flies. Even with these encouraging indications, it seemed to be asking quite a bit to recreate an entire D/V patterning system in frogs using fly genes.

When I was a graduate student, the embryonic cell dissociation was actually used as an example of how one had to be very careful when designing an experiment to avoid generating artifactual results due to experimental manipulations. It is somewhat ironic that this well-chosen example of an artifact has itself turned out to be a misinterpretation of a very informative experiment.

Pursuing this idea, Kimelman's group injected *sog* mRNA synthesized from the fly *sog* gene into the ventral region of frog embryos and obtained embryos with a partial double axis (Fig. 5.3; see also plate 2D, E for example of a tadpole with a duplicated neural axis). This experiment meant that the fly Sog protein (translated from the injected *sog* mRNA) could act as a potent neural-inducing substance in frogs! De Robertis and collaborators came to the same conclusion and also showed the converse—that vertebrate Chordin could function in flies.

These experiments and those of McGinnis demonstrated that the primary mechanisms for patterning both the D/V and A/P axes have been strictly preserved during the course of evolution. A noteworthy point with respect to D/V axis formation in vertebrates and invertebrates is that the observed similarities between these two classes of organisms are evident at a much earlier stage in development than for the A/P axis. As mentioned above, segment-identity genes function during the phylotypic stage of development midway through gastrulation (for a review of fly embryonic development see Fig. 3.1). In contrast, the similarities between the mechanisms guiding D/V axis formation in flies and frogs are observed much earlier, prior to cell movements, when the embryo is just activating expression of its own genes for the first time. The fact that D/V patterning shares similarities between vertebrates and invertebrates at such an early developmental stage raises the question of whether there may also be yet unknown aspects of early A/P patterning in common between vertebrates and invertebrates.

Neural Versus Nonneural Development: Déjà Vu, All Over Again

In Chapter 3, we discussed a proposed mechanism for the way in which *sog* functions to promote neural development in the neuroectoderm. The central premise of that hypothesis is that the diffusible Dpp protein produced in the nonneural ectoderm of the fly embryo leaks into the adjacent neural ectoderm where, in principle, it can activate its own expression though a positive feedback loop (see Figs. 3.9 and 3.10). In this scheme, Sog prevents the invasive combination of Dpp diffusion and autoactivation from spreading into the neuroectoderm thereby permitting cells to follow their preference to become neural. In support of this hypothesis, Dpp can invade the neuroectoderm of mutants lacking the *sog* gene and convert much of it into nonneural ecto-

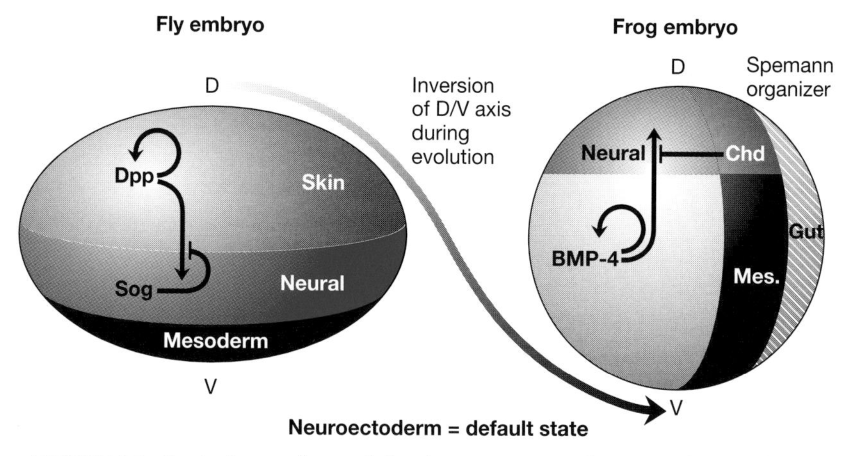

FIGURE 5.5. Equivalence of neural development in vertebrates and invertebrates.

derm. This conversion of neural ectoderm into nonneural ectoderm results from suppression of neural gene expression and from activtion of nonneural genes (e.g., *dpp*) in the neuroectoderm.

A remarkable indication of the preservation of D/V patterning during the evolution of vertebrates and invertebrates is that zebrafish mutants lacking the function of the *chordin* gene have defects nearly identical to those of *sog* mutants described above. In these mutants, expression of the *bmp4* gene spreads from the nonneural ectoderm into the neural ectoderm, indicating that BMP4, like its fly counterpart Dpp, is capable of diffusing and autoactivating. As a consequence of the spread of BMP4 signaling into the neuroectoderm in *chordin* mutant fish embryos, the formation of neural structures is compromised and the region of the embryo giving rise to nonneural structures is enlarged. Thus, in both vertebrates and invertebrates, a key function of "neural inducers" is to protect the neuroectoderm from invasion by BMP4 or Dpp, respectively. In this way, Chordin and Sog permit ectodermal cells to follow their default preference to become neuroectoderm (Fig. 5.5).

As we discuss further in Chapter 6, the commonalities between developmental mechanisms in vertebrates and invertebrates provide a basis for reconstructing the form of the most recent common ancestor of vertebrates and invertebrates. From what we have discussed in this chapter, it is clear that this ancestor was a highly structured creature which was subdivided into segments along the A/P axis and partitioned into basic tissue types along the D/V axis. In the near future it should be possible to draw a fairly detailed image of this wondrous creature that gave rise to most animal forms alive on earth today.

■ Summary of Early Vertebrate Development ■

The mechanisms by which vertebrate mothers polarize their eggs seem to be quite different from those employed by flies. For example, in flies, the mother polarizes the A/P axis by creating a graded distribution of the Bicoid morphogen and the D/V axis by generating a graded distribution of the Dorsal morphogen. In frogs, on the other hand, the mother polarizes the egg into animal and vegetal hemispheres, and an external agent (the sperm) defines the opposing position of the dorsal organizing center. Also, patterning of the A/P and D/V axes is coupled in frogs, whereas these axes are established by entirely independent mechanisms in the fly embryo. Although not enough is known about early mechanisms involved in patterning the A/P and D/V axes of other vertebrates, it seems likely that diverse mechanisms initiate axis patterning in different vertebrate species.

Despite the differences in creating initial asymmetries in the egg, an abundance of evidence indicates that there are deep similarities in how embryos use maternally provided information to generate segments along the A/P axis and to partition the D/V into primary tissue types such as neural versus nonneural ectoderm. During mid-gastrulation at the so-called phylotypic stage, embryos from across the animal kingdom share obvious features of segmentation. During this "phylotypic" stage, homeotic and Hox genes label segments according to their position along the A/P axis in flies and vertebrates, respectively. These segment-identity genes organize segment-specific developmental programs in corresponding regions of the fly and vertebrate embryo. The most dramatic demonstration that vertebrate and invertebrate embryos employ a common developmental mechanism for assigning segment identity was provided by the McGinnis experiment in which it was shown that mouse Hox genes could mimic the function of their fly counterparts during fly development. This seminal observation laid the groundwork for subsequent studies that have revealed a more detailed web of common mechanisms guiding early developmental decisions in all segmented animals.

Common mechanisms are also shared between vertebrates and invertebrates in subdividing the embryo into primary tissue types along the D/V axis. In both classes of organisms, the default state of ectoderm is neural, and, in the nonneural ectoderm, this neural preference is actively suppressed by cells sending a mutually inhibitory signal to one another. The molecular identity of this inhibitory signal is the same in vertebrates and invertebrates (i.e., BMP4 in vertebrates = Dpp in flies). In flies the neural suppressive activity of Dpp is blocked in the neuroectoderm by Sog. Similarly, Chordin, the vertebrate counterpart of Sog, blocks the neural suppressive activity of BMP4 and thereby functions indirectly to promote neural development. In both vertebrate and invertebrate embryos an important mechanism by which such neural-inducing substances act is to block a positive feedback loop created by the coupling of Dpp/BMP4 diffusion and autoactivation. As in the case of segment-identity genes along the A/P axis, these early-acting D/V patterning genes function when transplanted between vertebrates and invertebrates.

An important implication of the deep similarities in the primary patterning mechanisms driving vertebrate and invertebrate development is that the most recent common ancestor of all segmented animals must have been a highly evolved creature with well-defined segments and primary tissue types. This emerging image of our shared ancestor as a shrimp-like creature is several orders of magnitude more complex than the amoeboid slug-like organism that was the generally imagined form of this ancestor just 15 years ago. This realization is one of the most profound insights into evolution that one can extract from our current understanding of development.

6 Patterning Vertebrate Appendages and Eyes

In the previous chapter, we discussed patterning mechanisms for establishing basic positions and fundamental tissue types in early vertebrate embryos. We saw that the initiating asymmetries laid down in the egg by vertebrate mothers such as the frog seem to be quite different from those used by invertebrate mothers, but that vertebrate and invertebrate embryos respond to maternal information by very similar mechanisms to define segment identity along the anterior–posterior (A/P) axis and tissue-type identity along the dorsal–ventral (D/V) axis.

In this chapter, we examine the parallels between invertebrates and vertebrates further by considering the mechanisms for establishing polarity in developing appendages and sensory structures such as limb buds and eyes. Echoing themes of the previous chapter, the mechanisms for specifying the locations at which appendages develop appear to be different in vertebrates and invertebrates, but there are significant similarities in the way genes pattern the principal axes of appendages once their position has been determined. The specification of eyes in various species, which often appears very different, also relies on equivalent genes. The idea that appendages and eyes are patterned using similar mechanisms in diverse species runs counter to the prevailing view that vertebrates and invertebrates independently evolved appendages and sensory organs, and provides us with a new view of the most recent common ancestor of all segmented animals.

■ **Cast of Characters** ■

Terms

Anterior–posterior (A/P) axis The anterior (head)–posterior (tail) axis of an animal. In an appendage such as a human hand, the thumb is anterior and the little finger is posterior.

Aniridia A human disease that is associated with eye defects similar to those in *Small eye* deficient mice in which the vertebrate *pax-6* gene is mutant.

Apical ectodermal ridge (AER) A specialized group of cells at the junction between the dorsal and ventral surfaces of vertebrate limb buds that plays a role in patterning the dorsal–ventral axis of the limb.

Cambrian extinction A massive extinction of life forms during the early Cambrian period.

Dorsal–ventral (D/V) axis The dorsal (back)–ventral (belly) axis of an animal. In an appendage such as the human hand, the back of the hand is dorsal and the palm is ventral.

Lens cells The cells that give rise to the transparent portion of the eye that focuses light on the retina.

Limb buds Developing vertebrate appendages such as legs, wings, and arms.

Neural tube A cylindrical infolding of the neural ectoderm that gives rise to the central nervous system of a vertebrate.

Optic vesicle Specialized neural tube cells that bud out from the neural tube to give rise to the vertebrate eye.

Photoreceptors Light-sensitive neuronal cells in the eye that respond to light by producing an electrical impulse.

Proximal–distal (P/D) axis The proximal (e.g., shoulder)–distal (e.g., hand) axis of a vertebrate appendage extends from the body (proximal) to the tip of the appendage (distal).

Retina A hollow sack of cells derived from the out-pocketing of the optic vesicle that gives rise to the portion of the eye containing the light-sensitive photoreceptor cells.

Zone of polarizing activity (ZPA) A small region in the posterior portion of each limb bud that organizes the anterior–posterior axis of the limb, at least in part.

Genes

bone morphogenetic protein-4 (BMP4) The vertebrate counterpart of the Dpp signal in flies.

distalless A homeobox gene that plays an organizing role in defining the proximal–distal axis of appendages.

eyeless The fly counterpart of the vertebrate *pax-6* gene required for early eye development.

fibroblast growth factor (fgf) A gene encoding a secreted signaling factor (FGF) that is critical for initiating limb outgrowth and for defining the anterior–posterior polarity of the limb.

Notch A gene encoding a receptor, which is activated in the AER of vertebrate appendages and is required for outgrowth of limbs.

pax-6 The vertebrate counterpart of the fly *eyeless* gene that plays an essential role in initiating eye development in mice and humans.

Small eye A mutant form of the mouse *pax-6* gene that results in development of smaller than normal mouse eyes.

sonic hedgehog (shh) A gene encoding a vertebrate version of the Hedgehog morphogen (Shh) that is produced by the ZPA and has anterior–posterior organizing activity.

PATTERN FORMATION IN VERTEBRATE APPENDAGES

Development of both vertebrate and invertebrate appendages can be broken down into two distinct steps: (1) determining where appendages will form along the A/P and D/V axes of the animal and (2) patterning of the appendage itself. The first step seems to be accomplished by diverse mechanisms in different organisms consistent with the fact that the positions of limbs vary among species. In contrast, the second step employs similar mechanisms to pattern the primary axes common to all appendages, which are usually referred to as the A/P axis, the D/V axis, and the P/D axis. To illustrate how these axes are oriented with respect to a typical vertebrate appendage, we can consider the human arm. The A/P and D/V axes of appendages derive from the polarity of the animal itself. These axes are easy to visualize in the hand, which has a clearly discernable anterior (e.g., the thumb) and posterior (e.g., the little finger) as well as dorsal (e.g., the back of the hand) and ventral (e.g., the palm) polarity. The proximal end of the appendage is connected to the body wall of the animal (e.g., the shoulder), and the distal tip (e.g., the hand) is farthest from the body. In addition to these general features of limbs, there are, of course, characteristics that distinguish different appendages, such as the tiny forelimbs versus the enormous back legs of *Tyrannosaurus rex*.

Segment-identity Genes Determine the Positions of Limb Buds

As discussed in Chapter 5, vertebrate segment-identity genes define positions along the A/P axis by mechanisms that have remained largely unchanged since the evolutionary split between vertebrates and invertebrates. As in the case of invertebrates, segment-identity genes determine where appendages will form along the A/P axis. The segment-identity genes involved in defining the positions of vertebrate limbs do not generally correspond to those involved in specifying leg- or wing-bearing segments in flies, however. This last fact is consistent with the accepted view that there is no direct correspondence between vertebrate and invertebrate appendages. We revisit this very interesting topic below.

The mechanism by which *Hox* genes determine the position of limbs involves localized activation of a secreted signaling factor called fibroblast growth factor (FGF). FGF protein can diffuse several cell diameters from the point at which it is produced, bind to FGF receptors present on neighboring cells, and activate a signaling pathway in those recipient cells (see Fig. 1.6). FGF, like the Dpp signal, is used in many developmental decisions and often is used in situations where cells undergo coordinated movements, such as formation of the mesodermal cell layer in gastrulating fly and frog embryos. FGF also is the key signal for shaping the tracheae of the lungs in vertebrates and a corresponding set of breathing tubes in invertebrates. These roles of FGF provide additional examples of developmental mechanisms that have been preserved during evolution. The use of FGF in defining the position and initial outgrowth of limb buds, on the other hand, appears to be a vertebrate-specific innovation.

FGF is involved in initiating limb outgrowth and defining A/P polarity of the limb. Induction of limb-bud formation by FGF leads to the production of FGF in the limb bud itself through a positive feedback loop. This spread of FGF expression into the limb bud through a positive feedback loop is similar to that discussed previously for Dpp signaling in the ectoderm of early embryos (see Chapters 3 and 5). FGF produced by the limb bud, in combination with a *Hox* gene expressed only in the posterior part of limb buds, defines a small region in the posterior portion of each limb bud called the zone of polarizing activity (ZPA; see Fig. 6.1). In contrast to the role of FGF in determining where limb buds will grow, which is a vertebrate invention, the ZPA organizes the A/P axis of the limb by mechanisms that are very similar to those used in flies to pattern the A/P axis of appendages such as the wing. As an example, we consider how the ZPA organizes the A/P axis of the chick wing.

Signals Emanate from Organizing Centers in Developing Limbs

Classic transplantation experiments using ZPA grafts in vertebrates performed by Saunders and Gasseling in 1968, similar to those of Mangold and Spemann, led to the identification of an organizing factor present in early limb buds of gastrulating chick embryos.

■ ZPA Graft Experiments ■

Saunders and Gasseling searched for organizing centers in developing limbs by grafting little patches of limb bud derived from different positions into various locations in a host limb bud. These experiments revealed that a region in the posterior base of a wing bud, when transplanted to an anterior position in a host wing bud, resulted in the formation of a wing having twice the normal number of digits (Fig. 6.2B; see also Plate 3M for an example of ZPA graft experiment). Moreover, the extra digits in these wings were arranged as a mirror image of the normal digits. This phenotype is very similar to that of segment-polarity mutants in *Drosophila,* in which one-half of each segment is replaced by a mirror duplication of the other half (see Chapter 3, Fig. 3.3). The duplicated structures resulting from ZPA graft experiments were interpreted as evidence for a posteriorizing signal that when transplanted to an anterior location could reprogram cells to form posterior rather than anterior structures. As in the case of the Spemann organizer, cells comprising the extra digits were mostly formed by anterior cells surrounding the ZPA implant rather than by cells from the ZPA itself. For example, it was possible to kill ZPA cells by irradiation prior to transplantation and still obtain duplicated wings.

A. *Hox* genes activate autoregulatory FGF expression in limb buds

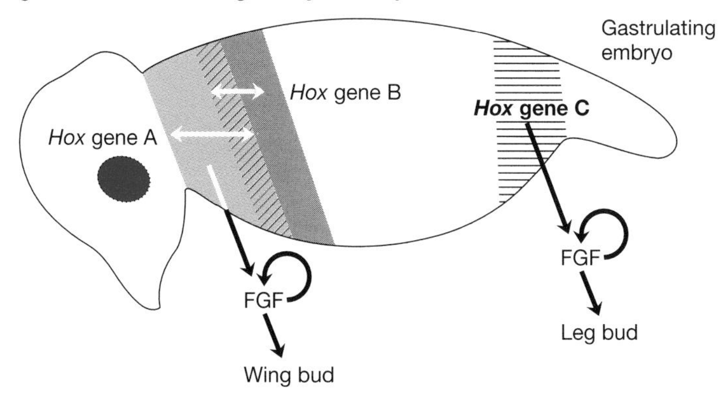

B. FGF and a second *Hox* gene activate Hh expression in ZPA

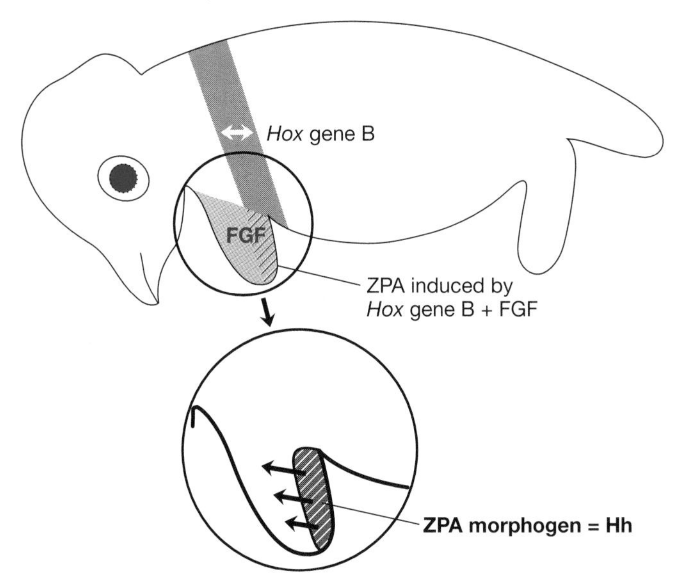

FIGURE 6.1. Initiation of limb-bud formation in vertebrates.

An important aspect of the ZPA transplantation experiments was that the ZPA acted in a dose-dependent fashion, suggesting that it was the source of a diffusible morphogen. Thus, if one transplanted only a small piece of the ZPA, one obtained just a partial duplication of digits (Fig. 6.2C). When such partial duplications were generated, the duplicated digits were always those which normally formed in more anterior positions. This result indicated that low levels of the ZPA morphogen specified anterior digits, whereas higher levels specified posterior digits.

Further analysis of A/P patterning in the vertebrate limb bud by Cliff Tabin's group and several other laboratories revealed the outlines of a genetic pathway that is strikingly similar to that acting in fly appendages. For example, they found that the ZPA factor sonic

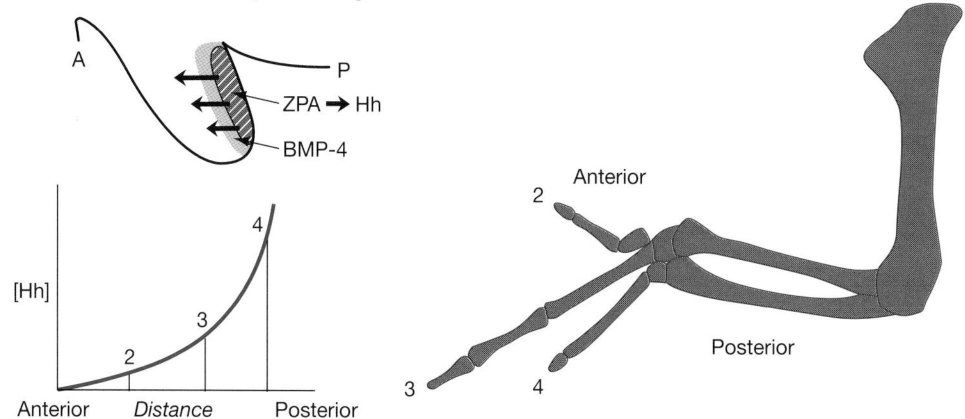

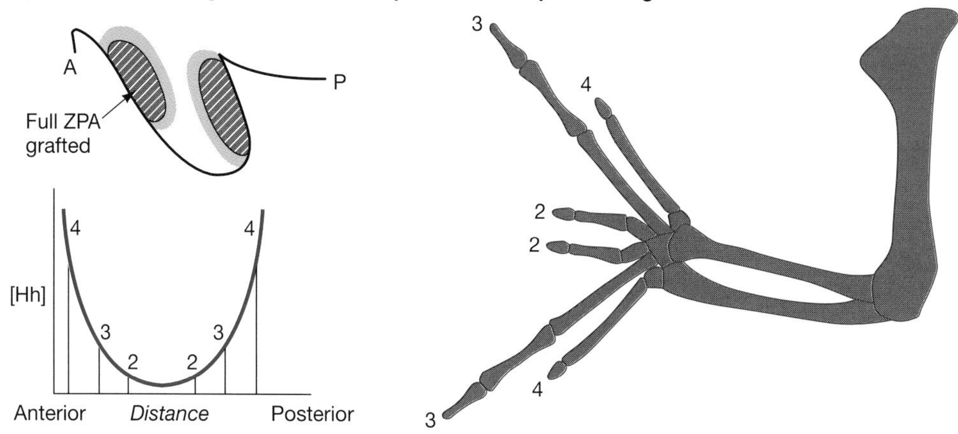

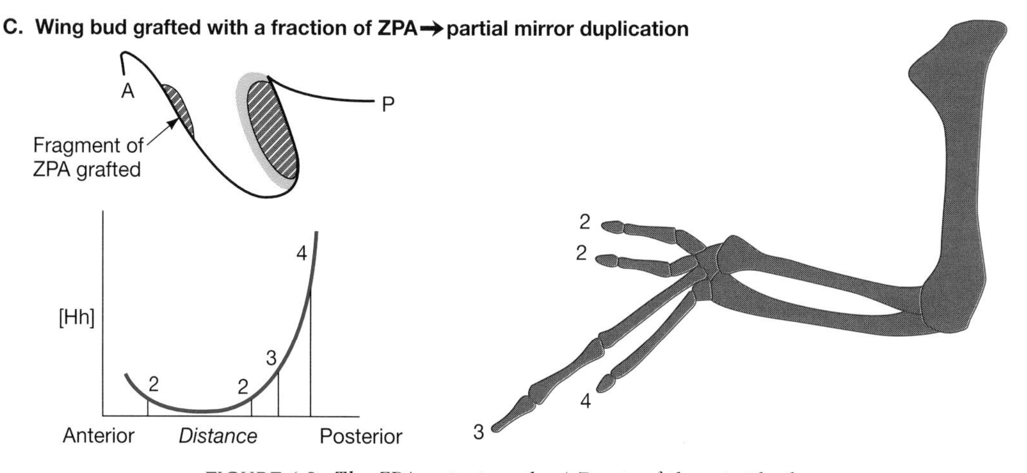

FIGURE 6.2. The ZPA organizes the A/P axis of the wing bud.

■ Hedgehog Is the Morphogen Produced by ZPA ■

A major breakthrough in the limb-patterning field was the identification of the morphogen produced by the ZPA in the laboratory of Cliff Tabin at Harvard University. Tabin found that a vertebrate version of the Hedgehog signal, which he called Sonic Hedgehog after a cartoon character, was expressed in the same small group of posterior cells that produced the ZPA activity (see Plate 3L). He and other workers then performed a series of elegant experiments demonstrating that Hedgehog was indeed the ZPA morphogen. For example, Tabin implanted plastic beads soaked in Hedgehog protein into the anterior part of limb buds and generated wings with a perfect set of mirror duplicated digits, just as in the classic ZPA transplantation experiments (see Plate 3N). Most importantly, he showed that the pattern of extra digits generated depended critically on the dose of Hedgehog protein (see Plate 3N). Beads soaked in concentrated solutions of Hedgehog caused full digit duplications, whereas beads soaked in more dilute concentrations generated partial duplications. As in the case of the ZPA grafting experiments, when partial duplications of the wing were observed, the duplicated structures were always the most anterior digits.

Hedgehog diffuses from the cells in which it is produced, binds and activates a Hedgehog receptor in those cells, and initiates expression of several of the same target genes activated by Hedgehog in fly appendages, such as bone morphogenetic protein-4 (BMP4), the vertebrate counterpart of the Dpp signal in flies. The domain of BMP4 expression is centered over Hedgehog-producing cells, but is significantly larger, indicating that Hedgehog protein can diffuse from the ZPA and activate BMP4 expression in a broader domain. By now it probably comes as no shock to the reader that vertebrate and invertebrate forms of Hedgehog can replace one another functionally in cross-species gene-transplantation experiments. These similarities in patterning the A/P axes of fly and chick wings, and other parallels in the development of the other two axes of the limb (D/V and P/D) discussed below, reveal that vertebrate and invertebrate appendages have more in common than previously thought.

Similarities between Patterning Vertebrate and Invertebrate Appendages

As mentioned above, all appendages have three primary patterning axes, A/P, D/V, and P/D. It is worth backtracking momentarily into fly development for a refresher on the basic features of D/V and P/D patterning in wing and leg imaginal primordia, as it has recently become apparent that there are notable similarities in the mechanisms by which these axes are patterned in fly and vertebrate appendages (Fig. 6.3).

The fly wing, like a human hand, has a dorsal and a ventral surface. A narrow strip of cells lying at the junction between the dorsal and ventral surfaces of the wing organizes the D/V axis in much the same way that the A/P organizer sets up the A/P axis. An important signaling

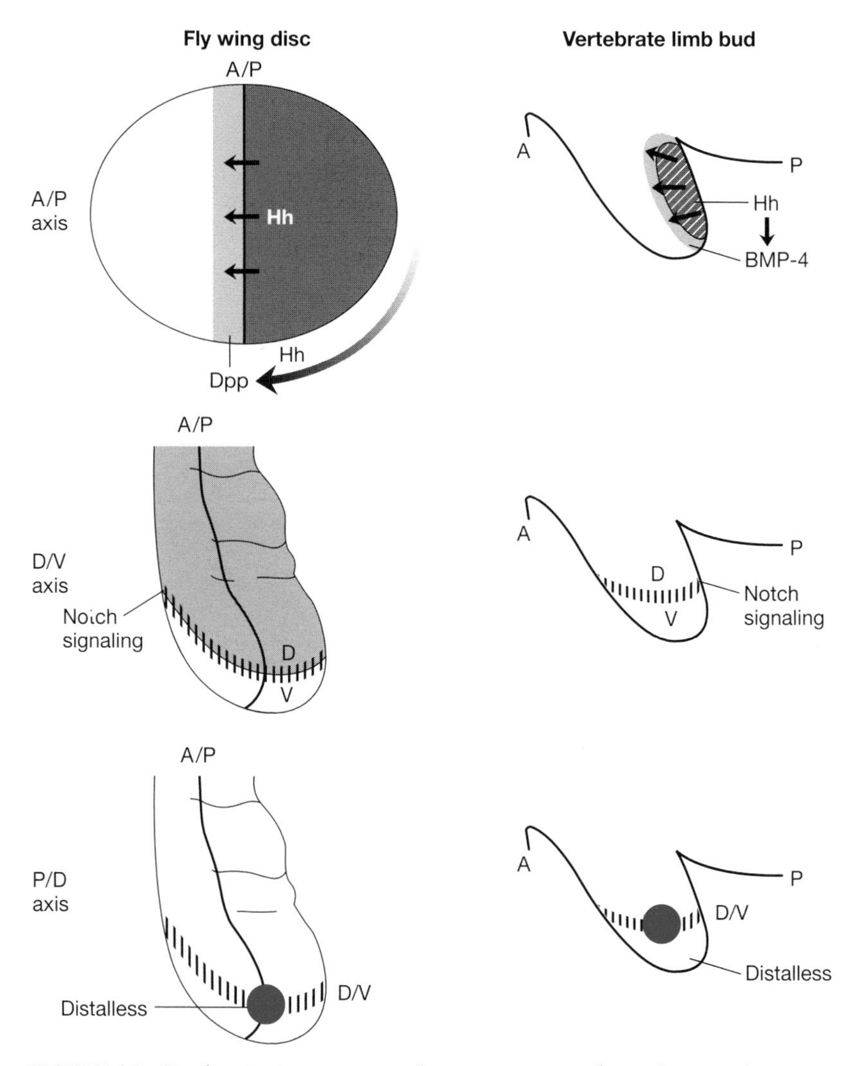

FIGURE 6.3. Similarities in patterning the primary axes of vertebrate and invertebrate appendages.

Cliff Tabin (1954–)

Cliff Tabin was born in Chicago, Illinois. He did his undergraduate studies at the University of Chicago and was a graduate student with Bob Weinberg at the Massachusetts Institute of Technology. His Ph.D. thesis work included the identification of a mutation that activated a cancer-causing gene, which was the first example of an activating mutation in a cancer-causing gene.

Tabin did a brief postdoctoral stint with Doug Melton, working on Hox genes in frogs, and then took a position as an independent postdoctoral fellow at Massachusetts General Hospital, where he initiated work on regeneration of newt limbs using molecular techniques. He then joined the faculty in the

Department of Genetics at Harvard Medical School, where he has made most of his groundbreaking discoveries on limb patterning in vertebrates.

One of Tabin's most important discoveries is that a vertebrate version of the fly Hedgehog morphogen, which he named Sonic Hedgehog (after the cartoon character), is the primary anterior–posterior organizing factor produced by the zone of polarizing activity (ZPA). Tabin credits this discovery in part to two critical factors defining the intellectual setting at that time. First, his lab consisted of a team of exceptionally bright young scientists, excited about working on vertebrate development at a molecular level, something that at the time was very new. The limb-patterning problem was very timely because his group and others were in the process of developing the new molecular tools and methods required for this analysis. Their studies also benefited enormously from an existing wealth of anatomical and embryological observations obtained in classic grafting experiments using chick embryos. This rich history provided a strong theoretical context for their work. The second aspect of the intellectual environment came from the close collaboration Tabin established with Phil Ingham (at the time at Oxford) and Andy McMahon (then at Roche Institute). Tabin recalls, "Not only did the close interactions accelerate the work, but it also meant that I felt a part of the discoveries of the role of Sonic in the neural tube (or spinal cord), while they participated intellectually in the discovery of the role in the limb bud."

Tabin emphasizes the importance of previous classic experiments in their recognizing that Hedgehog was likely to be the ZPA factor. For one thing, transplants had clearly shown that the posterior limb made some factor responsible for patterning the A/P axis. In addition, the location and timing of the production of that factor had been carefully mapped. Therefore, when they discovered Sonic Hedgehog, Tabin was able to compare its developmental expression profile to that of the ZPA signal, and found a close correspondence. Moreover, other transplants had shown that although most tissues did not induce limb-pattern duplications, several others besides the ZPA did. In particular, pieces of the notochord, floorplate of the neural tube, and Hensen's Node all caused such an effect. This was correctly interpreted as indicating that a common signaling pathway was used by these various organizing centers. Even before they demonstrated the biological properties of Sonic Hedgehog, Tabin was very excited by the fact that this gene was expressed in exactly those tissues shown to cause limb duplications. Tabin concluded that Sonic was very likely to be part of the ZPA signaling story, based on the expression profile of this gene alone.

In addition to these discoveries, Tabin also discovered the first two genes to be expressed in a left–right asymmetric pattern and showed that they are regulators of morphological left–right asymmetry (e.g., the heart forms on the left); and he performed the first biochemical experiments characterizing the Hedgehog receptor. As exciting as these important discoveries were for Tabin, the most exciting moment in his scientific life actually occurred much earlier. He recounts the following description of his maiden scientific voyage: "When I was a graduate student, one of my first projects was collaborating with Steve Goff in David Baltimore's lab on the biological characterization of the Abelson gene product. My role was simply to transfect v-Abl clones (i.e., introduce virus) into fibroblasts to see which were active and could transform cells (i.e., convert them into a cancer-like state). While my control transfections worked great, I never saw anything with a bunch of independent v-Abl clones. I had to report my data at a Baltimore lab group meeting, and I (a second-year student) was terrified that David would say I was incompetent and Steve would decide to collaborate with someone else. Instead, after hearing my controls, etc., David simply said that perhaps there was some reason that one could not transfect v-Abl (i.e., he trusted my data more than I did myself). That was a tremendous boost to my confidence and sort of a turning point in my early scientific development. With renewed confidence I designed a simple test to show that this was the case (co-transfect a marker, and show that the v-Abl DNA killed the ability to isolate colonies transformed with the marker). Hypothesizing that high levels of v-Abl (virus) were transiently being expressed and killing the recipient cells, I then came up with a way around the problem (transfect not just the v-Abl virus, but co-transfect it mixed with a helper virus so that active Abl virus could be produced and stably infect surrounding cells, even if the originally transfected cell died.) This worked. In retrospect it was a minor experiment verifying that a clone was biologically active, and a simple technical trick to get around an experimental obstacle. However, it was the first time I had confronted a scientific problem, and actually solved it; making a hypothesis, testing it, and being right. I can still vividly remember my excitement when I saw the Abl-transformed cells on the transfected plates. It is the one time in my career I can recall actually trembling when getting a new result. And that thrill of solving a scientific problem, and thereby learning something new (albeit minor in that case) about biology is really what has propelled me to continue as a scientist ever since."

Tabin's lab works on a very wide range of scientific problems, including exploring limb development in terms of early patterning of the P/D axis, specification of cartilage, migratory pathways of muscle precursor cells, defining determinants of left–right asymmetry and understanding how such signals create morphological asymmetry, looking at a variety of other embryological events including gut patterning and cell-type specification in somites, and also biochemical characterization of the Hedgehog pathway. Finally, Tabin has an Evo-Devo project on the genes underlying the morphological variation observed in Darwin's finches in the Galapagos Islands. He encourages people to define their own interests. "I give people in my lab complete freedom in terms of what they work on. I try to 'sell' them on various projects, but ultimately it is their own choice. Also it means I can usually let postdocs take the complete projects with them when they leave, as we have so much else going on. Importantly, giving people more freedom also means you get to feed off of their creativity and ideas to a greater extent too."

In considering important factors for discovery, Tabin acknowledges that there are many paths and that generalities are hard to make. In his own case, Tabin remarks, "I have benefited a lot from working in great systems just when they were getting off the ground (oncogenes, vertebrate development). I wish I could say I was insightful enough to have decided, in a careerist fashion, that limb development would be a hot area in the future and hence I should go into it. But that is not at all true. I started in limb development at the same time as Jeremy Brockes began doing limb regeneration, but before anyone else had worked on the problem on a genetic level. The field exploded when others subsequently entered the field because they cloned genes that happened to be expressed there, e.g., Hox gene (Dennis Duboule), FGF (Gail Martin), etc. But, I was not prophetic in this regard. I simply thought vertebrate development was the coolest question (the genetics of 'where do babies come from') and felt it would be fun to do molecular biology in the context of a classical experimental area like limb patterning."

pathway active in the strip of cells comprising the D/V organizer is known as the Notch pathway (Fig. 4.4). As its name indicates, mutants in the Notch gene have notches in the edge of the wing, giving it the appearance of having been nibbled on. Because the dorsal and ventral surfaces of the wing meet at the edge of the wing, this notching reflects a defect in communication between cells belonging to the dorsal and ventral surfaces. The Notch pathway also plays a role in a specialized group of cells at the junction between the dorsal and ventral surfaces of vertebrate limb buds. This specialized band of cells known as the apical ectodermal ridge (AER) plays a role in patterning the D/V axis of the limb akin to that of the ZPA along the A/P axis. Although the mechanisms used by vertebrates and invertebrates to restrict activation of the Notch pathway to narrow stripes of cells at the border between the dorsal and ventral surfaces of appendages are different, it is nonetheless striking that the Notch pathway ends up being activated in cells having the same organizing functions in both classes of animals.

There also is evidence for a common mechanism in defining the P/D axis of vertebrate and invertebrate appendages. The key patterning gene involved in initiating outgrowth along the P/D in flies is known as *distalless,* which derives its name from the lack of distally derived structures in fly mutants having reduced activity of this gene. Positional information generated along the A/P and D/V axes of the early fly embryo is used to define the location where *distalless* will be expressed in gastrulating embryos. Cells expressing *distalless* then organize outgrowth of appendages in these positions. An intriguing finding made in the laboratory of Sean Carroll at the University of Wisconsin was that *distalless* is expressed at the tips of growing appendages in other distantly related insects such as butterflies. For example, in the caterpillar, *distalless* is expressed as a pair of dots in every segment that will form legs, but not in segments that do not. This finding was generalized when a vertebrate counterpart of *distalless* was found that also is expressed at the distal tips of limb buds. The expression of this particular transcription factor at the distal tips of appendages throughout the animal world strongly suggests that Distalless controls some important general aspect of appendage outgrowth. A corollary to this hypothesis is that all appendages derive from some archetypal outgrowth in the most recent common ancestor of vertebrates and invertebrates.

Did the Ancestor of Vertebrates and Invertebrates Have Appendages?

Is it a coincidence that key genetic pathways involved in patterning the A/P axis (Hedgehog and Dpp), D/V axis (Notch), and P/D axis (Distalless) of vertebrate and invertebrate appendages are the same? Perhaps, there are important properties of each of these gene systems that make them optimally suited for carrying out particular developmental tasks. According to this hypothesis, two different lineages of animals could end up using the same genes to do the same jobs because of what one might overstate as inevitable natural selection. For several reasons, this hypothesis seems unlikely to be correct. First, each of the signaling pathways involved in appendage patterning is used in many other developmental events. These other developmental events often are specific to particular animals. For example, BMP4 derives its name, bone morphogenetic protein-4, from its ability to promote bone formation in vertebrates, a role it obviously does not play in boneless invertebrates. The multiple and diverse uses of individual signaling pathways (e.g., the Hedgehog, Dpp, and Notch pathways) during development also argue pointedly against the hypothesis that there is something innately better about one signaling pathway over another for a given developmental process. Similarly, it is not obvious that there is anything special about Distalless so that it should always be chosen among the hundreds of known transcription factors for defining the distal tip of appendages. In all apparent respects, Distalless is just another transcription factor that binds to a typical simple sequence of DNA base pairs and alters gene expression of target genes. Another important consideration is that the A/P, D/V, and P/D axes of growing appendages do not form independently, but rather interact with one another during development to create the final proportions of the limb. For example, *distalless* expression at the distal tip of fly appendages is induced at the intersection of the A/P and D/V organizers (Fig. 6.3). As a consequence of these types of interactions, signaling pathways involved in patterning the various axes are connected to one another by positive and negative feedback loops. On balance, the whole picture seems to be more than coincidence could possibly allow.

One way to account for the consistent parallels between appendage development in vertebrates and invertebrates is to hypothesize that all appendages use a complex integrated genetic circuit that evolved in the common ancestor of vertebrates and invertebrates to pattern some type of appendage or protrusion from the body wall. If the ancestor of vertebrates and invertebrates had invented such complex interactions between signaling pathways, one could imagine that all its descendants continued to use this genetic machine to perform the same or related functions. This argument is exactly the same one used to rationalize the preservation of function of segment-identity genes and genes involved in specifying neural-versus-nonneural ectoderm in early vertebrate and invertebrate embryos. The problem with this hypothesis is that it seems to conflict with a wealth of fossil evidence that early vertebrates had no appendages, certainly no bony outgrowths. There also is evolutionary evidence that vertebrates arose from echinoderms (e.g., starfish and sea anemones), which do not have the organized appendages of flies or current-day vertebrates.

One way to reconcile the existing information is to suppose that a common ancestor of vertebrates and invertebrates had appendages or simple body-wall protrusions and that the mechanisms used to create these structures have been preserved for one purpose or another during the evolution of both vertebrates and invertebrates. According to this view, there is no direct equation of any particular vertebrate appendages with a counterpart in invertebrates. However, the basic layout of all appendages would be constructed according to similar sets of blueprints. Consistent with the idea that the highly specialized appendages of modern-day vertebrates and invertebrates arose during evolution as independent modifications of a simple patterned protrusion present in an ancestor, the developing tentacle-like suction feet of starfish have recently been shown to express the *distalless* gene at their distal tips.

It also is possible that the most recent common ancestor of vertebrates and invertebrates had complex appendages and that early vertebrates which descended from this organism either lost these well-defined appendages entirely or retained only simplified versions of them that do not show up in the fossil record. According to this scenario, when vertebrates first evolved the prototypic structures that would give rise to modern-day limbs, they once again engaged the complex module of genetic pathways invented by its ancestor for building appendages. This would be akin to having blown out your modern hi-tech living room receiver and going upstairs to your bedroom to fetch your grandfather's old radio to listen to the news. You would be much more likely to trot upstairs to get this primitive, but working, antique than to build another radio from scratch, particularly if you are interested in hearing today's news. This last scenario carries with it a hidden constraint, however, which is that the complex genetic machine for appendage construction must have been continually used for some purpose during the intervening period when vertebrates did not have visible appendages. If this were not the case, the unused circuits would have fallen into disrepair because complex relationships between genes can only be preserved if there is a constant pressure of natural selection demanding that they remain intact (e.g., if your grandpa's radio were left unused in the attic for centuries, it would be very unlikely to work when your great-great-great grandson found it and plugged it in). Perhaps in primitive vertebrates, the appendage-making genetic device might have been used in the formation of gills or some other external body structure. As shown below, appendage formation is not the only ambiguous case where there are striking mechanistic similarities in the development of structures that were previously thought to have evolved independently.

EYE DEVELOPMENT IN VERTEBRATES

The neuroectoderm of vertebrates, which forms dorsally, runs nearly the entire length of the embryo. The posterior portions of the neuroectoderm give rise to the spinal cord and the anterior region to the brain. The nervous system becomes internalized within the embryo by virtue of the neuroectoderm folding in as a coherent sheet through a process called invagination (Fig. 6.4). This concerted motion of cells is very similar to that discussed previously in Chapter 3 with respect to formation of the mesoderm in fly embryos (Fig. 3.1). The result of the infolding of the neural ectoderm is the formation of a tube called the neural tube, which lies directly beneath and separated from, the nonneural ectoderm (Fig. 6.4). The overlying nonneural ectoderm primarily gives rise to skin.

The eyes of vertebrates derive from an anterior region of the neural tube at a mid-level in the brain. These specialized neural tube cells, which are referred to as the optic vesicle, bud out from the neural tube

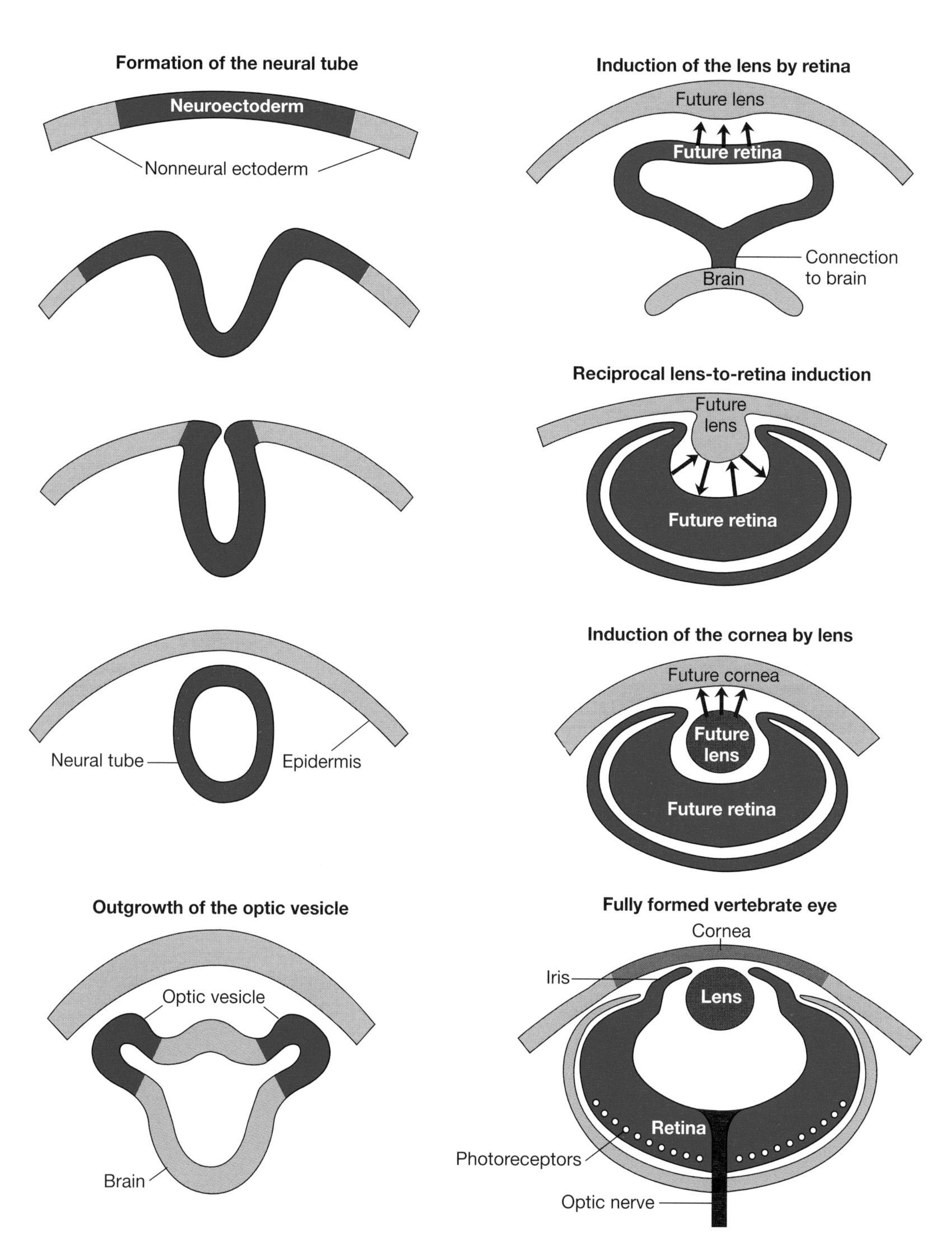

FIGURE 6.4. Eye formation in vertebrates. (Based in part on summary figure in Harris 1997.)

by a process that is topologically opposite to that of invagination (Fig. 6.4). The out-pocketing of optic vesicle cells protrudes from the rest of the brain and forms a little hollow sack that will become the future retina. As the optic vesicle extends away from the rest of the brain, it ultimately contacts the overlying ectoderm and sends an inductive signal that causes the ectoderm to thicken and begin forming the future lens of the eye. Once the future retina has induced the overlying cells to develop as lens, the lens sends back a second signal to trigger further development of the retina and an encapsulating layer of cells called the pigment epithelium. Cells within the retinal layer of the eye proliferate to generate a large number of neurons as well as other cells required for insulating neurons from one another. These cells communicate with one another to determine which of the various retinal cells they will become (e.g., photoreceptor cells sensitive to color versus light intensity). Thus, final cell fate determination in the vertebrate retina, as in the fly eye imaginal disc, depends on cell–cell interactions rather than on predetermined specification of cells at the time they are generated.

From the above description, it would seem vertebrate eye development is very different from eye development in invertebrates such as flies, which have compound eyes (Fig. 6.5). It also has been noted that the eyes of other invertebrates such as the squid, which resemble those of vertebrates in final form, develop by a very different sequence of inductive events (see below). However, there are more similarities in these various developmental processes than meet the eye.

Genes Required for Making Eyes

One of the most exciting insights into vertebrate eye development came from identification of the gene that is disrupted in a mouse mutant known as *Small eye*. Some of the most severely affected *Small eye* mutant individuals entirely lack eyes. There also is a human disease known as aniridia, which is associated with eye defects similar to those in *Small eye* mutant mice. It came as quite a surprise when it was discovered that the same gene—*pax6*—was disrupted in mouse *Small eye* mutants and human aniridia patients, and that this gene was none other than the vertebrate version of the *eyeless* gene described in Chapter 4, which is required to initiate eye development in fruit flies.

The vertebrate *pax6* gene is expressed in the early neuroectoderm in a stripe running perpendicular to the body axis at the level of the eyes. This *pax6* expression precedes any morphological hint of eye development such as out-pocketing of the optic vesicle. The expression of *pax6* in the eye primordium and the loss of eyes in mouse and human mutants that have reduced *pax6* activity argue strongly that *pax6,* like its fly counterpart, plays a critical role in specifying the eye primordium. It is important to note, however, that *pax6* is also expressed in other parts of the brain and that total loss of *pax6* activity leads to widespread defects

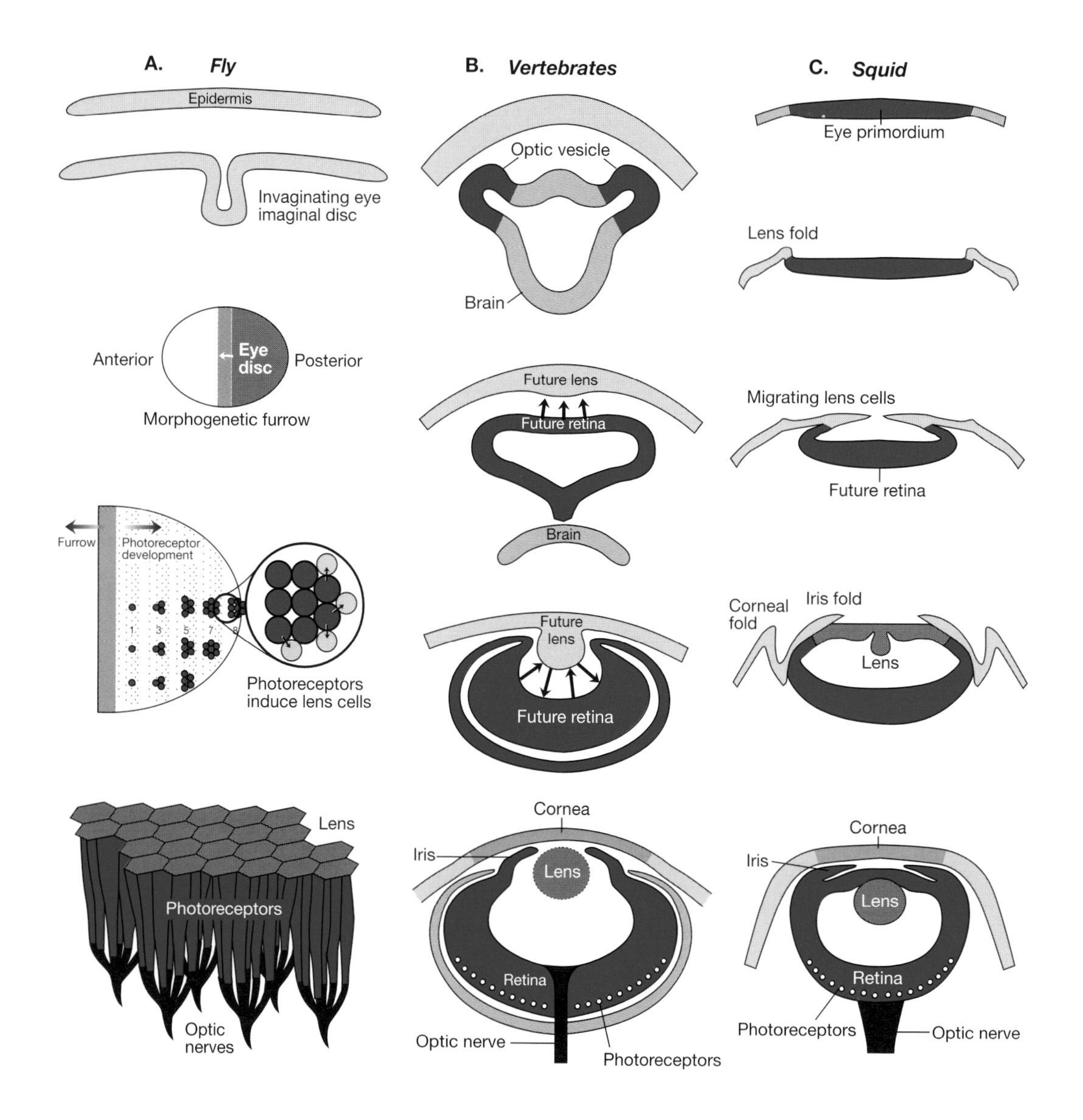

FIGURE 6.5. Comparison of eye development in different species. (Based in part on summary figure in Harris 1997.)

in these regions of the brain as well as to the loss of eyes. Thus, whereas *pax6* is clearly essential for initiating vertebrate eye formation, it has other roles in brain development as well.

It also has been found that vertebrate counterparts of several other genes involved in early fly eye specification, such as *sine oculus,* are involved in early vertebrate eye formation. The regulatory interactions among these genes, however, appear not to be the same in flies and vertebrates. The potential role of the ancestral version of the *pax6/eyeless* gene in defining eyes is discussed further below.

Cross-species Function of Eye Genes

Given the success of cross-species function experiments with segment-identity genes, D/V patterning genes, and limb-patterning genes, it should almost be expected that the same kind of result would be obtained with eye development. As mentioned in Chapter 4, when the fly *eyeless* gene is misexpressed during adult development, freakish flies are generated with eyes forming in bizarre locations such as the middle of their wings. Similarly, when vertebrate *pax6* was misexpressed in developing flies, well-formed eyes were generated in similar inappropriate positions (see Plate 3H). The ability of the *pax6* gene to function in such cross-species experiments, and the similar eye defects observed in vertebrates and flies lacking the respective function of the *pax6* and *eyeless* genes, strongly suggest that the most recent common ancestor of flies and vertebrates used a *pax6*-like gene in some aspect of eye development or function. In the next section, we discuss what this role may have been.

Evolution of Eyes

Like the evolution of appendages, it has generally been assumed that eyes have evolved independently in several animal lineages. This conclusion is based primarily on morphological and developmental considerations. For example, the morphology, organization, and development of the compound fly eye is about as different as could be imagined from that of vertebrates (Fig. 6.5). In flies, the eye primordium (the eye imaginal disc) is set aside as a small sack of epidermal cells, which is then patterned from posterior to anterior by the passage of the furrow (Fig. 6.5A). The furrow initiates a series of cell–cell interactions among eye disc cells that results in their organization into a repeated pattern of clusters that will become the individual facets of the compound eye (see Fig. 4.5 for a refresher). In contrast, the unitary vertebrate retina derives from an out-pocketing of the brain, which then induces the overlying epidermis to form the lens (Fig. 6.4, Fig. 6.5B). As mentioned above, the lens primordium then sends a signal back to the retina to initiate the final stages of retinal development, detaches from the epidermis, and induces the epidermis to form the cornea. These involved series of developmental events and their final outcome in flies and vertebrates bear no apparent similarities.

The squid, an invertebrate whose ancestors diverged from flies at around the same time as the vertebrate lineage (e.g., >500 million years ago), also is thought to have evolved eyes independently from that of vertebrates. Although the final structure of the squid eye is superficially quite similar to that of vertebrate eyes, it develops very differently (Fig. 6.5C). In the squid, the retina derives from the ectoderm rather than the neural tube, and the lens and cornea are generated from thickenings of epidermal cells flanking the retina, which then migrate over the retina in successive waves. Thus, the origins and types of inductive interactions involved in creating vertebrate versus squid eyes appear to be quite distinct. Because of these differences in vertebrate versus squid eye development, Walter Gehring's laboratory asked whether a squid version of *pax6* was expressed in the eye primordium of squid embryos. They successfully isolated the squid *pax6* gene and showed that indeed it was expressed in the developing eye primordium. Thus, the ancestor of squids and vertebrates also was likely to have used the *pax6* gene in some aspect of light detection.

How can we resolve the apparent conflict between the evolutionary arguments that eyes evolved independently in vertebrates, flies, and squids and the observation that *pax6* is expressed in the primordia of all of these eyes and is required for eye development (at least in vertebrates and flies)? This question is very similar to that raised above regarding the origins of appendages and can be answered in much the same way. One idea has been put forward by Charles Zuker, at the University of California, San Diego.

■ The Zuker Hypothesis ■

Zuker studies the mechanism by which photoreceptors in the eye detect light and convert that sensation into a pattern of neural impulses. His group and others have shown that nearly all the key molecular mechanisms acting to create a neural copy of a visual scene are the same in vertebrates and invertebrates. On the basis of this fact, he suggested that the common ancestor of vertebrates and invertebrates was able to sense light and did so using the same sensing mechanism that is operating in modern vertebrates and invertebrates. Because many components are involved in the light detection pathway, it is possible that a single regulator of genes encoding this battery of proteins evolved (i.e., the ancestor of the vertebrate *pax6* and fly *eyeless* genes). This is a very reasonable proposal, because even in very primitive organisms such as bacteria, genes involved in a common pathway are often activated in a concerted fashion. This coordinated activation of a set of genes is usually accomplished by a single transcription factor which is turned on in response to appropriate conditions. According to the Zuker hypothesis, the ancestor of vertebrates and invertebrates may not have had a specialized eye, but could detect light. Subsequently, during the independent evolution of eyes in vertebrate and invertebrate lineages, the mechanism for activating light-sensing genes was retained because expression of these genes in the new light-sensing organs was necessary. Since at all stages of the evolution of eyes in different lineages these same light-sensing genes would be required, the key activator of these genes (i.e., *pax6*) was also preserved as a genetic switch.

There are some problems with the Zuker hypothesis; most notably that one might expect that *pax6* would be involved in the activation of light-sensing genes in the retina of nearly fully formed eyes and not necessarily in the earliest phases of initiating eye development. This objection can be countered by imagining that *pax6,* being in a very central regulatory position in the ancestor of vertebrates and invertebrates, gradually took over more and more duties in evolving eyes until it became a major regulator of all eye genes. It is also possible, as in the case of appendage development, that the ancestor of vertebrates and invertebrates had well-developed eyes, whose development was controlled by the precursor of the *pax6* and *eyeless* genes, and that during evolution, different genes came under the control of *pax6*, resulting ultimately in the evolution of very different morphological rearrangements during development and, in some cases (e.g., flies versus vertebrates), in dramatically different final eye morphology. This last hypothesis could be compared to a king who lived a thousand years and during that time held unbroken dominion over a series of kingdoms spanning ancient feudal societies to a modern jet-age world. If he were a truly gifted leader, it is possible to imagine that he always would have the right people executing his will at any given time. Maybe it was just too hard to overthrow the dominion of *pax6* in the eye and so it has kept its throne for all of these years.

Whatever the actual basis is for *pax6* being involved in early eye development in diverse animals, one thing seems clear; the ancestor of these animals could at least detect light, and *pax6* had something to do with this process. However, as mentioned above and in Chapter 4, it is necessary to remember that *pax6* is also important for other aspects of brain development. Furthermore, as mentioned in Chapter 4, there are several reasons it seems unlikely that the *pax6/eyeless* genes act as "master" genes for eye development. One reason is that misexpression of *eyeless* converts some, but not all, cells of the developing fly into eyes. Another indication that *eyeless* is part of a ruling coalition rather than a supreme dictator is that other genes are also required for eye development (e.g., *sine oculus, eye gone,* and *dachshund*) and that misexpression of several of these genes can similarly induce the formation of eyes in abnormal locations (see Chapter 4).

So where does all of this analysis leave us? I think at least two major points can be extracted from this story. First, eyes in all organisms probably did originate from a common ancestral light-detecting organ of some kind. Second, although the interactions between the key regulators of eye development are complex and may have changed during evolution, the same cast of genes has always been involved in one way or another. The important conclusion one can draw from these two points is that the most recent common ancestor of vertebrates and invertebrates had eyes or light-sensing organs of some kind, and the formation of these structures depended on the same set of genes that now guide eye development in a myriad of its varied descendants.

THE MOST RECENT ANCESTOR OF VERTEBRATES AND INVERTEBRATES

Nearly two centuries of experimentation and the enormous flurry of recent work, which has been only been touched on in the previous chapters, can be distilled into two very important but simple lessons about development and evolution. It is difficult to convey the great implications of these points in the history of the science of biology without some hint of melodrama. Thus, let us now consider these points and their profound implications.

The most visual of these two points is that we can now reconstruct a reasonably detailed image of the most recent common ancestor of vertebrates and invertebrates (Fig. 6.6). This wondrous creature who begat us all certainly had a well-defined head and tail with repeated segments in between, a belly and back, and basic tissue types such as nerve, muscle, and skin. It is likely that this creature also had some type of appendages or outgrowths from its body wall and a light-sensitive organ which served as the precursor to current-day eyes. It also seems clear from a variety of experiments similar to those described above that there are profound similarities between the formation of the rudimentary heart of invertebrates and the early stages of vertebrate heart development. In addition, the mechanisms by which neurons connect during the early stages of wiring the nervous system, trachea, or lungs branch, germ cells are produced, and basic immune system functions are based on common molecular machinery. Taking all of these similarities into account, our ancestor must have been a bilaterally symmetric animal that looked something like a shrimp. As this picture of our most recent common ancestor comes into ever-sharpening focus, it is apparent that this creature was the product of a great deal of previous evolution and must have lived at the same time as many other organisms, some of which may have eaten it routinely for a snack. A big question then is, Why was our ancestor the sole complex animal survivor of its time? Why did the progeny of the multitude of other animals that must have coexisted with our ancestor die out? We may never know the answer to this mystery, since these other organisms left no known trace either in the fossil record or as genetic material carried by their descendants.

Steven Jay Gould has popularized a similar massive extinction of life forms that occurred during the early Cambrian period. In his excellent book, *Wonderful Life: The Burgess Shale and the Nature of History* (1989), Gould tells the story of the Burgess shale fossil beds in which soft-bodied animals of the early Cambrian were preserved in exquisite detail owing to particularly favorable conditions of fossilization. This story of the early Cambrian period (which occurred long after the evolution and dominion of our sole surviving ancestor) is the first period in evolution in which large animals are adequately represented in the fossil record. The most interesting point of the Burgess Shale story is that among the many different basic categories of animals known as phyla which evolved in the late pre-Cambrian period, only a few survived the Cambrian extinction to give rise to modern-day animals. As an analogy to the explosion of life forms and its collapse during the Cambrian, Gould considers a bush that grows thick with branches during an ideal summer and then is severely pruned at the end of the growing season so that only a limited number of branches

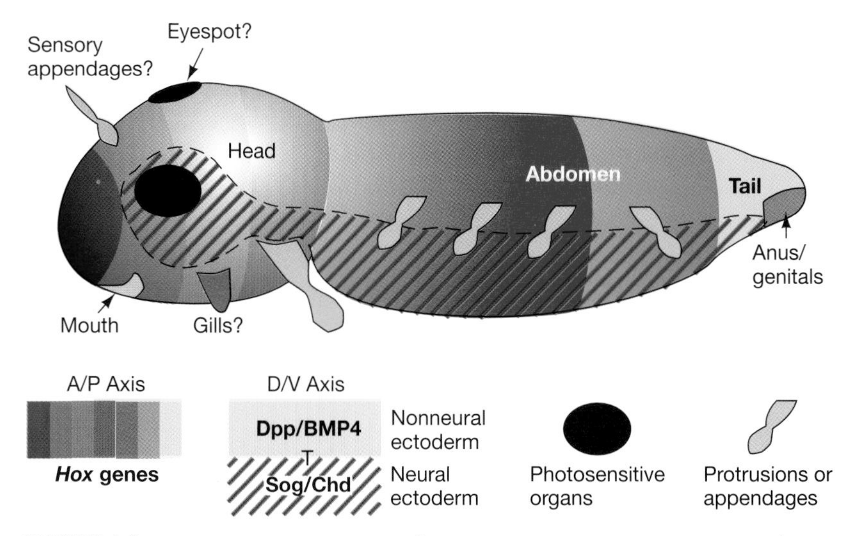

FIGURE 6.6. A possible representation of the most recent common ancestor of vertebrates and invertebrates.

are left to serve as shoots for growth in the following year. Following one of Gould's favorite themes, he argues that it may well be random as to which branches survive and which ones are trimmed. Whether or not the survivors of this cataclysm had something special to offer in the face of environmental adversity or were only the lucky ones who escaped the shrapnel of a comet impact, the main point is that the Cambrian extinction is one of several clear examples in which only a fraction of the organisms alive at a given time survived some type of devastating worldwide crisis. Luckily for us, one of the few survivors of that extinction was a primitive ancestor of vertebrates. If that nondescript creature had disappeared along with many of its comrades, perhaps some green juice-oozing creature with an exoskeleton would be writing this book now—or perhaps intelligent life, as we consider ourselves to be, would have never evolved at all.

Given that the Burgess Shale provides a clear example of a great diversification of life and a subsequent brutal pruning of that bush of life, it is reasonable to propose that a similar type of story took place during a previous event when our most recent common ancestor emerged as the sole triumphant survivor of its time. If we adopt Gould's view of the Cambrian extinction, we might wonder whether this animal was special or just lucky. We also should again thank our lucky stars that this animal survived, because complex animal life might have otherwise perished altogether, perhaps never to have been replaced. Another obvious question is, What kind of crisis wiped out all but one animal form? Was it a comet like the one that is suspected of bringing about the demise of the dinosaurs? Was it some dramatic climactic change like an increase in the level of oxygen? Or could it have been some biological innovation, like the invention of homeotic genes, that gave our ancestor and its descendants a great advantage over all other forms of life? According to this last scenario, there may have been no cataclysmic event at all, but rather life forms descended from our ancestor diversified with time and systematically displaced all other forms of life because of the advantage they all shared. We may never know the answers to these fascinating questions. Then again, who would have ever thought we could build as detailed a picture of our ancestor as we have in the last decade? I suspect that there are still several exciting installments of this story to come, so stay tuned.

The second major implication of the most recent common ancestor of vertebrates and invertebrates being a highly structured creature rather than some type of simple amoeboid form of life is perhaps less lofty than the first, but is more relevant to concrete matters such as our health. The fact that we and model experimental animals such as flies and worms have inherited the same molecular devices for accomplishing the equivalent basic biological processes means that we can get very valuable information from these model systems and apply what we learn to humans. In addition to the clear examples of shared genetic systems for patterning embryos, appendages, and eyes, there also are a growing number of examples of genes involved in human diseases such as cancer and Alzheimer's disease that also play important roles in fly development. Studies of flies and worms may also have important implications for longevity, since there are mutants in both worms and flies that have lifespans 30–60% longer than normal. The commercial implications of the deep similarities between humans, flies, and worms have not gone unnoticed. There are several new biotechnology companies organized around using flies or worms to make fundamental discoveries that might lead to marketable products for humans. This is also good news for researchers studying flies and worms, since we are all hopeful that the major federal funding institutions for science will continue to make basic research on model genetic organisms a high priority.

In many ways, the realization that we share deep similarities in fundamental developmental processes with creatures as lowly regarded as invertebrates is the last of a series of assaults on the sanctity and special station of humans. This systematic erosion of our image of ourselves as unique in the universe was unleashed by Copernicus and Galileo when they demonstrated that the sun, not the earth, was the center of our solar system, and has progressively grown less palatable. Darwin dealt another big blow to our self-image when he concluded in his masterpiece, *On the Origin of Species* (1859), that monkeys are our cousins and that all life forms including humans may have originated from a single common ancestor. The discovery that the genetic code and basic biochemical processes are common to all life forms, made in the middle of this century, confirmed Darwin's visionary hypothesis and demoted us firmly yet another notch. And now, it turns out that we are made in the same creepy mold as flies and worms. Can it get any worse? Our bruised human egos aside, I think that we are finally beginning to see where we fit into the larger scheme of nature, and I personally find that comforting rather than disappointing. We are, after all, the descendants of millions of generations of survivors. I do not see any tragedy in not being uniquely fashioned by the hands of an orchestrating god, but rather see our being sculpted from the likeness of a fly as just one chapter in the amazing story by which against all odds, we are still in the game after having traveled a very, very, long and winding road. This perspective is cause for great celebration and for doing everything possible to preserve the unbroken chain of life yet one more cycle so that future generations of wild-and-crazy creatures will be permitted to succeed us. Who knows, maybe one of our descendants will conceive of a frightening science fiction story where they get transformed into a primitive dull-witted human.

■ Summary ■

Vertebrate appendages, like those of invertebrates, are patterned along the three interrelated A/P, D/V, and P/D axes. The positions of vertebrate limb buds with respect to the primary body A/P axis are determined by segment-identity genes, which control expression of the secreted factor FGF. FGF initiates outgrowth of the limb bud and defines the orientation of the A/P axis of the limb primordium, and thereby the position of the primary A/P organizing center of the limb (the zone of polarizing activity or ZPA). The key patterning gene expressed in the ZPA is the Hedgehog morphogen, which is produced in a posterior domain of the limb primordium. Hedgehog protein diffuses from the ZPA and activates expression of target genes in a threshold-dependent fashion. One important target gene activated by moderate levels of Hedgehog is BMP4, the vertebrate counterpart of the Dpp morphogen in flies.

Despite the fact that appendages are thought to have evolved independently in vertebrates and invertebrates, patterning of the A/P axis in vertebrate limbs is quite similar to that of fly appendages such as the wing. In both systems, a source of Hedgehog is provided in the posterior region of the appendage and diffuses anteriorly to activate genes in a threshold-dependent fashion. An important target of Hedgehog in vertebrates and invertebrates is the morphogen BMP4/Dpp. Consistent with vertebrate and invertebrate Hedgehog proteins having equivalent functions, fly Hedgehog can mimic the effect of vertebrate Hedgehog as a ZPA signal. There also are notable similarities in how the D/V and P/D axes are established in vertebrates and invertebrates. The Notch signaling pathway plays an important role at the D/V border of vertebrate limbs in a structure known as the apical ectodermal ridge (AER) and in defining the corresponding boundary of fly appendages (e.g., at the interface between dorsal and ventral surfaces of the wing). In addition, the distal tip of vertebrate and invertebrate appendages is associated with the universal expression of a transcription factor called Distalless, which is essential for initiating P/D polarity in fly legs. These similarities in patterning the three organizing axes of vertebrate and invertebrate appendages suggest that a network of genes controlling early steps in patterning appendages or protrusions from the body wall existed in the most recent common ancestor of vertebrates and invertebrates and these genes have continued to exert similar functions throughout the evolution of vertebrate and invertebrate lineages.

Eyes appear superficially to develop by very different mechanisms in vertebrates and invertebrates. Surprisingly, however, formation of eyes in mice and humans depends on the function of a common transcription factor encoded by the *pax6* gene, corresponding to the fly *eyeless* gene, which is required for initiation of eye development in flies. *pax6* functions much like *eyeless* because misexpression of vertebrate *pax6* in flies induces the formation of fly eyes in inappropriate locations. As in the case of appendage development, the underlying similarities in molecular mechanisms for initiating eye development in diverse species suggest that the most recent common ancestor of vertebrates and invertebrates possessed some type of light-sensing organ, the formation of which depended on the function of a *pax6/eyeless* gene.

The unanticipated underlying similarities between the establishment of primary body axes in vertebrate and invertebrate embryos (Chapter 5) and the similarities in mechanisms for patterning appendages and eyes in these two anciently diverged forms of life strongly suggest that the most recent common ancestor of current-day vertebrates and invertebrates had a well-developed body plan, appendages, and light-sensing organs. In addition, a variety of other work suggests that our most recent common ancestor had also invented early elements of a heart, trachea, reproductive cells, neural wiring, and basic immunity. These revelations permit us to reconstruct an image of this common ancestor, which most likely resembled a shrimp-like creature. Because this ancestor must have lived in a world filled with other organisms, a major question is, Why was this the only creature to give rise to surviving descendants? Was there some great cataclysm that eliminated all other life forms, or did our ancestor invent some revolutionary biological property that gave it a great advantage over all other life forms? Perhaps the future will hold the answer to this tantalizing mystery.

7 Establishing the Primary Axes of Plant Embryos

We now turn to the development of the other major branch of multicellular organisms on earth, plants. In this chapter, we examine early events in establishing the primary axes of the plant referred to as the apical–basal (A/B) axis (i.e., shoot versus root) and radial axis (i.e., tissue layers) (see Fig. 7.1). In Chapter 8, we discuss the formation of flowers and leaves, which in many respects can be considered as plant appendages. Because most plants are essentially rotationally symmetric, positional information is imparted by the two perpendicular A/B and radial axes. The A/B axis is subdivided into domains forming the root, stem, and shoot, and the radial axis is typically partitioned into three primary germ layers giving rise to the epidermis, cortex, and vascular tissue.

■ Cast of Characters ■

Terms

Apical shoot meristem A small group of self-renewing cells located at the apical tip of a plant from which all above-ground structures of the plant derive.

Apical–basal axis (A/B) The vertical axis of a plant extending from the shoot (apical end) to the root (basal end).

Auxin A plant hormone that can stimulate proliferation of plant cells and influences establishment of the A/B axis by favoring basal cell fates (e.g., roots).

Carpel The female reproductive organ onto which the pollen is deposited to begin the life cycle of the flowering plant.

Carrier An individual having one mutant and one good copy of a given gene.

Cell wall A rigid protective layer of plant cells that prevents cell migration.

Chloroplast An intracellular structure in plants that performs photosynthesis and contains the light-absorbing chlorophyll molecule.

Cloning (of an organism) The process of creating an exact copy of an organism.

Cortex (L2 layer) The middle of the three germ layers partitioning the radial axis of a plant, which provides rigidity and substance to the plant.

Cotyledon Seed leaves of a plant embryo. Embryos of dicot plants have two cotyledons, and those of monocots have one.

Cytokinin A plant hormone that can stimulate proliferation of plant cells and influences establishment of the A/B axis by favoring apical cell fates (e.g., shoots).

Dormancy The intervening stage between embryonic and adult development.

Epidermis (L1 layer) The outer of the three germ layers partitioning the radial axis of a plant, which protects the plant from the outside world.

Ethylene A hormone that promotes growth in cell width over length. Ethylene competes with gibberellic acid (G.A.), which, conversely, promotes growth in cell length over width.

Genetic screen A systematic hunt for mutations.

Gibberellic acid (GA) A hormone that promotes growth in cell length over width. Gibberellic acid competes with ethylene, which, conversely, promotes growth in cell width over length.

Globular embryo A spherical morphologically undifferentiated mass of embryonic cells that forms after several divisions of the fertilized egg cell.

Glutamate An amino acid (i.e., a building block of proteins) that is also widely used as a signal in the nervous system of diverse animals.

Glutamate receptor A receptor protein that can be activated by binding the amino acid glutamate, which functions as a signal in the nervous system.

Heart-stage embryos The first stage of plant embryonic development when the embryo becomes visibly polarized and in which the primordia of cotyledons can be distinguished as the lobes of a heart.

Homeobox The DNA-binding region of a subtype of transcription factors such as those encoded by the homeotic genes and the plant *stm* gene.

Homeotic genes Genes that determine floral organ identity in plants.

Hormone A circulating chemical signal in plants or animals that is typically produced by a small set of specialized cells, travels in body fluids, and triggers a response in distant target cells.

Morphogen A secreted signal that elicits different cellular responses at different concentrations.

Morphogenesis The process by which the developing organism attains its final shape. In plants, this is accomplished by controlling the orientation of the plane of cell division and by regulating the direction of cell expansion following division.

Mutagen A chemical compound that causes mutations.

Mutant An organism with a mutated gene causing an identifiable defect.

Mutation An alteration in the base sequence of a gene.

Ovary (floral) The portion of the female organ (carpel) in which the egg-containing ovules develop.

Plasmodesmata Large pores connecting plant cells to their neighbors through which large molecules can diffuse directly from one cell to another.

Pollen grain The product of the male organ of a plant (stamen) that houses the sperm.

Pollen tube A long tube through which the sperm nucleus descends to the egg within the ovary portion of the carpel.

Radial axis The width dimension of a plant, which is subdivided into three tissue layers (outer layer = L1 = epidermis, middle layer = L2 = cortex, and inner layer = L3 = vasculature).

Stamen The male reproductive organ in a flowering plant that produces pollen.

Stigma An exposed portion of the carpel onto which the pollen is deposited.

Stromatolites Organized colonies of bacterial cells that date back more than three billion years and still exist today.

Suspensor An umbilical cord-like structure that connects the plant embryo to nutrients stored within the seed.

Unicellular (*organism*) A single-celled organism such as a bacterium or yeast cell.

Vascular tissue (L3 layer) The inner of the three germ layers partitioning the radial axis of a plant, which serves as a transport system for water and nutrients.

Genes

clavata1 (CLV1) A gene encoding the receptor protein Clavata1 (activated by the Clavata3 signal), which is required to restrict the number of proliferating meristem cells and is expressed in the L2 and L3 layers of the apical shoot meristem.

clavata3 (CLV3) A gene encoding a signal, Clavata3 (activating the Clavata1 receptor), which is required to restrict the number of proliferating meristem cells and is expressed exclusively in the L1 layer of the apical shoot meristem.

knolle An embryonic patterning gene required for epidermal development and for assembly and orientation of a platform in the cell upon which new cell walls are built during cell division.

shootmeristemless (STM) A gene encoding a homeobox type protein that is required for determining apical shoot meristem identity in heart-stage embryos.

wuschel (WUS) A gene encoding a homeobox type protein that is required for determining apical shoot meristem identity in globular-stage embryos and activating expression of *STM.*

PATTERNING IN PLANTS

Much less is known about the nature of genes patterning the primary axes of the early plant embryo than about the genetic hierarchy that establishes positional identities in the fly embryo. Of the handful of identified genes involved in patterning the A/B axis of plant embryos, however, two genes encode transcription factors of the homeobox class. As mentioned in previous chapters, homeobox transcription factors play a prominent role in determining regional cell identities in animal embryos. For example, in the fly embryo, key patterning homeobox proteins include the maternal morphogen Bicoid, pair-rule and segment-polarity genes, and the homeotic/*HOX* (segment-identity) genes. Although we are only beginning to appreciate the parallels between early plant and animal development, it is already evident that there are several striking similarities in the principles of organization and the types of genes used to implement patterning in these two most visible kingdoms of life.

Plant Shape Is Determined by Regulated and Oriented Cell Division

The most conspicuous difference between plant and animal cells is that plant cells are surrounded by rigid cell walls. Because cells must be able to change shape rapidly and to regulate adhesive interactions with their neighbors and environment in order to move past one another, encapsulation by rigid cell walls prevents cell migration. The absence of cell migration has profound implications on plant development because plants cannot undergo reorganization by gastrulation, which as you may recall, is more important than marriage or death to animals. Cells encased in walls must stay where they were born, like little cinder blocks, and thus, developing plants rely primarily on regulating the direction of growth to change shape. To expand in one direction versus another, plants can either orient the plane of cell division or alter the proportions and total volume of individual cells (Fig. 7.1). Although animal cells also control the orientation of cell division and can regulate cell size in certain situations, these mechanisms play relatively minor roles during gastrulation compared to the complex rearrangements of cells accomplished by cell migration. Growing plant embryos and mature plants have two major centers of dividing cells, one at the tip of the root and one at the apex of the shoot. In the case of the root, elongation of cells along the axis of the root also plays a very significant role in growth. The levels of two competing plant hormones, gibberellic acid (GA) and ethylene, determine whether cells expand along their length (GA) or their width (ethylene) (Fig. 7.1). Following cell division, root cells increase their volume 50 times by expanding lengthwise but not widthwise. Thus, the root is a good example of a structure that attains its final shape as a result of both oriented cell division and cell expansion.

Apical-basal Patterning in Plant Embryos

The life cycle of a flowering plant begins when pollen produced by the stamen is deposited on the exposed portion of the carpel (called the stigma). The sperm carried in the pollen grain sheds its protective shell and sends a process down through the hollow tube of the stigma, forming a long tube (the pollen tube) through which the sperm nucleus descends to the egg within the ovary portion of the carpel. As in animal development, the sperm injects its nucleus containing one full copy of the genome (1× DNA) into the egg and this nucleus fuses with the 1× DNA nucleus of the egg, resulting in a 2× DNA fertilized egg to generate the embryo (Fig. 7.2).

The poles of the apical–basal (A/B) axis in a developing plant embryo are defined by the opposing positions of the apical shoot and basal root primordia. The early embryo is visibly polarized at the time of the first division of the fertilized egg within the developing seed (Fig. 7.2). It is unclear whether the asymmetry of the fertilized egg is induced by con-

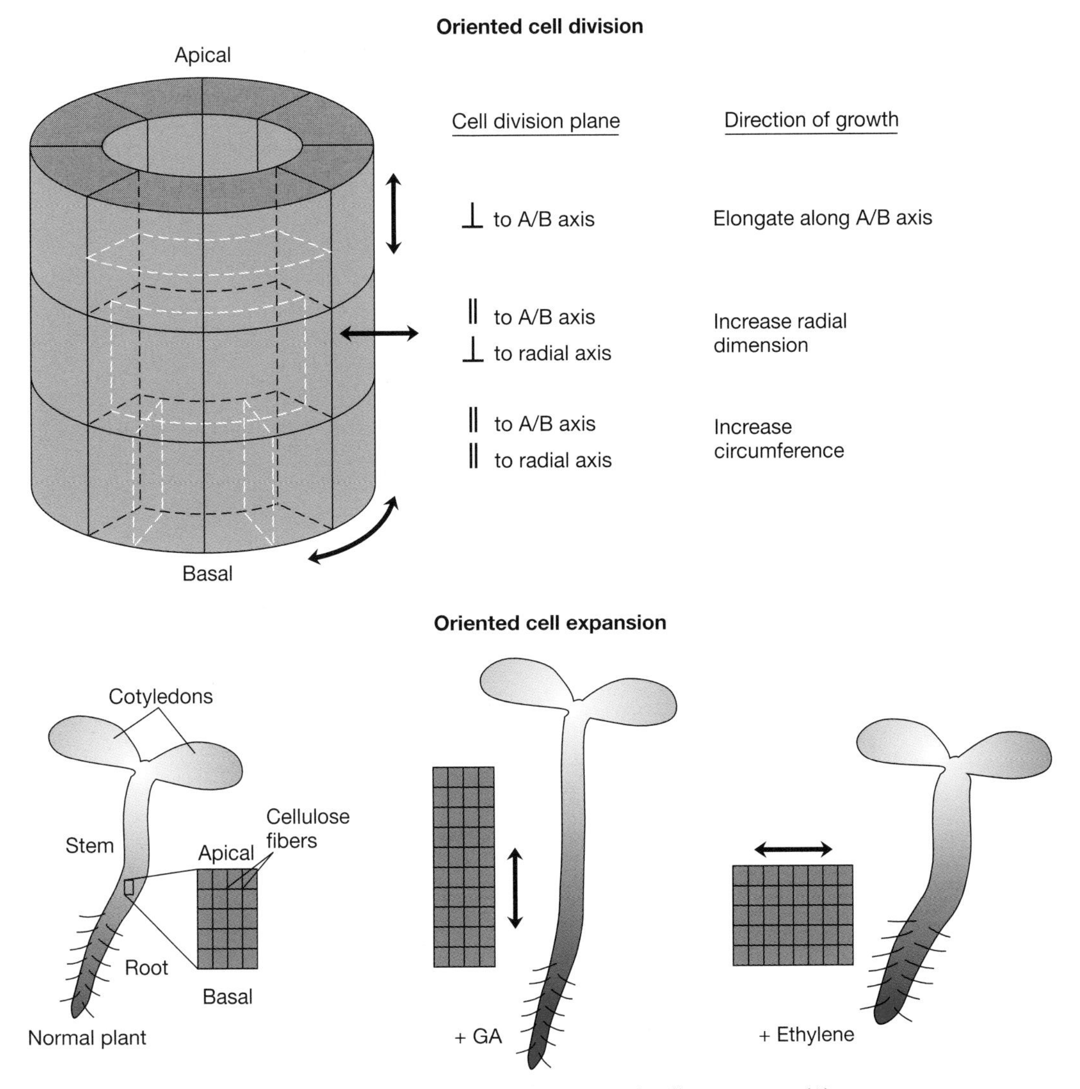

FIGURE 7.1. Plants change shape by oriented cell division and cell expansion. *(A)* Oriented cell division (plane of division is indicated by dotted white lines). *(B)* Oriented cell expansion.

tact with the sperm, which enters the ovary in a fixed orientation, or whether the egg itself is polarized as a consequence of residing in the asymmetric environment of ovarian cells surrounding the egg. Whatever the source of initial polarity, the first division of the egg cell is asymmetric, giving rise to a small compact cell and a larger elongated counterpart. The small cell divides many times to generate the embryo itself, whereas the elongated cell divides a few times to generate an umbilical cord-like structure called the suspensor, which connects the embryo to the nutrients stored within the seed. After several cycles of cell division, the embryo becomes a spherical mass of cells, known as the globular embryo. The globular embryo looks something like a lollipop on a stick

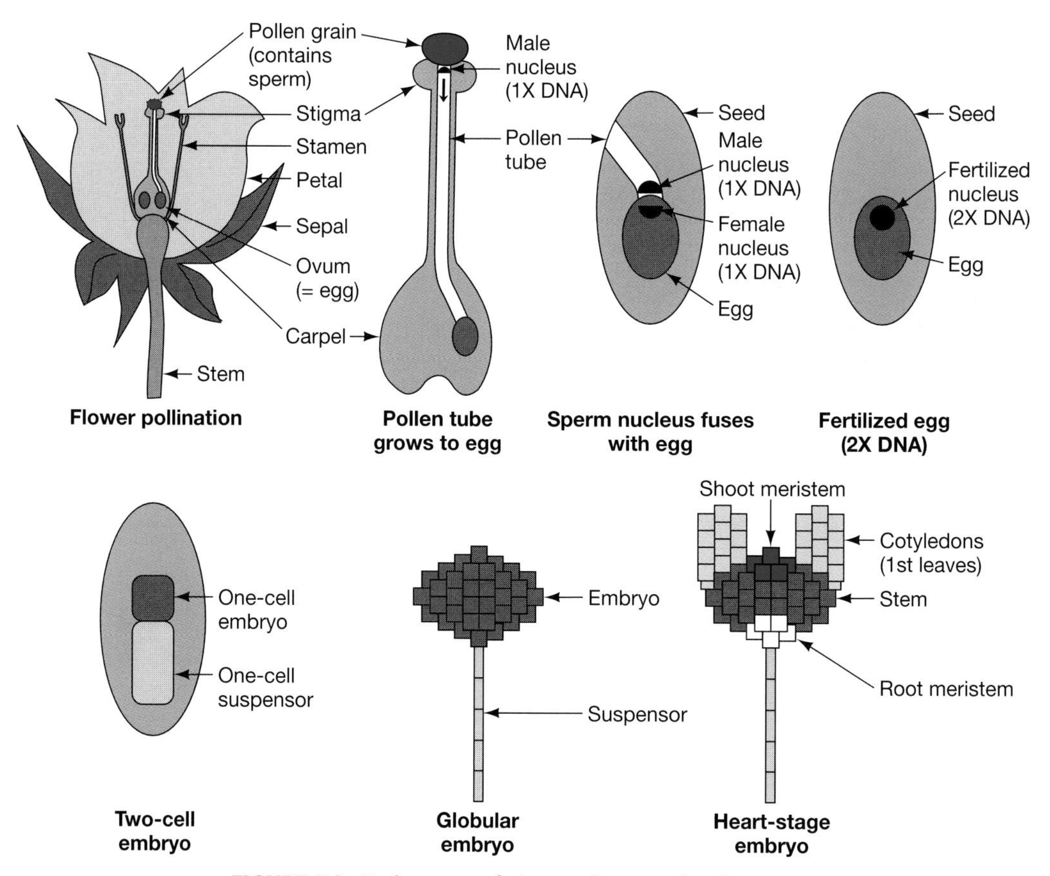

FIGURE 7.2. Early stages of plant embryonic development.

(the suspensor). Like the blastoderm embryo of a fruit fly or the blastula stage of a vertebrate embryo, the globular plant embryo manifests no overt morphological differences between cells, yet as discussed below, key patterning events are initiated prior to and during this phase. Several cell divisions later, the plant embryo begins to show clear signs of morphological asymmetry as it takes on a heart shape. The pointed end of the heart-stage embryo, which will become the root, is attached to the suspensor, and the shoot primordium forms at the other end between the two rounded lobes of the heart in a mound-like structure referred to as the apical shoot meristem. The lobes of the heart then elongate further to become the first leaf-like structures of the plant referred to as cotyledons. Between the root and shoot primordia is a central domain of the embryo which will give rise to the first section of the stem.

Radial Patterning in Plant Embryos

As the plant embryo develops, it generates three discrete tissue layers along the radial dimension referred to as the epidermis (or L1 = outer layer), cortical layer (or L2 = middle layer), and vascular tissue (or L3

= inner layer). Like all other aspects of morphological change in a plant, radial growth is accomplished by controlling the orientation of the plane of cell division as well as the shape of individual cells. For a plant embryo to grow in length versus girth, the plane of cell division must lie perpendicular to both the surface of the embryo and the A/B axis (Fig. 7.1), whereas to grow in the radial dimension, the plane of cell division must be parallel to the surface of the embryo. After the first few rounds of cell division, three layers of cells are generated. To accommodate expansion of the inner two layers (cortical and vascular tissues), the outer layer of the plant (epidermis) must also grow in girth. To increase in circumference, the plane of cell division must be oriented perpendicular to the surface of the plant and parallel to the A/B axis (i.e., perpendicular to the axes required for growing lengthwise and radially). Thus, morphogenesis along both the A/B axis and radial dimension in plants is accomplished chiefly by controlling the orientation of the plane of cell division and by regulating the direction of cell expansion following division. As mentioned previously, these mechanisms for creating plant shape are very different from those involving complex choreographed cellular movements during the gastrulation of animal embryos.

Axis and Polarity Mutants in Plants

After completing his postdoctoral work with Christiane Nüsslein-Volhard and Eric Wieschaus on the famous screen for mutations disrupting fly embryonic development, Gerd Jürgens went off on his own (to Tübingen, Germany) to conduct a similar genetic screen for mutations affecting the development of mustard plant embryos (the mustard plant is the best-studied plant model organism, an equivalent of the fruit fly). Although there were some important methodological differences between the plant and fly genetic screens, the idea was much the same, which was to treat plants with a mutagen, recover a large number of mutant strains propagating mutations as carriers (i.e., *m*/+ individuals), and examine mutant plant embryos (i.e., *m*/*m* individuals) for defects at the heart stage (see Plate 3F–H for actual examples of normal and mutant heart-stage embryos). As in the case of the earlier fly screen, the function of a given plant gene was deduced from the types of defects observed in mutants disrupting the function of that gene (see Chapter 3 for a refresher on genetic screens). A decade after the publication of the first seminal paper by Wieschaus and Nüsslein-Volhard outlining the genetic control of embryonic development in the fruit fly embryo, the Jürgens lab published a similar type of paper in which they reported several distinct classes of mutants affecting specific regions of the plant embryo (Fig. 7.3). Since then, other laboratories, such as those of Elliot Meyerowitz at California Institute of Technology and Kathryn Barton at the University of Wisconsin, have also identified important mutants defective in embryonic patterning. One interesting class of mutants, which are reminiscent of the gap mutants in flies (see Chapter 3), lack large regions in the embryo such as apical structures, the stem, or the root. For example, in *gurke* mutants, apical structures such as shoot and cotyledon are missing (Fig. 7.3; see also Plate 2G), and in *monopteros* mutants, the basal root primordia are missing (Fig. 7.3; see also plate 2H). In another mutant (*gnom*), the embryo consists of only the stem portion of the embryo (i.e., it lacks both root and shoot). Another group of mutants identified by Jürgens and other investigators have defects in patterning various radial elements of the embryo. For example, the *scarecrow* mutant, which has been studied in the laboratory of Philip Benfey (New York University), lacks the endodermal layer in roots, whereas the *knolle* mutant does not form normal epidermis. In another mutant called *sabre* characterized by Phil Benfey, cells only expand radially rather than in the longitudinal dimension, which is the predominant axis of expansion in normal roots. The radial expansion defect in *sabre* mutants can be greatly ameliorated by reducing the level of ethylene acting on roots (recall that ethylene normally promotes radial cell growth). Thus, the *sabre* gene may normally play a role in counterbalancing the effect of ethylene.

Among embryonic mutants exhibiting regionally specific defects (Fig. 7.3), the best-studied group is that affecting formation of the most apical structure of the plant embryo, the shoot apical primordium (or meristem) (Fig. 7.4). The shoot apical meristem forms the growing tip of the plant and is the site at which upward growth is initiated throughout the life of the plant. Removal of the shoot meristem stunts the vertical growth of a plant and stimulates branching, a fact used often by gardeners who wish to limit the height of shrubs and promote lateral growth to make them bushy. One class of shoot mutants, represented by the *wuschel (WUS)* and *shootmeristemless (STM)* mutants, lack shoot meristems and therefore do not develop beyond the embryonic stage because they cannot elaborate a growing shoot (Fig. 7.3; see also Plate 3J, K). Another group of mutants including the *clavata1 (CLV1)* and *clavata3 (CLV3)* mutants have the opposite defect, namely great overgrowth of the shoot meristem (Fig. 7.3; see also Plate 4R, S). This latter category of mutants is discussed in greater detail below in the section on cell–cell signaling.

Homeobox Genes Pattern the Apical Region of Plant Embryos

Genes involved in patterning the A/B axis of plant embryos are just now beginning to be identified. Among this currently limited group of genes are those encoding transcription factors. For example, *WUS* and *STM,* two key genes required for determining shoot meristem identity, encode homeobox-type transcription factors distantly related to the homeotic/*Hox* genes that determine segment identity in animal embryos.

The field of embryonic plant development is still in its infancy, thus, patterning genes have not yet been characterized that are ex-

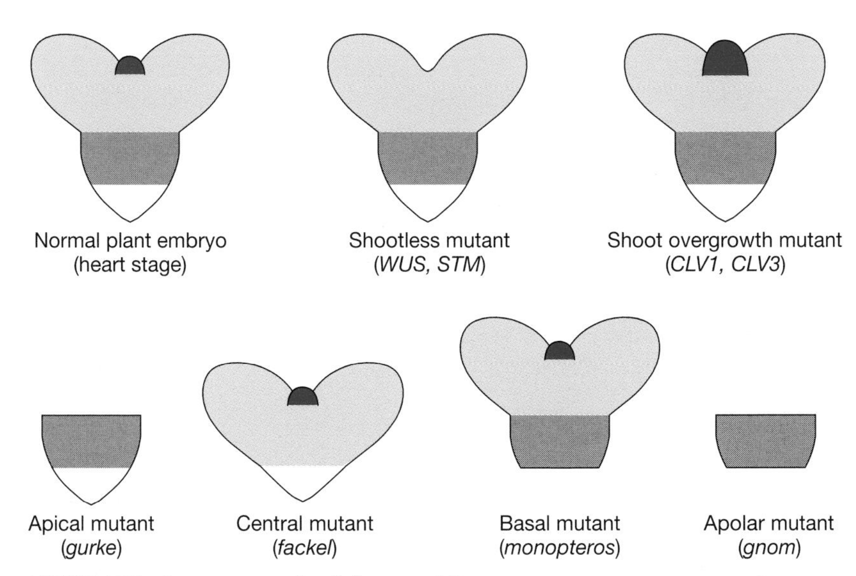

FIGURE 7.3. Domain-specific defects in A/B patterning mutants suggest that patterning genes function regionally.

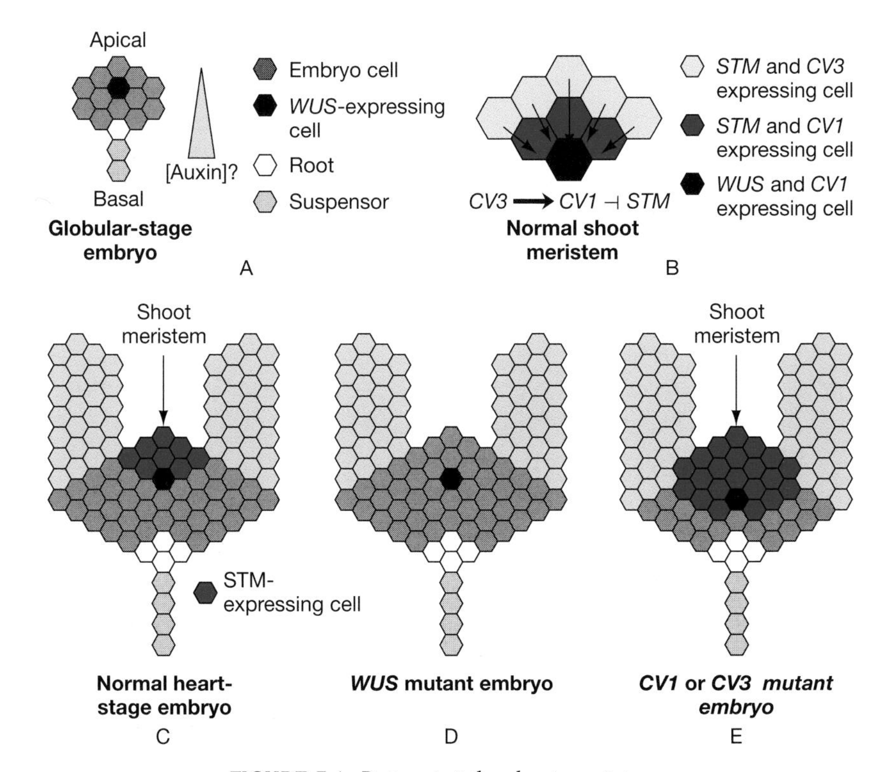

FIGURE 7.4. Patterning the shoot meristem.

Gerd Jürgens (1949–)

When Gerd Jürgens set up his laboratory, he embarked on a comprehensive genetic hunt for mutants disrupting the formation of early mustard plant embryo. Using principles and strategies similar to those used in the hunt for fruit fly mutants which he, along with Christiane Nüsslein-Volhard and Eric Wieschaus conducted a decade earlier, he launched a new era of analysis of early plant development.

Gerd Jürgens started out as biology student at the venerable University of Göttingen, then earned a masters degree at the University of Berlin. He then went on to do his graduate studies with the leading developmental neurobiologist Jose Campos-Ortega at the University of Freiburg, where he worked on the genetic control of cell–cell communication in the developing fruit fly eye. Being at three different places as a student was rather unusual in Germany at the time, because almost everybody did his undergraduate and graduate studies at the same place, thus pointing to Jürgens's intellectual wanderlust. Upon completing his graduate studies, Jürgens joined Christiane Nüsslein-Volhard and Eric Wieschaus at the EMBL in Heidelberg as a postdoctoral fellow and played a key role in the great hunt for mutations disrupting development of the fruit fly embryo.

After dabbling a little further in flies at the Max Planck Institute in Tübingen, he struck out on his own at the University of Tübingen, where he boldly began his groundbreaking studies of plant embryonic development, confident that the previous success of the fly screen would carry over to plants. Jürgens recalls that when his group undertook this genetic screen, "there were no models for embryo pattern formation (I actually borrowed the term from the fly field)." He designed a clever genetic screen in which he reasoned that embryonic pattern mutants in plants would be able to complete embryogenesis, as had been the reasoning for the fly mutant hunts. His team therefore began by collecting mutant lines that produced abnormal seedlings. After having done this, they sorted them into those with general problems, those that didn't turn green (a very large class), and, finally, a small class with specific defects in the apical–basal or radial pattern. One of the most exciting moments Jürgens remembers is "when I did my first mutagenesis of *Arabidopsis* seed without prior experience and found lots of distinct mutants—the scheme worked!" They then examined seedling mutants for defects in earlier periods of development, and found that, as hoped, the seedling defects could be traced back to problems arising during early embryonic development—as early as the very first division. The Jürgens group published a seminal paper describing this screen in which they identified a variety of mutants with defects restricted to particular domains of the embryo. These various mutant classes were very reminiscent of the types of mutants that he and his colleagues had found previously in the fly screen.

The Jürgens laboratory has gone on to clone several genes that were disrupted in several of their most interesting embryonic patterning mutants. These studies led Jürgens to propose, among other things, that key patterning events in plants revolve around orienting planes of early cell division and assembly of cell walls. These types of mutant defects are not common among animal patterning mutants and may reflect the profoundly different mechanisms by which plants and animal grow and assume various shapes.

Jürgens views scientific discovery as consisting of "phases of exploration, focusing, and readjustment." He also notes that it often requires strength and persistence to harvest the rewards of one's efforts. He favors "problem-oriented rather than method-driven" approaches and "flexibility in the analysis of well-chosen problems" in his work. Ultimately, Jürgens credits success to "a mix of intuition, reflection and chance(!)" So far, all of these ingredients have served him well.

■ ***WUS* and *STM*** ■

The *WUS* gene has been isolated and analyzed by the Jürgens group, and the *STM* gene has been characterized in Kathryn Barton's laboratory. In mutants lacking either *WUS* or *STM* activity, shoots are absent because the primordium for the shoot (i.e., the apical meristem) fails to form (Fig. 7.3; see also Plate 3J, K). The *WUS* and *STM* genes are both expressed in localized patterns in the apical region of the embryo, the *WUS* gene being activated at an earlier developmental stage than *STM* (Fig. 7.4; see also Plate 3I). Consistent with the timing of gene expression, the *WUS* gene acts before *STM* and is required to activate normal expression of *STM* in cells that will form the shoot meristem. It is not known what activates the *WUS* gene in its restricted pattern in the 16-cell embryo (well before the globular-stage embryo); however, it may respond to a particular level of a morphogen such as the plant hormone auxin (see next section). Activation of *STM* expression in heart-stage embryos (see Fig. 7.2) by *WUS* requires communication between cells, because the *WUS* gene is expressed exclusively in the L3 layer and the *STM* gene is expressed predominantly in the L1 and L2 layers (Fig. 7.4; see also Plate 3I). This communication between cells in different radial layers is thought to instruct more superficial cells, which ultimately will give rise to the shoot meristem, to divide asymmetrically so that they assume the correct apical position in the embryo to form the shoot meristem. *STM* plays an important role in defining the shoot meristem cell fate, since misexpression of the *STM* gene in other areas of the developing plant, such as leaves, leads to the production of freakish little plants with shoots coming out of leaves—a sight that competes with fly eyes popping out of wings (see Plate 4X)! These observations indicate that the *STM* gene normally functions to define the apical meristem cell fate and that it can impose this fate on a variety of other cell types.

pressed in localized patterns corresponding to the other natural subdivisions of the plant embryo such as the primordia of the root, stem, or cotyledon. Nonetheless, it seems very likely that such genes exist, given that several known patterning mutants display regionally restricted embryonic defects. In addition, many genes have been identified that are expressed in various subdomains of the embryo. It remains to be determined, however, whether these latter genes play roles in patterning the early embryo or, instead, simply respond to earlier-acting patterning genes. Identification of embryonic patterning genes will be greatly aided by a cooperative effort among many plant laboratories to determine the full DNA sequence and expression pattern of all genes in the mustard plant genome.

Pattern formation in plants is ultimately manifested by changes in cell growth and cell shape, thus there is a strong expectation that genes acting early during plant development to determine cell fates (e.g., transcription factors such as STM) will control the expression of genes involved in positioning the cleavage plane in dividing cells or genes involved in remodeling the cell walls during cellular expansion. Indeed, the early-acting patterning gene *knolle,* which has been characterized in detail by the Jürgens group, plays such a role in assembling, and presumably orienting, a platform in the cell upon which new cell walls are built during cell division. Patterning in the radial dimension also seems to be mediated by genes involved in controlling the mechanics of cell division, as revealed by a mustard mutant known as *keule,* in which the outer epidermis (L1 layer) assumes characteristics of inner vascular tissue (L3 layer). These patterning defects in *keule* mutants result from a defect in erecting new cell walls.

Cell–Cell Signaling Is Required for A/B Axis Formation

Although genes involved in initiating cell–cell signaling in plant embryos remain to be identified, two classic plant hormones called auxin and cytokinin are good candidates for primary morphogens involved in establishing the A/B axis. Auxin and cytokinin were originally identified on the basis of their ability to stimulate proliferation of plant cells. In addition to this general function in promoting cell division, and hence growth, a classic series of elegant experiments performed in the mid-1950s by Carlos Miller working in Folke Skoog's laboratory showed that the ratio of auxin to cytokinin determines which types of embryonic cells will form. Cells grown in the presence of low levels of auxin and high levels of cytokinin develop as shoots, whereas reciprocal ratios of high auxin to cytokinin favor root development. These hormones can be used sequentially to generate patterned embryos from individual plant cells.

■ The Miller-Skoog Experiment ■

In one of the first successful full organism cloning experiments, Miller and Skoog induced a single cell to divide in low equal concentrations of both hormones. They transferred the resulting mass of undifferentiated cells onto an agar plate containing high levels of cytokinin and a low concentration of auxin, which triggered development of an apical shoot. This polarized growth of cells was transferred to another agar plate containing low levels of cytokinin and a high concentration of auxin, whereupon it formed a root in a position opposite to that of the shoot and organized itself into an embryo, which then developed into a mature plant.

In accord with the possibility that auxin and cytokinin play an important role in determining the A/B axis in normal plant embryos, recent data reveal that auxin is directionally transported in plant embryos. Additionally, in another developmental situation where auxin is proposed to play a key role in radial patterning of the root, it has been empirically determined by Göran Sanberg's group that there is a graded distribution of auxin in roots that correlates with its effects in promoting the differentiation of cell types in different tissue layers. Thus, early patterning information, possibly laid down in the egg itself, may lead to the localized production of auxin in cells lying at the junction of the embryo and suspensor. This critical hormone may then be transported up to the shoot by an active process leading to the formation of a graded distribution of auxin, which is high in the root and low in the shoot. Consistent with there being a source of auxin in the root tip, the *monopteros* mutant (Fig. 7.4) fails to form normal roots and has morphological defects at the junction between the root and suspensor. This mutant is suspected of being defective in responding to auxin.

Another example of cell–cell communication in plants involves genes functioning to confine the size of the apical shoot meristem (Fig. 7.4). The shoot meristem is required to generate all above-ground parts of the mature plant, illustrating the need to maintain a group of perpetually dividing meristem cells and to balance this need with the incorporation of cells into emerging organ primordia. Perhaps the most dramatic example of the growth potential of this relatively small group of cells is the giant sequoia tree, which manages to maintain a shoot meristem whose size remains unchanged for thousands of years and is responsible for the steady growth of one of the largest organisms on earth. As described above, *STM* is a key gene involved in defining the shoot meristem cell fate. Cells in the shoot meristem divide continuously during the life of the plant to assure that there will always be sufficient numbers of cells from which new sections of stem, branches, and flowers can be generated. Because shoot meristem cells are perpetually dividing, they must be kept in check to prevent them from dividing wildly and overtaking the plant like a tumor. *clavata1 (CLV1)* and *clavata3 (CLV3)* are two genes required to restrict the number of proliferating meristem cells. Mutants lacking function of either of the *CLV* genes have overgrown meristems that can be as much as 1000 times larger than those of normal plants (Fig. 7.3; see also Plate 4R, S). The greatly enlarged meristems in these mutants lead to gross abnormalities in the structure of the final plants. Overgrowth of the meristem in *CLV* mutants is most likely the result of a failure to confine the expression of the *STM* gene to a small group of meristem cells, as the number of cells expressing *STM* is significantly increased in either *CLV1* or *CLV3* mutants. The hypothesis that the *CLV1* and *CLV3* genes function to limit the effect of *STM* is further supported by the observation that *CLV1* or *CLV3* mutants can develop into normal plants if the gene dose of *STM* is reduced by 50% as in the *m*/+ condition. This result indicates that when the levels of *STM* fall below a certain critical point, the *CLV* genes are no longer required to limit the extent of the shoot meristem. Isolation of the *CLV1* and *CLV3* genes in the Meyerowitz laboratory strongly suggests that these genes are part of a common signaling system, since *CLV1* encodes a receptor type of protein and *CLV3* encodes a likely secreted signal. Interestingly, the CLV1 receptor is expressed only in the L2 and L3 layers of the meristem (see Plate 4P), whereas the CLV3 ligand is expressed exclusively in the L1 layer (see Plate 4Q). These observations suggest that CLV3 is secreted from the epidermis (i.e., the L1 layer) and diffuses to nearby internal cells (i.e., the L2 and L3 layers) during a time when cell walls are still immature and relatively porous.

In addition to the standard secreted form of cell–cell signaling such as that mediated by the *CLV3* and *CLV1* genes, it appears to be possible for large molecules such as transcription factors to diffuse directly from one plant cell to another through large pores called plasmodesmata, which connect plant cells to their neighbors. These plant-specific channels link the inside compartments of adjacent cells and permit large molecules to pass between cells without having to be secreted from one cell and received on the surface of a neighboring cell. Plasmodesmata do not allow all proteins to diffuse between cells in an indiscriminate fashion, but rather function as regulated gates. There is evidence that the STM protein can move between cells through plasmodesmata, indicating that this transcription factor may act as a signal in the sense that it is synthesized in one cell and affects gene expression in surrounding cells. Thus, the size of the shoot meristem appears to be determined by the counterbalancing extracellular diffusion of the CLV1 signal and the intracellular movement of the STM protein. The ability of large proteins such as STM to pass between cells through plasmodesmata is another important difference between plants and animals, since, in general, transcription factors cannot move between animal cells. Animal cells must therefore rely on extracellular signals to communicate with one another.

DORMANCY: A DEVELOPMENTAL STAGE BETWEEN EMBRYONIC AND ADULT DEVELOPMENT

Although certain animals also can enter into quiescent states in which development is arrested, such developmental detours are not generally integral parts of an animal's life cycle. In contrast, dormancy is nearly a universal component of plant development.

In the following chapter we discuss the development of leaves and flowers in mature plants. There is, however, an intervening quiescent stage known as dormancy in plants, interposed between embryonic development (which occurs in the seed) and adult development (which takes place upon germination of the seedling). Maturation of the embryo into a seedling is initiated upon contact with a suitable growth environment such as wet soil. As even the most amateur gardener is aware, seeds can be kept for a very long time (e.g., years) without germinating. This ability to remain in a dormant state is critical, because it would not do a plant embryo much good to keep developing in the seed and use up all of the stored nutrients before it is in the ground. Dormancy is carefully regulated, as evidenced by many annual plants whose seeds and embryos mature in the summer or fall, but do not germinate until spring.

One indication that dormancy is a well-defined developmental stage is that mutants have been identified which disrupt this process. Some mutants fail to arrest following embryonic development while others are unable to break dormancy. We will not consider this interesting aspect of plant development further here; however, it is clear that dormancy is a critical stage in the life cycle of plants.

PARALLELS BETWEEN PLANT AND ANIMAL EMBRYONIC DEVELOPMENT

In addition to important differences between plant and animal development, such as plant cells having rigid cell walls that are unable to migrate and the continuous growth of mature plants as opposed to the fixed size of adult animals, there are also several unanticipated similarities. Although we do not yet have sufficient information to make detailed comparisons between plant and animal embryonic development, some striking parallels are already apparent, particularly given the significant mechanistic differences in how plant and animal embryos undergo morphological changes. One notable similarity is that typical animal and plant embryos are organized into three fundamental germ layers (i.e., ectoderm, mesoderm, and endoderm in animals, and epidermis, cortex, and vasculature in plants). Although this similar basic organization could reflect a need for embryos to be organized into a small number of tissue layers, there does not seem to be anything particularly magical about three. For example, in animals, there are several different mechanisms by which the three primary tissue layers are generated during embryogenesis. In some cases, the mesoderm invaginates before the endoderm (e.g., flies), whereas in other organisms, these tissues invaginate in a concerted fashion (e.g., frogs). In addition, during eye development, distinct series of inductive events and morphological transformations generate the varied and complex arrangements of cell layers characterizing the eyes of diverse species (see Fig. 6.5), indicating once more that it is possible to generate multiple tissue layers by various combinations of invagination and delamination of cells.

Another noteworthy similarity between patterning of early animal and plant embryos is the conspicuous role played by homeobox-type transcription factors. Again, this similarity could be coincidence or could reflect some special property of homeobox proteins that makes them particularly well suited for patterning purposes. This explanation seems unlikely, however, given that there are several other classes of structurally unrelated transcription factors involved in controlling developmental decisions in both plants and animals. Alternatively, the homeobox genes may have played a key developmental role in a common ancestor of plants and animals. In support of this latter possibility, a gene distantly related to plant and animal homeobox genes determines which of two different mating types will be adopted by individual yeast cells.

An intriguing recent finding is the identification of a receptor protein in plants that is clearly related to a receptor in the nervous system of animals for the signal glutamate. Glutamate is among the most widely used signals in our nervous system. The glutamate receptor is a member of a large family of molecules that relay many different kinds of signals in animals cells. Gloria Coruzzi at New York University and Julian Schroeder at UCSD made a very interesting discovery when they identified a cell-surface receptor that is the mustard plant counterpart to the animal glutamate receptor. Although the function of the plant glutamate receptor is unclear, it may be involved in some aspect of responding to light. It is well known, and not surprising, that many types of molecules involved in basic biochemical processes are common to plants and animals. Such molecules typically carry out functions common to all cells, including bacteria and fungi, and therefore have been inherited from an early form of single-cell (unicellular) ancestor of plants and animals that probably evolved more than 3 billion years ago. What is unexpected about a receptor molecule being common to plants and animals is that it begs the question, What was the function of such a protein in a unicellular organism that had no obvious need to communicate with other cells? Perhaps the glutamate receptor evolved independently in plants and animals from a molecule that functioned in a unicellular ancestor to bind the essential amino acid glutamate and transport it into the cell to be used as a building block for synthesizing proteins. This ancestral molecule may not have had any function in receiving signals, but because it could bind glutamate, it was used in two different branches of multicellular life as the starting point from which to evolve a receptor. It also is possible that an ancestral unicellular organism was capable of sending and receiving signals that could have been used for purposes such as building organized colonies of cells such as the ancient stromatolites (>3 billion years old). Stromatolites still exist today in isolated bodies of water (e.g., Shark Bay in Western Australia) that are sheltered from seaweeds and animals (for example, in water that has double the salinity of normal seawater) and can grow to be over a meter tall. Remnants of these ancient stromatolite colonies, which formed giant reefs, are still visible today as cliffs hundreds of feet high.

Finally, another titillating parallel between plant and animal cell communication pointed out by Elliot Meyerowitz is that the receptor encoded by the plant gene *CLV1* (described above) is a member of a large family of receptors present in plants, which also includes receptors involved in recognizing plant pathogens. In response to pathogens such as bacteria, plant cells produce a cocktail of toxins to kill infected cells and stop the spread of infection. The family of plant immunity receptors including CLV1 is distantly related to a family of receptors found in animal cells. This family of receptors includes a member in flies that functions during formation of the egg in the mother to concentrate the Dorsal morphogen in ventral cells of early embryos. Intriguingly, this same fly receptor is involved in protecting larvae from bacterial infection, and when activated, leads to the production of bactericidal proteins to limit infection. In addition, the same signaling pathway is present in humans where it mediates an acute-phase immune response. Did the same type of receptor get selected independently in plant and animal lineages to carry out both immune and developmental patterning functions? Alternatively, was this receptor used in an ancient unicellular organism to combat other microorganisms?

WHAT IS THE BASIS FOR THE SIMILARITIES IN PLANT AND ANIMAL DEVELOPMENT?

Each of the examples of similarities between plant and animal embryonic development discussed above could be coincidence or could be rationalized by hypothesizing that there is something special about a certain process or molecule that makes it particularly useful for a given developmental process. As emphasized in Chapters 3–6, inheritance from a multicellular common ancestor is another possible reason for shared developmental mechanisms and does seem to be the basis for many of the deep similarities between vertebrate and invertebrate development. In the case of plants versus animals, however, the prevailing view is that the common ancestor of plants and animals was a unicellular organism, possibly an amoeboid creature of some kind. Perhaps the foremost reason that plants and animals are thought to be descended from separate unicellular ancestors is that all plants and their nearest relatives, photosynthetic algae, carry an intracellular structure called a chloroplast. Chloroplasts perform photosynthesis and contain the light-absorbing chlorophyll molecules that give plants their green color. Because no known animal cells contain chloroplasts although many unicellular species of plants and algae do, it has been argued that the common ancestor of plants and photosynthetic algae was a single-cell organism with chloroplasts and that a separate unicellular organism without chloroplasts gave rise to animals.

Assuming that the systematists are right in asserting that plants and animals evolved independently into multicellular organisms, it becomes all the more interesting to consider the basis for the similarities in development of plant and animal embryos such as the subdivision of the embryo into three primary germ layers, the involvement of homeobox genes in specifying regions of the embryo, and the common use of the glutamate and immune-related signaling systems in plants and animals. As shown in the following chapter, there also are remarkable parallels between appendage development in animals and leaf/flower development in plants. Why should two lineages of life independently evolve such similar developmental strategies and use the same types of molecules for related purposes? Perhaps the hypothetical unicellular ancestor of plants and animals performed functions that made it particularly well suited for evolving into multicellular forms. Alternatively, there may be fewer solutions to evolving multicellular life than one might expect. One of the most exciting aspects of experimental biology today is that fundamental questions such as the origin of plants and animals remain to be answered. We are still in the embryonic stages of knowledge—the story is still developing.

■ Summary ■

An important property of plant cells, which distinguishes them from animal cells, is that they are encased by rigid cell walls. Because cell walls prevent cells from migrating, plant embryos cannot rely on cell migration as animal embryos do during gastrulation to effect complex morphological reorganization. The primary mechanisms for changing shape available to developing plants are oriented cell division and asymmetric cell growth.

Plant embryos are visibly polarized along the shoot–root axis, referred to as the apical (shoot)–basal (root), or A/B, axis from the time of the first division of the fertilized embryo. The product of this first cell division is a small cell that will give rise to the embryo, and a larger cell which connects the embryo to the yolk. The plant embryo divides several times to generate a globular shape, which corresponds to the earliest stage that patterning genes are known to act. After another few rounds of cell divisions, the embryo acquires a polarized heart shape, in which the lobes of the heart form apically and the point of the heart basally where the roots will develop. In addition to the A/B axis, different cell types develop along the radial axis of plant embryos (e.g., an outer epidermal layer, an intermediate cortical layer, and an inner vascular layer).

Several mutants affecting early embryonic patterning have been identified in various mutant screens. In several of these mutants, large sections are missing along the A/B axis of heart-stage embryos. This class of plant mutants is reminiscent of the gap mutants in flies. There also are mutants that affect patterning along the radial axis (e.g., specific radial layers are deleted or malformed). Among the embryonic patterning genes that have been isolated by molecular cloning, two encode transcription factors of the homeobox subclass (*WUS* and *STM*) that are required for formation of the apical shoot region of the embryo. Consistent with their requirement apically, *WUS* and *STM* are expressed in apical regions of the globular or heart-stage embryos, respectively.

Cell–cell signaling is clearly important during early plant development, even though plant embryos are surrounded by cell walls. It is likely that some developmental signals are transmitted from cell to cell through large channels called plasmodesmata, which connect adjacent cells. The earliest signals thought to be involved in A/B patterning are two plant hormones called cytokinin and auxin. Classic experiments performed by Miller and Skoog revealed that high ratios of auxin to cytokinin initiate development of roots, whereas high ratios of cytokinin to auxin favor shoot development. A variety of evidence suggests that auxin may be synthesized by cells in the root and then transported apically to create a graded concentration of auxin which would be highest in the root and lowest in the shoot. Cell–cell signaling also appears to be important for determining the identities of apical embryonic cells. For example, two genes (*CLV1* and *CLV3*) that normally function to limit the size of the apical region encode, respectively, a signal and a receptor. Because the cells expressing the ligand (CLV3) are different from those expressing the receptor (CLV1), it is likely that the signal must pass from one cell layer to another.

Although plants and animals are thought to have evolved independently from unicellular ancestors, there are several striking similarities between embryonic development in these two great kingdoms of life. For example, plant and animal embryos are both subdivided into three primary tissue layers. In addition, related genes perform similar functions in plants and animals. Thus, homeobox class transcription factors play important roles in determining identities of cells in localized regions of the embryo, structurally related genes are involved in cell–cell signaling in early embryos and then later for immunity in both plants and animals, and a receptor identified in plants corresponds to a receptor in animals that plays a prominent role in nervous system function. Whether these similarities are mere coincidence or reflect the use of many of these same molecular systems in a unicellular ancestor of plants and animals remains an open and intriguing question.

8 Patterning Plant Appendages

In this chapter, we discuss the formation of flowers and leaves, which are secondary structures in plants that are comparable in several respects to appendages in animals. In Chapter 7, we discussed formation of the apical shoot meristem during embryonic development of mustard plants. The central region of the shoot meristem continues to grow and give rise to sections of stem throughout the life of the plant. Once the plant begins to mature, secondary meristems arise along the periphery of the shoot meristem, which give rise to branches, leaves, and flowers. These secondary meristems share several properties with the primary shoot meristem, including the function of a common set of genes. Unlike the primary shoot meristem, which continues to grow and produce secondary meristems, however, secondary floral meristems produce only a single flower. It is worth distinguishing between "determinate" and "indeterminate" forms of development at this point, as these are the two primary modes of adult plant development. The primary shoot meristem is an example of indeterminate development in that it continues to generate branches, leaves, or flowers throughout the life of the plant. In contrast, under normal circumstances, floral meristems follow a determinate form of development in that they generate the primordium for a flower and then, after a short time, cells stop dividing and differentiate into floral organs (e.g., sepals, petals, stamens, and carpels). As we will see, there are mutants in which the floral meristems fail to undergo determinate development and instead continue to generate floral organ primordia. Unlike most mutants, which are often freakish and unappealing, indeterminate floral mutants such as roses can be very beautiful and have been selected by intensive breeding. There also are mutants in which the shoot meristem fails to maintain indeterminacy and instead terminates in a flower. Let us now consider the first step in flower development, the formation of a floral meristem.

■ Cast of Characters ■

Terms

ABC model A model of flower development to explain the behavior of three different classes of floral patterning mutants (A, B, and C mutants) which proposes that A, B, and C genes function in a pairwise fashion to specify the four floral organ identities.

Apical shoot meristem A small group of self-renewing cells located at the apical tip of a plant from which all above-ground structures of the plant derive.

Apical–basal axis (A/B) The vertical axis of a plant extending from the shoot (apical end) to the root (basal end).

Carpel The female reproductive organ onto which the pollen is deposited to begin the life cycle of the flowering plant.

Cotyledon Seed leaves of a plant embryo. Embryos of dicot plants have two cotyledons whereas those of monocots have only one.

Determinate development A mode of meristem development in which cells proliferate for a limited period to generate the primordium for a structure (e.g., a flower) and then stop dividing to differentiate.

Dicot plants Flowering plants that have embryos with two seed leaves or cotyledons.

Dorsal surface of leaf The surface of the leaf forming nearest the shoot.

Floral meristem A small group of cells forming at the flank of the apical shoot meristem that generate the primordium for a flower and differentiate into floral organs.

Floral organs The concentrically organized sepals, petals, stamens, and carpels of a flower.

Fruit The seed-containing structure that develops from the basal portion of the carpel following fertilization of the eggs.

Globular embryo A spherical morphologically undifferentiated mass of embryonic cells that forms after several divisions of the fertilized plant egg.

Heart-stage embryos The first stage of plant embryonic development when the embryo becomes visibly polarized and in which the primordia of cotyledons can be distinguished as the lobes of a heart.

Homeotic genes Genes that determine regional cellular identities, such as *homeotic/Hox* genes in animal embryos and floral organ-identity genes in plants.

Indeterminate development A self-regenerating mode of meristem development (e.g., the apical shoot meristem) in which cells proliferate continuously during the life of the plant to provide new cells for growth of the plant (e.g., the central stem) and for formation of secondary meristems (e.g., primordia giving rise to branches and leaves or to flowers).

Leaf primordium The default state of a secondary apical meristem.

MADS-box genes Genes encoding a class of transcription factors present in plants and animals. For example, in plants, MADS-box genes define the identities of floral organs and direct fruit development.

Margin of leaf The edge of the leaf that forms at the junction between the dorsal and ventral surfaces.

Medial–lateral axis of leaf (M/L) The axis of the leaf running perpendicular to the proximal–distal (P/D) axis, which is marked by structures such as veins that branch in particular locations.

Monocot plants Flowering plants that have embryos with a single cotyledon.

Ovary (floral) The portion of the female organ (carpel) in which the egg-containing ovules develop.

Petals The second whorl of floral organs that form just inside the sepals and are the most prominent structure of the flower (e.g., the red petals of a rose).

Proximal–distal axis of leaf (P/D) The axis running from the stem (proximal) to the tip (distal) of the leaf.

Secondary meristem A small group of cells that arise during maturation of the plant along the periphery of the apical shoot meristem and give rise to branches, leaves, or flowers.

Sepals The outer whorl of floral organs, resembling leaves, that encloses the flower.

Shattering The process by which pod-type fruits break open and release their seeds.

Stamen The male reproductive organ in a flowering plant that produces pollen.

Valve The fleshy sectors of a fruit that get eaten in edible fruits.

Valve border The narrow stripes of cells running along the edges of the sectors of valves.

Ventral surface of leaf The surface of the leaf forming farthest from the shoot.

Whorls The four concentric rings of floral organs (e.g., sepals, petals, stamens, and carpels).

Genes

Agamous (AG) A C-function floral homeotic gene expressed in the center of the floral meristem in cells giving rise to stamens and carpels.

Apetala1 (AP1) An A-function floral homeotic gene expressed in an outer ring of floral meristem in cells giving rise to sepals and petals.

Apetala3 (AP3) A B-function floral homeotic gene expressed in a central ring of floral meristem cells that overlaps the domains of cells expressing AG and AP1 and gives rise to petals and stamens.

Apterous A fly homeobox gene that is required for formation of dorsal wing cells.

Cauliflower (CAL) A gene closely related to AP1 that functions together with AP1 to prevent floral meristems from developing as primary indeterminate shoot meristems.

Cup-shaped2 (CUC2) A gene expressed in a stripe of cells bisecting the apex of the globular embryo which is required to split the cotyledon primordium into two separated parts.

Cup-shaped1 (CUC1) A gene required to split the cotyledon primordium into two separated parts.

Fruitful (FUL) A gene expressed in the region giving rise to the primordium of the fleshy fruit valve that is required for formation of the valve.

FT A gene encoding a protein related in structure to that of TFL that promotes initiation of floral development by opposing the activity of *TFL*.

Leafy (LFY) A gene expressed in early developing floral meristems that is required to initiate development of the floral meristem.

Notch A gene encoding a receptor that is required for outgrowth of the wing and formation of the wing margin in flies and for formation of a specialized group of cells (the AER) at the junction between the dorsal and ventral surfaces of vertebrate limb buds.

Phantastica A snapdragon gene required for outgrowth of the leaf and for formation of the leaf margin at the junction between the dorsal and ventral surfaces of the leaf.

***Shatterproof* genes** A pair of highly related genes expressed specifically in valve border cells that are required for formation of valve borders.

Terminal flower (TFL) A gene encoding a likely inhibitory signal that suppresses floral development in secondary meristems by repressing expression of *AP1* and *LFY*. Flowering is initiated by genes such as *FT*, which oppose the action of *TFL*.

THE ABCS OF FLOWER DEVELOPMENT

The Emergence of Secondary Floral Meristems from the Shoot Meristem

As the seedling grows, the apical shoot meristem enlarges. Once it gets to a certain size, secondary meristems, which will give rise to flowers or leaves, begin to form around its periphery. It is likely that these secondary meristems produce signals that inhibit the formation of other secondary meristems. This hypothetical inhibitor would account for the fact that secondary meristems form in an evenly spaced pattern around the perimeter of the shoot meristem. Presumably, new secondary shoots form when the primary meristem grows sufficiently to separate existing secondary meristems by enough distance so that the inhibitor falls below the concentration necessary to block initiation of meristem development at points between two meristems. The primary shoot meristem also can develop as a secondary floral meristem and is normally prevented from doing so by another type of inhibitory signal, which is produced by cells lying just under the meristem. The gene *terminal flower (TFL)* encodes a likely inhibitory signal produced by these central shoot meristem cells (Fig. 8.1). In *TFL* mutants, the shoot meristem seems to be unable to restrain flower development and is consumed in the process of developing into a flower rather than continuing to serve as a regenerative growth center at the apex of the plant. The product of the *TFL* gene may be a diffusible signal, since it is similar to animal signals used in the nervous system. Consistent with central meristem cells being the source of an inhibitory signal, the *TFL* gene is expressed selectively in the center of the shoot meristem throughout the life of the plant. The effect of TFL appears to be counteracted by the product of another gene called FT. Interestingly, FT encodes a protein related in structure to that of TFL. Perhaps FT competes for binding to the TFL receptor but is unable to activate it. Alternatively, TFL and FT may bind to the same receptor but elicit op-

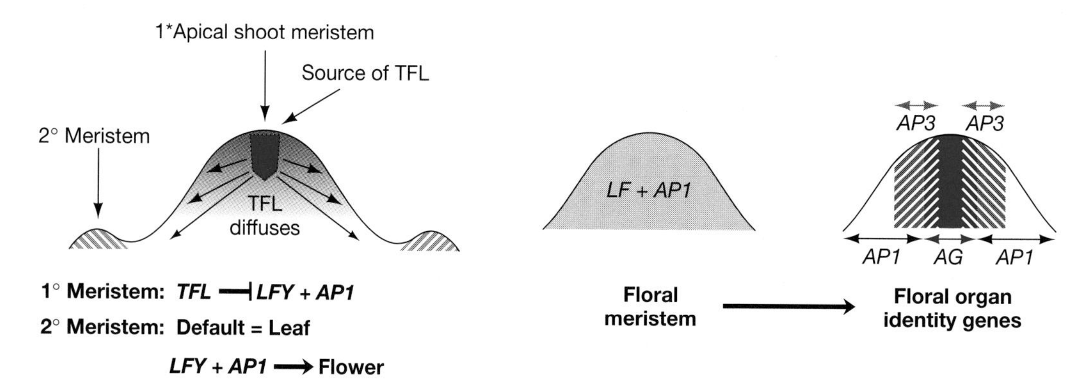

FIGURE 8.1. Overview of floral meristem development.

posite responses. The antagonistic roles of the *TFL* and *FT* genes underscore the general principle that development is generally guided by counterbalancing forces.

Several genes referred to as floral meristem identity genes are required to initiate formation of the floral meristem. Two of these genes, *leafy (LFY)* and *apetala1 (AP1)*, are expressed early in developing floral meristems. In the absence of both *LFY* and *AP1*, flowers are replaced by leaves and an adjoining shoot meristem which forms a branch. The result of this transformation of secondary meristem identity is the production of a highly branched plant with leaves but no flowers. The reason that these mutant plants are highly branched is that secondary meristems that would normally become flowers instead behave as shoot meristems. Shoot meristems are typically indeterminate, and thus continuously elaborate organs (e.g., leaves) on their flanks, whereas the determinate flower meristem produces a fixed number of organs.

Another variation on the theme of converting secondary meristems into shoot meristems is the cauliflower mutant, in which flowers are converted into a seemingly endless proliferation of shoot meristems (see Plate 4F). The defect in cauliflower mutants, which has been characterized in Martin Yanofsky's lab by Sherry Kempin, is the result of inactivating AP1 and a second very closely related gene called cauliflower (CAL). When both AP1 and CAL are absent, secondary meristems simultaneously express LFY and TFL, something that normally does not occur since LFY and TFL are expressed in flower and shoot meristems, respectively.

The fact that leaves are generated in the same locations that normally would be occupied by flowers in double mutants lacking the *LFY* and *AP1* genes indicates that the mechanism for generating secondary meristems at the flank of the apical shoot meristem is independent from that specifying floral meristem identity. Experiments conducted in the laboratories of Detlef Weigel (Salk Institute) and Martin Yanofsky (UC, San Diego) demonstrated that *LFY* and *AP1* can confer floral meristem identity. When these groups misexpressed either *LFY* or *AP1* in shoot meristems, they found that they could convert the shoot into a flower. Cumulatively, these experiments reveal that genes involved in meristem formation per se (e.g., *STM* and *CLV*) function independently of genes that determine meristem identity (e.g., *LFY* and *AP1*). The independent action of genes involved in assigning positional information and genes specifying organ identity is reminiscent of segmentation in the fruit fly embryo, where genes such as pair-rule and segment-polarity genes partition the A/P axis into discrete repeated segmental units which are then separately labeled with particular segment identities by homeotic genes.

Homeotic Genes in Plants

The now blossoming field of flower development was launched in the laboratories of Elliot Meyerowitz at Cal Tech and Enrico Coen at the John Innes Center. Meyerowitz was originally a fly guy who for some perplexing reason decided to turn his talents to another organism, the mustard plant. Over the last decade, his lab has conducted a series of classic experiments that have defined the genetic basis for flower development. As mentioned previously (see Fig. 7.1), flowers consist of four organ types (Fig. 8.2; see also Plate 4A, B), which are arranged in a series of concentric rings (or "whorls" in the flowery jargon of the green contingent). The outermost whorl, which consists of leaf-like structures called sepals, encloses and protects the flower bud. The next whorl is occupied by petals, the often pigmented organ we most associate with flowers and an inspiration for poets and romantics. In the inner two whorls reside the male and female reproductive organs called stamens and carpels (in the center), respectively.

Although floral mutants have been known for centuries, a major conceptual breakthrough in understanding flower development was made by Meyerowitz and Coen, when they realized that floral mutants in which organs developed with incorrect identities could be grouped into three major categories, which have been dubbed the A, B, and C mutant groups. Mutants in each of these three groups have organs in two adjacent whorls of the flower that are transformed into other organs (Fig. 8.2). In mutants lacking function of genes in the A group, sepals and petals (outer organs) are transformed, respectively, into

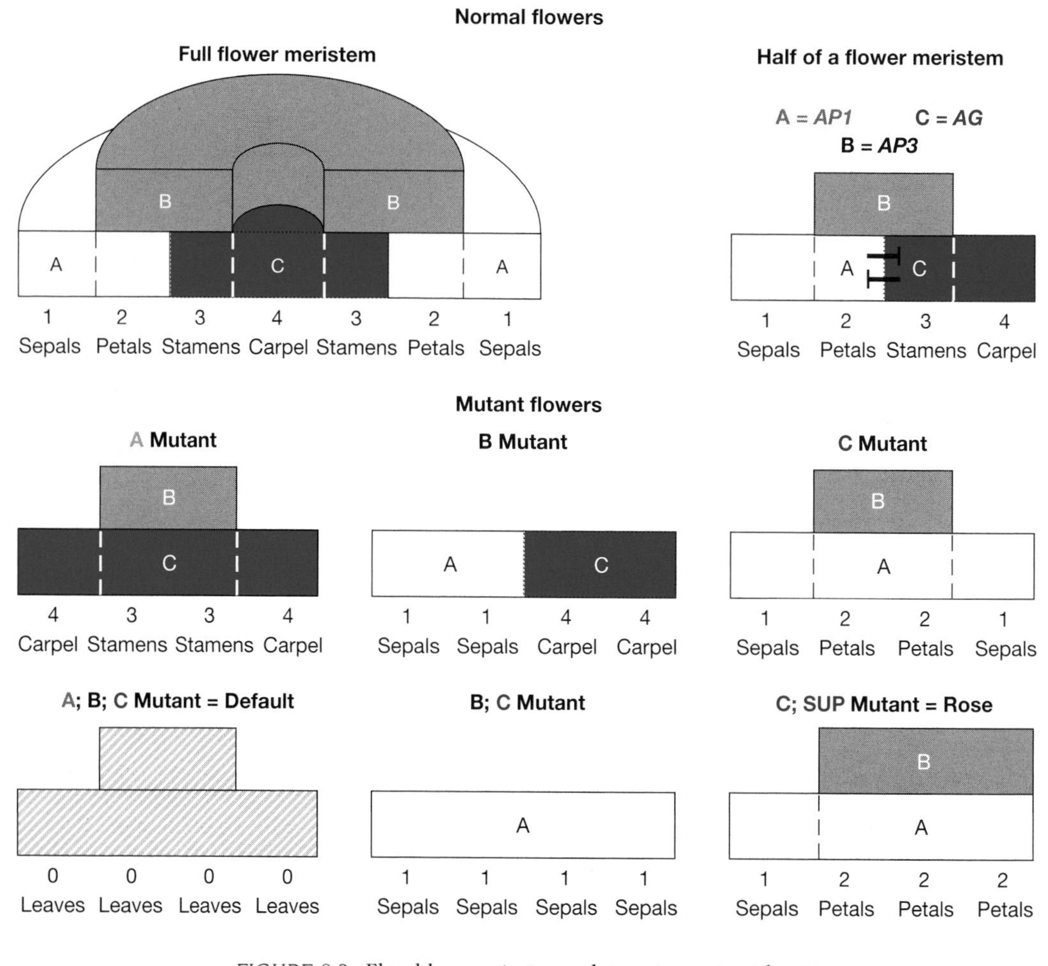

FIGURE 8.2. Floral homeotic genes determine organ identity.

carpels and stamens (inner organs) (Fig. 8.2; see also Plate 4E). In group C mutants, the opposite type of transformation is observed in that stamens and carpels (inner organs) are transformed, respectively, into petals and sepals (outer organs) (Fig. 8.2; see also Plate 4M). In group B mutants, the middle whorls of organs, petals and stamens, take on the identities of the extreme outer whorl (sepals) and extreme inner whorl (carpels), respectively (Fig. 8.2; see also Plate 4I). Meyerowitz and Coen independently proposed a very simple and elegant model to explain the behavior of these mutants in which the A, B, and C genes function in a pairwise fashion to specify floral organ identity (Fig. 8.2). In this so-called ABC model, sepals are defined by the sole action of the A group genes, petals are specified by the combination of A and B group functions, stamens are determined by the sum of B and C group functions, and carpels are specified solely by C group activity. An important feature of the ABC model is that the activities of the A, B, and C group genes must be confined spatially such that A function is present only in the outer two whorls (i.e., sepals and petals), B function is restricted to the middle two whorls (i.e., petals and stamens), and C function is limited to the inner two whorls (i.e., stamens and carpels). As shown in the next section, this prediction has been borne out by cloning of A, B, and C group genes and determination of the expression patterns of these genes. One mechanism by which the A and C group functions are restricted to complementary domains of

In the late 1980s, Enrico Coen and Elliot Meyerowitz independently proposed the ABC model for floral development. Coen's work leading to this model was on homeotic mutants of the snapdragon, which he did in collaboration with his colleague Rosemary Carpenter. He focused on understanding the earliest steps in morphogenesis by identifying and analyzing key plant genes that switch meristems from making shoots to making flowers. His insightful analysis of these odd mutants led to the now-accepted ABC model for genetic control of the floral ground plan.

Enrico Coen was born in Liverpool, UK. He attended King's College at the University of Cambridge as an undergraduate where he focused on genetics. He remained at the University of Cambridge for his graduate studies where he worked on fruit fly genetics for his doctoral thesis. He then took a position at the Genetics Department at John Innes Center, Norwich, UK, where he is now a Group Leader.

Given the ability of the ABC model to account for all single and double mutant combinations of floral homeotic mutants, it is not surprising that one of Coen's most exciting moments was the realization that a simple model could account for various mutants affecting floral organ identity. In his recent book, *The Art of Genes* (1999), Coen recalls, "Goethe's perspective (on floral mutants) only came to experimental fruition in the twentieth century, as mutations affecting development started to be investigated in detail. The unraveling of the ABC model is a good example of how the outlook underwent a change. Given its basic simplicity, it seems quite remarkable that the ABC model for flower development was only proposed in the late 1980s, even though the experimental approach that lay behind it, the production and classification of mutants, had been well established for many decades before this. The advance had more to do with a change in the way the flowers were being looked at than in the development of a new technology. I remember, when we had first obtained one of the class of mutants (carpel, stamen, stamen, carpel), going home in the evening after having spent some time looking at its flowers. It was clear that the outer whorl of sepals had been replaced by female organs, but it was less obvious what had happened to the next whorl, where petals normally form. It seemed that these organs were narrow and strap-like with abnormal structures at the ends. As I considered various models at home, it occurred to me that if the strange strap-like structures were due to a transformation of petals toward male organs, the stamens, a simple model could account for the various classes of mutant we knew about. The next morning, I rushed into the greenhouse to look at the mutant flowers again. To my delight, the strap-like organs did indeed have some tell-tale features of stamens that I had overlooked the previous day. Later on we obtained some much clearer examples of this type of mutation where there could be little doubt that stamens had replaced petals, but the earlier anticipation of the result has remained with me as a striking example of how observations and descriptions are influenced by what you are looking for. In the 1980s we had started to look at flowers in a different way. At the back of our mind we had the notion that genes might act in combination to confer distinctions in identity. And one of the most important contributions to this new outlook on flowers came from studies on quite a different organism: the fruit fly, *Drosophila melanogaster.*"

Coen describes his scientific style as a mixture of many things, "from worrying about the details of a strange flower, to thinking about very general aspects of development and biology. Indeed I think one of the most important aspects of science is to be able to move easily between different levels, much as someone looking at a painting may first look closely at some detail, then stand back to look at the whole and then move in again to examine further particulars. In this way, the picture is comprehended in a dynamic way at a number of levels. In my view, this continual movement between ways of looking at something is one of the most important ingredients of science."

Coen is the recipient of many awards including the Science for Art Prize (1996), the EMBO Medal (1996), and the Linnean Gold Medal (1997) and was elected as a Fellow of the distinguished Royal Society in 1998.

Elliot Meyerowitz (1951–)

Generally regarded as founders of modern floral developmental biology, Elliot Meyerowitz and Enrico Coen independently conceived the ABC model of floral development in the late 1980s—Meyerowitz working with the mustard plant and Coen with the snapdragon. In addition to his research contributions, Meyerowitz has also trained many of the trailblazers in plant development.

Elliot Meyerowitz was born in Washington, DC. He did his undergraduate studies at Columbia University and then entered graduate school at Yale where in Doug Kankel's laboratory he performed a genetic analysis of a mutant disrupting eye development in fruit flies. For his postdoctoral work, he joined David Hogness's laboratory at Stanford University and studied how the expression of *glue* genes (genes encoding proteins that stick pupae to hard surfaces) was controlled. He then moved south to Pasadena in 1980 to take a faculty position at California Institute of Technology, where he has remained.

When Meyerowitz set up his own laboratory as a faculty member at Caltech, he began by identifying regulatory sequences controlling the activity of *glue* genes. After establishing this line of research in his lab, he became interested in plant development and began searching for a plant species that would be particularly well suited for a modern genetic analysis of development. He settled on a small mustard plant known as *Arabidopsis thaliana* because it has a relatively short generation time (3 months) and a small genome (which makes gene cloning easier). As he completed his groundwork characterization of *Arabidopsis,* the first generation of plant molecular biologists joined the Meyerowitz lab. Among the first such postdocs were Marty Yanofsky, Hong Ma, and somewhat later, Detlef Weigel. Meyerowitz, now steeped in the biology and history of floral development, launched his new research direction in earnest. He quickly settled on homeotic mutations as a focal point for his group's studies, as these mutants were well known and one of the most likely categories of floral mutants to disrupt pattern formation. Meyerowitz recollects his thinking at the time he embarked on his analysis of floral homeotic mutants: "I'm sure that my thinking on it was greatly influenced by *Drosophila* work; the parallel between studies of the genetics of plant homeotic mutants and of fly ones is straightforward. Another influence was the long literature (mostly 19th century) on plant floral abnormalities (plant teratology), which gave a thorough list of the sorts of homeotic abnormalities (inherited or not) that had been found in many plant families, and thus a preview of the mutant phenotypes that we could expect."

Meyerowitz and his graduate student John Bowman soon noticed that floral homeotic mutants all shared one thing in common: Two adjacent whorls of the flower developed inappropriate floral organs. Bowman then began combining the single mutants together in pairs to determine which mutant would prevail in various combinations. Remarkably, Meyerowitz realized, the results from all of these pairwise combinations of mutants could be explained by a simple model, which became known as the ABC model (see also Enrico Coen biobox). He recalls, "I think one could reasonably have believed that there would be some answer (in the form of a model) that would come from the type of genetic analysis that we did, though the nature of the model and its generality were unanticipated. What led us to the experiments and interpretations were the phenotypes of the mutations that we got—the plants told us what to think about." When Meyerowitz conceived the ABC model, the clinching data were not yet in. "I remember coming up with the ABC model and explaining it to the people in the lab, but at the time all of the evidence wasn't in, and I don't think any of us believed it would turn out to be true—it was just another hypothesis that needed experimental testing."

In addition to adequately accounting for all of the genetic data the Meyerowitz group had generated, the ABC model also made very specific predictions about the expression patterns of different types of floral homeotic genes. For example, the model predicted that there would be an A function gene which was expressed in the two outer whorls of the flower (e.g., in the sepals and petals), a B function gene that would be expressed in the second and third whorls (petals and stamens), and a C function gene that would be expressed in the inner two whorls (stamens and carpels). This detailed prediction was indeed borne out as floral homeotic genes were cloned, such as the first one, *agamous,* which Marty Yanofsky found to be expressed in the inner two whorls of the developing floral primordium as expected. Similarly, Yanofsky subsequently showed that *AP1,* an A function gene, was expressed in the outer two whorls of the floral primordium, and the Meyerowitz group found that *AP3,* a B function gene, was expressed in whorls 2 and 3.

Meyerowitz has always favored genetic approaches in his work. He notes, "I've always tried to keep a general problem in mind (how cells know where they are in a developing tissue, and how they talk to their neighbors, and consequently form a pattern of cell types). This is the problem that pretty much all of the experiments I've done, flies and plants, are directed toward. My personal preference is strongly for genetic approaches, as inference from genetic results is to me the most exciting sort of science and the most fun as an intellectual enterprise. There are many ways to do science and to make scientific progress, this is just the one I enjoy." Meyerowitz takes a rather utilitarian view of science in that he measures success by what practical impact it has. "An experiment, method, model system or model that is useful to others is what it is all about. What is important is utility."

cells is by mutual repression between genes in these two groups (see below).

Localized Expression of Patterning Genes in Developing Flowers

An important characteristic of homeotic genes in animals is that they are expressed in localized patterns corresponding to the segments whose identity they specify. The ABC mutants are aptly referred to as floral homeotic mutants since the term "homeotic" was actually first applied to flower mutants by William Bateson in 1874. The analogy with animal homeotic mutants is further justified because the transformations of floral organ identity in A, B, or C mutants are similar in many respects to the segmental transformations observed in animal homeotic mutants. In addition, like animal homeotic genes, key A, B, and C genes are expressed in restricted patterns corresponding to the domains in which they exert function. For example, when Martin Yanofsky cloned the first floral homeotic gene called *agamous (AG)* in Elliot Meyerowitz's laboratory, he found that it is expressed only in the center of the floral meristem in the primordia of the third and fourth whorls, as expected for a C class organ identity gene (Fig. 8.2; see also Plate 4K). Shortly after starting his own research group, Yanofsky showed that *apetala1 (AP1),* an A function gene, is expressed in an outer ring of floral meristem cells corresponding to the domain of A gene function in the first and second whorls (Fig. 8.2; see also Plate 4C). He found that expression of *AP1* is strictly complementary to that of *AG.* The complementary patterns of *AG* and *AP1* expression result in part from mutual repression between A and C function genes as revealed by expression of *AP1* in a C class mutant, which expands to occupy the entire floral meristem (see plate 4O). Finally, the Meyerowitz group found that *apetala3 (AP3),* a B group gene, is expressed in a ring of cells which gives rise to the second and third whorls, respectively, and overlaps the domains of cells expressing *AG* and *AP1* (Fig. 8.2; see also Plate 4G).

Like homeobox proteins, MADS-box proteins are named after a DNA-binding domain which is common to all members in this family.

Interestingly, animals also have MADS-box genes, and in one well-studied case, an intimate relationship has been found between a homeobox-containing gene and a MADS-box gene that is required for initiation of heart development in vertebrates and invertebrates. It remains to be determined whether homeobox genes, which control early specification of the shoot meristem, also play a role subsequently in conjunction with MADS-box genes to specify floral organ identity.

AG, AP1, and AP3, and the products of several other genes involved in determining floral identity are structurally related proteins in the so-called MADS-box family of transcription factors.

The fact that several key homeotic genes encode related transcription factors is another parallel between the floral ABC genes and the animal homeotic/*Hox* genes, which, as you may recall from Chapters 3 and 5, are all members of the homeobox group of transcription factors.

How to Make a Rose

In addition to defining the identity of inner whorl organs, the C class organ-identity genes (e.g., *AG*) are also needed to prevent indeterminate growth of the floral meristem. Recall that a significant difference between a normal floral meristem and the shoot meristem is that flowers undergo finite determinate growth, whereas the center of the shoot meristem is in a constant regenerative state and spawns secondary meristems at its flank throughout the life of the plant. In many flowering plants such as mustard, indeterminate growth is the default state of a meristem because class C mutants, in addition to having their inner organs transformed into outer organs, undergo indeterminate growth. The combination of organ-identity transformation and indeterminate growth in group C mutants results in the production of flowers that lack sexual organs and have many outer whorls of petals and then sepals in the center (Fig. 8.2; see also Plate 4M). This type of mutant flower is not terribly attractive since it has a big glob of green stuff (sepals) in its center. When a C group mutant is combined with misexpression of a B group gene, however, the beauty of a rose is created, since the unattractive inner whorl of sepals, present in a C group mutant, now develops as petals (e.g., combined A+ B activity specifies petals). One mutant condition resulting in B gene function spreading into the inner whorl is the *superman (SUP)* mutant, in which excess male reproductive organs (stamens) form at the expense of the female organs (carpels). The result of a C; *sup* double mutant is an indeterminate flower with concentrically repeated whorls of petals. Roses have been created by successive breeding schemes that have taken place over many years. It now is possible, in principle, to do an end run around the protracted process of breeding and convert the scrawny wild cousin of the domesticated rose into a voluptuous bouquet of petals by selectively mutating two genes (e.g., *AG-*; *SUP-*). Similar combinations of mutations are most likely responsible for many commercially appreciated flowers such as camellias, carnations, and chrysanthemums. One consequence of our detailed knowledge of floral organ formation is that commercial flower growers will soon be able to create a myriad of spectacular new varieties of flowers by controlling the activity of the floral organ-identity genes during development. For example, elimination of both *AG* and *SUP* function should convert any flower into a rose-like structure. Good news for Valentine's day vendors!

Leaf Versus Flower Development: Goethe's Hypothesis Revisited

In a treatise on plant development in 1790, in which Johann Wolfgang von Goethe (p. 44) attempted to explain the nature of what we now call homeotic floral mutants, he proposed that all floral organs (i.e., sepals, petals, stamens, and carpels) were modified leaves. Goethe's grasp of the fact that leaf development is the default developmental fate of the floral meristem was truly visionary given that the inner organs of flowers bear no obvious morphological similarities to leaves. According to Goethe's hypothesis and the modern ABC model of flower development, leaves should form in all whorls in A;B;C triple mutants, which lack organ identity in all whorls of the floral meristem. As predicted, A;B;C triple mutants produce an indeterminate structure consisting of concentric whorls of leaves approximating a leafy version of a rose (Fig. 8.2; see also Plate 4N). So, it turns out that the poet was as versed at reading leaves as he was in writing flowery prose.

A great strength of the ABC model is that it also accurately predicts where leaves or other organs should form in various combinations of double mutants. For example, in B;C double mutants, all whorls develop as sepals (Fig. 8.2; see also Plate 4J). Recall that sepals are normally specified by A function alone. The reason that A function is present in all whorls of B;C double mutants is that the C function normally represses A function in the inner whorls. In the absence of this repression, A function is distributed throughout the floral meristem. Because the B function is also missing in these mutants, all whorls form sepals.

Fruit Organ-identity Genes Subdivide the Carpel into Distinct Regions

The ovary comprises the bottom portion of the flower carpel and houses the eggs, which when fertilized by sperm, develop into embryos within seeds. After fertilization, the carpel develops into the fruit of the plant. The development of a fruit from the carpel is a separate patterning process unto itself, the end point of which is the release of seeds into the environment to begin life on their own. In many fruits, seed dispersal is accomplished by the fruits ripening, falling to the ground, and rotting. In other plants of the legume family such as peas, nuts, or mustard, the outer portion of the fruit forms a pod which holds the seeds. In order for the seeds to be released, the pod must break open at an appropriate time. Fruits of the pod type consist of two major parts: valves, the fleshy sectors that get eaten in edible fruits, and valve border cells, narrow stripes of cells running between the sectors of valves (Fig. 8.3). When a pod-type fruit ripens, the valve border cells separate from each other, thereby dissolving the sutures between the sectors of valve and releasing the seeds inside the fruit. This process of seed release is known as shattering.

Martin Yanofsky's laboratory has pioneered a genetic analysis of fruit development in mustard plants. His group identified several key genes required for subdividing the fruit primordium into domains corresponding to the valve versus valve border. These genes, like many of the homeotic genes specifying floral organ identity, turn out to be members of the MADS-box family of transcription factors. Since these genes seem to function analogously to the floral organ-identity genes in specifying the primary parts of the fruit (e.g., valve versus valve border), they can be considered fruit organ-identity genes. As mentioned above, the first MADs-box genes involved in fruit development are C-function genes such as *AG,* which specify the carpel cell fate. If *AG* is misexpressed in the outer whorl of a floral meristem, sepals are transformed into carpels. Following fertilization of their eggs, these abnormally positioned carpels develop into fruit-like organs. For example, when the

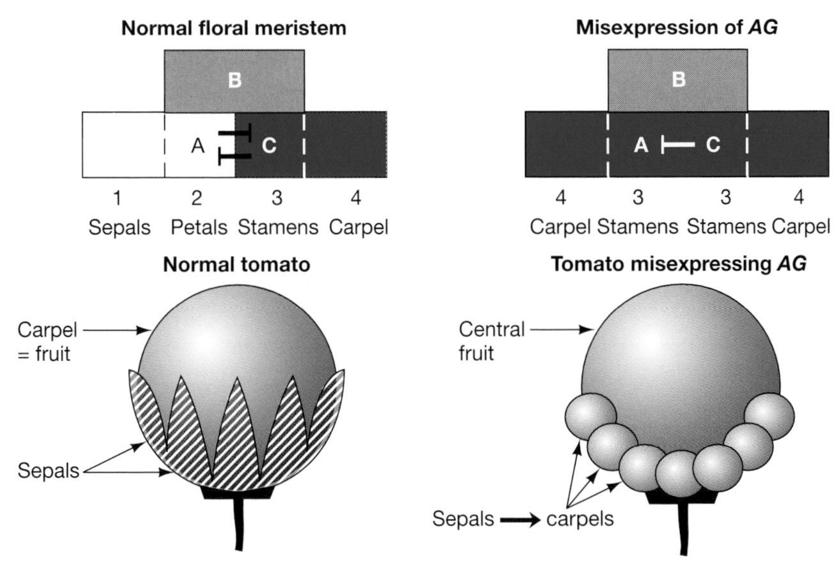

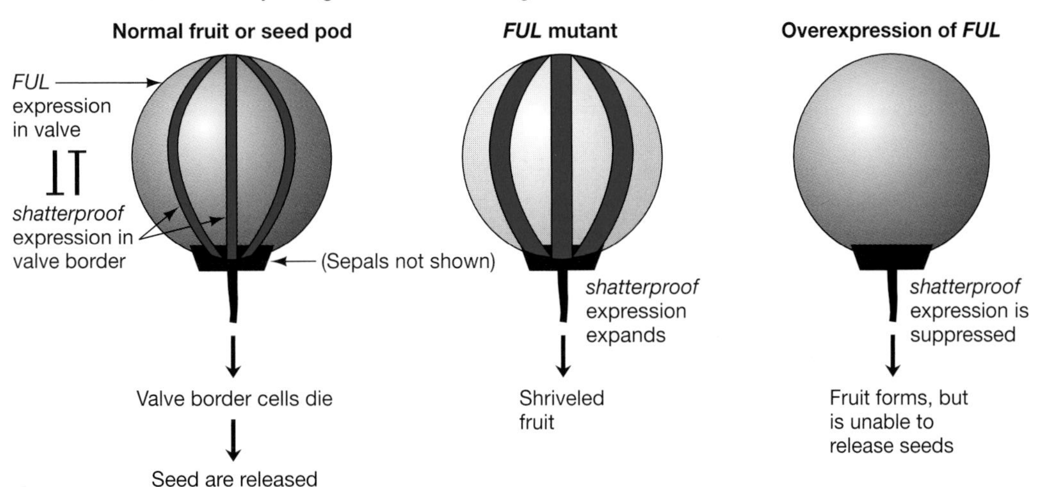

FIGURE 8.3. Fruit development. (*A*) Misexpression of *AG* can convert sepals into fruit. (*B*) The *fruitful* and *shatterproof* genes define fruit organs.

Yanofsky group engineered a tomato plant in which the *AG* gene was misexpressed in the sepal primordium (i.e., the outer leaf-like structures that normally surround a mature tomato), the result was a tomato plant with rings of little tomato-like fruits surrounding the central large fruit (Fig. 8.3; see also Plate 4T, U). One reason that *AG* can convert sepals into carpels is that expression of *AG* in the outer whorl suppresses expression of *AP1,* an A function gene specifying sepal identity.

Subdivision of the fruit primordium into separate domains corresponding to the valve and valve border involves two sets of MADS-box genes. The *fruitful (FUL)* gene is expressed in the region giving rise to the valve and the two *shatterproof* genes are expressed specifically in valve border cells (Fig. 8.3B). The *FUL* gene is required for formation of the fruit because mutants lacking *FUL* activity generate tiny shriveled fruits. The *FUL* gene also is capable of imposing valve development on valve border cells, since misexpression of *FUL* results in the outer portion of fruits forming one large unsegmented valve. As expected on the basis of their nearly identical expression patterns, the *shatterproof* genes are required for the formation of valve borders since these structures are missing in *shatterproof* mutants. Because valve borders are absent in *shatterproof* mutants, the valves do not separate from each other following ripening, which prevents shattering and hence seed release. As described further in Chapter 9, the failure of *shatterproof* mutants to release their seeds is of great agricultural and commercial value, since it provides a means for preventing the substantial seed loss caused by premature seed dispersal, and translates into harvesting much higher percentages of seeds in the field.

As in the case of the A and C function genes in the flower, the precise complementary expression patterns of the *FUL* and *shatterproof* genes arise from mutual repression. This mutual repression is evident in mutants lacking the function of the *FUL* gene in which expression of the *shatterproof* genes spreads into the region normally occupied by valve cells. Reciprocally, in *shatterproof* mutants, the activity of *FUL*

Martin Yanofsky (1956–)

Martin Yanofsky was born in Cleveland, Ohio, and then moved to Palo Alto, California, where he spent his childhood. He was an undergraduate at the University of California, San Diego, and then did his doctoral studies at the University of Washington in Gene Nester's laboratory, where he made seminal contributions to understanding the mechanism by which a group of bacteria known as agrobacteria infect plants by transferring some of their bacterial genes into host plant cells. These studies led to the development of methods for transferring genes into plants with high efficiency and have contributed significantly to the rapid advances in understanding a wide range of problems in plant biology.

Following his illustrious graduate career, Yanofsky joined the budding Meyerowitz plant group and began his work on cloning the first floral homeotic gene, *agamous,* in the experimental mustard plant *Arabidopsis thaliana.* He recalls feeling very fortunate to be at the right place at the right time to do these exciting experiments. "I was lucky to have been part of the Meyerowitz lab during the early studies on flower development, a lab I consider to be one of the best in the world. We had been working on the fantasy that most of these key regulatory genes would be members of a multigene family, and that if we could just get our hands on one of them we'd quickly get the rest. Soon after we cloned *agamous,* we showed that it was indeed part of an extended gene family (the MADS box family of transcription factors), and I was convinced that we had found our gold mine. I think it's fair to say that I've spent the past ten years mining that gold and I think we've only scratched the surface."

When Yanofsky started his own laboratory, he made use of the fact that other floral homeotic genes encoded MADS box proteins similar to *agamous* to clone a large number of members of this gene family. The first two of these genes that he characterized independently were *apetala1 (AP1)* and *cauliflower (CAL)*, which act at an earlier stage of floral development when secondary meristems adopt either the leaf or floral fate. Sherry Kempin in his lab found that *AP1* and *CAL* encode closely related and functionally redundant genes required for flower formation. Mutants that lacked both of these genes proliferated meristems in positions that would normally be occupied by single flowers, producing the "cauliflower" phenotype, similar to the dinner-table cauliflower. Yanofsky's initial analysis of additional MADS box genes also led to the identification of genes expressed in different parts of the developing fruit, such as *fruitful* (expressed in the fruit valve) and the *shatterproof* genes (expressed along the valve margin). After several years of a challenging genetic analysis, Yanofsky's group succeeded in getting mutants in these genes and found that they function to define the domains of the fruit in which they are expressed. These experiments led Yanofsky into an entirely unexplored new territory, which his group is currently exploiting very effectively. Regarding his early forays into this new area, he recalls, "Our subsequent studies on fruit development came out of our desire to carve out our own niche, and we realized that fruits were important and interesting, and yet very little was known about the molecular genetics of fruit development." Although Yanofsky does not think his discoveries in fruit development were entirely unanticipated, he notes that there are significant potential practical implications of this new understanding of fruit development. "One thing that strikes me is that we can manipulate many aspects of fruit development in very predicable ways by manipulating just a handful of genes, even though we still know relatively little about this complex process."

In addition to the great excitement in cloning the *agamous* gene and finding that it was expressed as expected for a C-function gene in the inner two whorls of the developing flower primordium, Yanofsky remembers the thrill of his first significant scientific success, when he was a graduate student. At the time, very little was known about the mechanism by which *Agrobacterium* transfers a portion of its genes into the genome of higher plants. He and his colleagues developed a simple assay to look for molecular changes in the *Agrobacterium* genome immediately after it is induced to undergo gene transfer. In one quick series of experiments, they were able to define the exact nucleotide at which transfer begins, to show that it is a single-stranded molecular that is transferred, and to identify the gene that initiates this process. Since these scientifically dramatic moments, Yanofsky sees progress in his work as occurring by many smaller but still intoxicating steps "Every six to twelve months I think we have one of those special moments where we take a significant step forward, and then there's just a lot of work in between. I think many of us live for those occasional moments of insights."

Yanofsky credits a component of his success to having a good system such as *Arabidopsis* to work in "since the critical mass of scientists working on this system ensures that many tools will be developed that can be applied to the problems we are interested in." In reflecting on how his own style of scientific investigation may have contributed to his advances, he notes "I think I tend to rely on my gut instincts for what problems I feel are the most interesting, and what is the quickest method of getting to the answer." The key is "Curiosity, tools, and a lot of hard work." He also acknowledges that "There are many ways of doing science and it's hard to say that one is better than the next. In the end, any approach can best be measured by productivity over many years."

expands to include the valve margin region. This example once again illustrates the generality of mutual repression as a refinement mechanism for cleanly subdividing domains of cells into two nonoverlapping subdomains.

THE XYZs OF LEAF DEVELOPMENT

As described above, analysis of double and triple floral organ-identity mutants has confirmed Goethe's hypothesis that the default state of floral organ development is leaf. In principle, all that is required for leaf formation is to avoid activating expression of floral meristem identity genes (i.e., *LFY* and *AP1*) in a secondary meristem. Typically, plants go through a vegetative phase of growth in which secondary meristems form leaves and lateral shoots. During this period of vegetative growth, secondary meristems express the *TFL* gene, which functions as it does in the primary meristem to suppress floral development. *TFL* is thought to exert its inhibitory effect on floral development primarily by suppressing expression of *LFY* and *AP1*. In response to various environmental factors, plants undergo a transition from vegetative growth to flowering, at which time they activate expression of *LFY* and *AP1*. The transition to flowering is promoted by genes such as FT, which antagonize the action of *TFL*, and result in activation of *LFY* and *AP1*. Various environmental cues such as light, temperature, or season induce the transition to flowering in various plants. It is likely that these diverse environmental stimuli define a combinatorial code to activate flowering-time genes such as *FT* to initiate flowering at the appropriate time. Control of flowering is one of the most important decisions a plant makes, since the reproductive strategy of the plant depends on its being able to read its environment for optimal conditions of seed production and dispersal. Such reproductive strategies vary enormously in different plants. For example, some plants such as trees or the agave only flower once in a hundred years, whereas others start flowering as soon as possible. Understanding the basis for these differences in flowering time is one of the most exciting areas of current plant research.

Once leaf development is initiated, cells in the leaf primordium begin to proliferate and organize themselves into a blade-like structure with two distinct surfaces, which I refer to as the dorsal and ventral surfaces. The developing leaf, like an animal appendage, also has a proximal–distal (P/D) axis (the proximal end joins the stem). As the leaf primordium forms, the P/D axis runs parallel to the A/B axis of the plant (i.e., the leaf starts off lying flat up against the stem). Because the leaf primordium can be unambiguously oriented with respect to the stem, the two surfaces are defined by their proximity to the stem. I refer to the surface of the leaf nearest the shoot as the dorsal surface and the opposing surface, forming farthest from the main shoot, as the ventral surface. As the leaf grows, the dorsal and ventral surfaces of the leaf must expand in proportion. Coordinated growth of the leaf requires interactions between the dorsal and ventral surfaces, a process which in many respects parallels appendage outgrowth in animals (see below). In addition to the P/D and D/V axes, leaves have an axis akin to the A/P axis in animal appendages that is perpendicular to the P/D axis. I refer to this axis as the medio–lateral (M/L) axis. Leaves have veins that form and branch in particular locations along the M/L axis. For example, the ma-

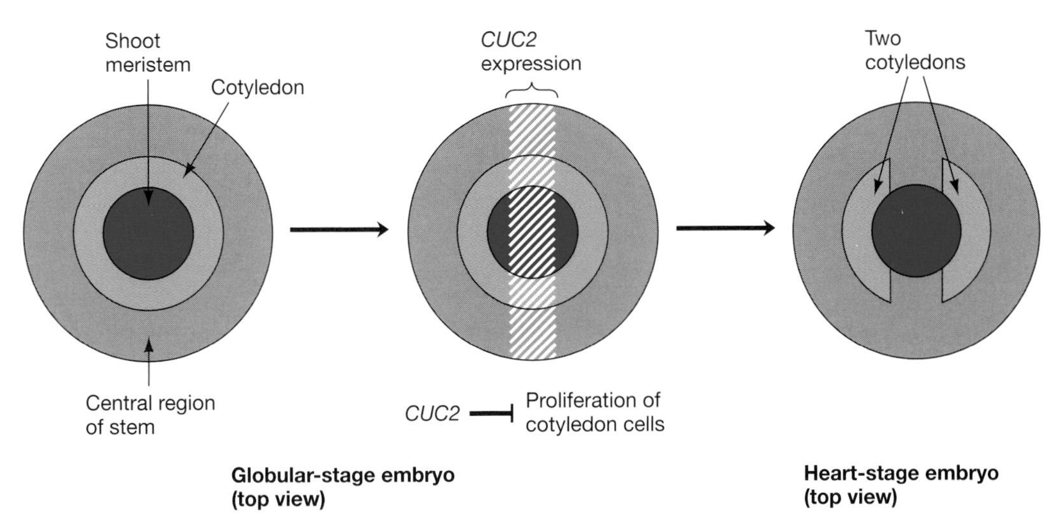

FIGURE 8.4. Splitting the cotyledon primordium into two domains in dicots.

jor leaf vein typically runs up the center of the leaf (i.e., up the middle of the M/L axis). In addition, some leaves (e.g., eucalyptus leaves) are asymmetric with respect to the M/L axis or have distinct patterns of pigmentation along the M/L axis. Although we do not consider the M/L axis further here, since little is known about how it is established, analysis of this patterning system is likely to become an area of future interest.

Carving the Two Cotyledons from a Goblet

In considering the origins of leaf development, we need to return briefly to embryogenesis when development of cotyledons, the first leaf-like structures, is initiated. Aside from their leaf-like appearance, one indication that cotyledons are indeed seedling versions of leaves is that in mutants that are unable to suspend embryonic development during dormancy, the cotyledons develop as mature leaves. Flowering plants are subdivided into two large groups based on whether they form one cotyledon (monocot plants) or two cotyledons (dicot plants). Because embryogenesis has been best studied in the mustard plant, we consider the formation of the two cotyledons in this representative dicot plant.

As you may recall from Chapter 7, the primordia of the cotyledons become morphologically visible by the heart stage of embryonic development. The initiating events for cotyledon formation, which precede this morphological development, occur during the globular embryo stage when the embryo appears to be rotationally symmetric. As discussed previously, the subapical position of the cotyledons is thought to be determined by genes functioning to define regional identities along the A/B axis of the embryo. Information arising from A/B patterning defines a ring of cells at the appropriate position along the A/B axis. How then, are two sites of cotyledon formation selected from a rotationally symmetric structure? The identification of two mutants, called *cup-shaped cotyledon1 (CUC1)* and *cup-shaped cotyledon2 (CUC2)* studied in the laboratory of Masao Tasaka (Nara, Japan) helped address this question. As their name implies, double mutants lacking function of both *CUC* genes generate seedlings with a goblet of cotyledon material encircling the shoot meristem (Fig. 8.4; see also Plate 2M–P). The formation of a cup-shaped cotyledon in *CUC* mutants suggests that the cotyledon primordium is normally initiated as a ring-like structure at the rotationally symmetric globular embryo stage, and that the *CUC* genes play some role in suppressing cotyledon formation at two opposing sites to split the cotyledon primordium in half. Consistent with this hypothesis, Tasaka found that the *CUC2* gene is expressed in a stripe of cells which bisects the apex of the globular embryo (Fig. 8.4; see also Plate 2L). Where the stripe of *CUC2* expression intersects the ring of cotyledon primordium, cell growth is suppressed, carving the cotyledon into two primordia. These two domains of cells proliferate to generate the two cotyledons and remain separated by nonproliferating *CUC2*-expressing cells, which ultimately become part of the stem. How *CUC2* expression is activated in a stripe of cells running parallel to the A/B axis in dicots is one of the key questions regarding early cotyledon formation. It will also be interesting to know whether monocots similarly express *CUC* genes, but in a different pattern, to define the position of the single cotyledon in these plants.

Apposition of Dorsal and Ventral Cells Induces Outgrowth of Leaves

One of the most striking similarities between developmental mechanisms in plants and animals is how outgrowth of plant leaves and fly animal appendages depends on interactions between the dorsal and ventral surfaces along the margin of these structures. As you may recall from Chapter 4, when cells on the dorsal surface of the fly wing confront cells on the ventral surface, they send a signal via the Notch receptor to initiate formation of the wing margin, which is a necessary condition for outgrowth of the wing. This signaling event is required for wing outgrowth, and thus mutants such as *apterous,* which are unable to specify the dorsal surface, lack wings. In addition, mutants partially lacking function of the Notch pathway display defects in the marginal region of the wing manifested by the formation of long narrow wings consisting only of central wing structures. Remarkably, the same types of interactions between dorsal and ventral cells of developing leaves seem to be required for leaf outgrowth and formation of the marginal region of the wing.

A snapdragon mutant called *phantastica,* which has been characterized by Richard Waites and Andrew Hudson (University of Edinburgh), most vividly illustrates the requirement for dorsal–ventral communication between leaf cells. *phantastica* mutants, like *apterous* mutants in flies, are defective for formation of the dorsal surface of leaves. There are various degrees of severity of the *phantastica* muta-

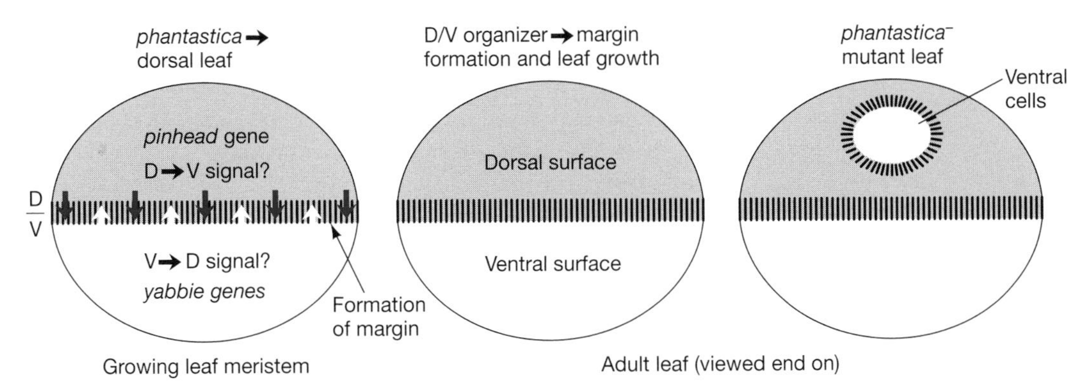

FIGURE 8.5. D/V signaling is required for margin formation and outgrowth of leaves.

tion. In severely affected individuals there is no outgrowth of the leaf at all. The only structure visible in such leaves is a little nub of a leaf stem. Less severely affected leaves are smaller and proportionally narrower than usual (see Plate 4V, W), and are cupped upward as a consequence of there being fewer cells on the dorsal surface than on the ventral surface. These moderately affected leaves often contain little islands of tissue resembling ventral cells imbedded in the mutant dorsal surface (Fig. 8.5). At the border between these islands of ventral-like cells and their surrounding dorsal neighbors, Waites and Hudson observed the formation of marginal structures. This inductive event in the leaf bears an uncanny similarity to the fly wing in which a margin is induced at the interface between islands of *apterous*⁻ mutant cells and normal cells on dorsal surface of the wing (see Fig. 4.4). Another interesting parallel between plant and fly mutants affecting leaf margin formation is that in partial loss-of-function *phantastica* mutants, narrow leaves are generated that comprise only central leaf structures. As in the fly wing, there are genes that are expressed in dorsal versus ventral domains of developing leaves. An early-acting transcription factor called *pinhead* is expressed on the future dorsal surface of leaf primordia in mustard embryos, and later during leaf growth, members of a group of related genes, referred to as *yabbie* genes, are expressed exclusively on the ventral surface of leaf primordia, where they appear to promote ventral cell fates. Thus, in both plant and animal appendages, juxtaposition of the dorsal and ventral surfaces induces formation of an organizing margin at the interface between the two surfaces, which controls outgrowth of the appendage and patterns the appendage in the vicinity of the margin.

GENERALITIES OF DEVELOPMENTAL STRATEGIES IN PLANTS AND ANIMALS

Perhaps the most striking similarity between plant and animal development is the iterative subdivision of developing embryos and appendages into a small number of domains. The general strategy by which this simple subdivision is accomplished in developing animals is by diffusible signals activating genes at one of a few thresholds. Although it remains to be demonstrated that morphogens act in similar ways in plants, this seems highly likely, particularly in light of the organizing activities of plant hormones such as auxin and cytokinin. It is also notable that particular subtypes of transcription factors seem to play prominent roles in defining regional cell identities. Thus, homeobox genes play an important role in specifying the apical tip of plant embryos. Similarly, during flower development, regionally expressed MADS-box genes function as homeotic genes to define organ identities in flowers and fruits in much the same way that homeobox-containing homeotic/*Hox* genes act during animal embryogenesis and appendage formation. The use of long- and short-range signaling is another significant similarity between plant and animal development. In plants, hormones such as auxin and cytokinin may diffuse over long distances to function as morphogens. Likewise, graded distributions of Bicoid and Dorsal initiate patterning in early fly embryos and Dpp/BMP4 function as long-range morphogens to pattern the A/P axis of animal appendages. Shorter-range signaling also plays an important role in resolving borders between domains of cells. For example, in plants, CLV-mediated signaling restricts the domain occupied by the shoot meristem, and short-range signaling between the dorsal and ventral surfaces of the leaf is required to initiate outgrowth and patterning. Similarly, the Hh signal acts locally to pattern a narrow group of cells along the A/P border of fly appendages, and short-range signaling through the Notch receptor is necessary to trigger outgrowth and patterning at the interface of the dorsal and ventral surfaces of the fly wing. As mentioned in Chapter 7, another striking finding is that many structurally related genes in plants and animals perform similar functions. Thus, in addition to the homeobox transcription factors mentioned above, plants and animals use related receptors for both developmental patterning and immunity, and a receptor used widely in the nervous system of animals is also present in plants.

Because plants and animals are believed to have diverged from a unicellular ancestor, the parallels between the developmental mechanisms in plants and animals most likely result from the independent evolution of common patterning strategies. I must confess to being quite surprised by the degree of similarity in plant and animal development, particularly given that we are only beginning to understand many of these developmental events. Although it is true that I have selected among many possible topics to discuss in this book, and that I have made a point of highlighting common developmental strategies, this has not been very hard to do. One way to think of the similarities in developmental strategies in plants and animals is that there may be a fairly limited number of molecular mechanisms by which cells can communicate with one another to coordinate their actions during development. As a final analogy to electronics, imagine a space alien who comes to earth and tries to figure out how our various electrical devices function. This creature might be surprised at first by the remarkable similarities between the innards of a TV and a microwave oven. When this alien, excited by its initial discovery, calmed down a bit, it might realize that all it had discovered was that primitive humans use a small number of circuit elements such as resistors, capacitors, transistors, and induction loops in their circuits and that they build their myriad of machines and gadgets using different combinations of these few basic circuit elements. Perhaps the unicellular ancestor of plants and animals possessed a fairly small number of circuit elements that were assembled in different combinations during evolution to generate the remarkable diversity of plants and animals currently on earth. Whether complex organisms, like Legoland, are indeed mammoth assemblies of simple units piled on top of one another over the vast expanses of evolutionary time is one of the big questions for us to address in the future.

■ Summary ■

Flowers and leaves derive from secondary meristems that form around the perimeter of the primary shoot meristem. The even spacing of these secondary meristems is likely to be the result of inhibitory signals emanating from these meristems. The primary meristem also can develop as a secondary meristem, but is prevented from doing so by factors such as TFL, which is produced by primary meristem cells. The effect of TFL is counterbalanced by that of FT, which functions to limit the size of the primary meristem. In many plants, such as mustard, cells in primary shoot meristem divide continuously to generate a series of secondary meristems throughout the life of the plant. This type of iterative process is referred to as indeterminate development. Secondary meristems giving rise to flowers and leaves, however, undergo cell division to grow to the appropriate size and then stop dividing in a process referred to as determinate development.

Floral meristem identity genes such as *LFY* and *AP1*, which are expressed in early developing secondary meristems, are required to initiate floral development. Once a secondary meristem has been specified as a floral meristem, expression of floral organ-identity genes (or floral homeotic genes) is initiated in one of three primary radially organized territories. The outermost ring of meristem cells, which gives rise to leaf-like sepals or petals, expresses A function genes such as *AP1*. An inner ring of meristem cells, which gives rise to stamens or carpels, expresses C function genes such as *AG*, and a central ring of cells that partly overlaps both the outer and inner rings of cells expresses B function genes such as *AP3*. The *AP1*, *AG*, and *AP3* genes all encode transcription factors in the MADS-box family. Genes in these three territories interact to define the four floral organs according to simple rules set forth in what has come to be known as the ABC model. According to the ABC model, the outermost floral organ (sepals) is specified by A function genes only. The second ring (or whorl) of floral organs (petals) is specified by having both A and B functions active, the third whorl (stamens) by having both B and C functions active, and the inner whorl (carpels) by having C function alone. The ABC model accurately predicts the defects observed in all single and double combinations of floral identity mutants. Plants lacking the functions of all floral identity genes (e.g., A; B; C triple mutants) make flower-like structures in which all organs develop as leaf-like structures. This result confirms a hypothesis made by the famous poet and amateur botanist Johann Goethe in 1790, who concluded from examining various floral mutants that the default state of floral organ development was leaf.

Following fertilization of the eggs housed within the carpel, the carpel develops into the fruit of the plant. In fruits of the legume family, the outer portion of the fruit is a pod, which encapsulates the seeds. Once the fruit is fully developed, the pod opens in a process referred to as shattering, and the seeds are released into the environment. Two types of cells make up the pod. The majority of cells contribute to the separate parts of the pod itself, called the valves, and the narrow strip of cells forming between the valves are referred to as valve border cells. When the fruit ripens, cells along the valve border die and the valves become detached from an internal supporting structure along their edges, which results in shattering. Two types of MADS-box genes play an important role in distinguishing the fate of valve cells from valve border cells. The *FUL* gene is expressed in the future valve cells and plays a key role in specifying valve cell fates, and two highly related *shatterproof* genes, which are expressed in the valve margin, are required to specify the fates of valve border cells. In *shatterproof* mutants, cells that would ordinarily form the valve border develop as valve cells, resulting in the formation of a pod made up of a single fused valve. Because this valve is unable to split open (i.e., there are no valve border cells), the resulting fruit cannot spill its seeds. This type of mutant is likely to be of significant commercial value since a sizable percentage of seeds from plants such as canola are typically lost in the field due to premature or uncontrolled shattering.

Leaves form as the default state of secondary meristems and have two sides referred to as the dorsal and ventral surfaces, which meet at the margin of the leaf. Outgrowth and patterning of the leaf margin relies on signals passing between the dorsal and ventral surfaces of the leaf primordium. This inductive interaction between the two surfaces of the developing leaf is remarkably similar to that required in flies for formation of the wing margin and outgrowth of the wing. Although plants and animals are thought to have descended from different single-cell ancestors, there are many similarities in the developmental strategies used by these two kingdoms of life. In addition, many of the same families of genes perform similar functions during plant and animal development (e.g., homeobox genes and signaling genes involved in patterning and immunity). Understanding the basis for these similarities is one of the great remaining challenges for developmental biologists.

9 The Future of Biology and Man

It is hoped that, by this point, the reader is convinced that we have determined many of the important principles guiding animal and plant development. A central theme regarding animal development has been that basic developmental mechanisms in vertebrates and invertebrates are shared because all segmented animals inherited a variety of genetic devices from a common ancestor that had already invented these fundamental patterning processes. As discussed in the previous two chapters, there also are surprising similarities between the mechanisms used to pattern plants and animals. In the latter case, however, it is thought that these two major kingdoms of life evolved independently from a unicellular ancestor into multicellular organisms. The parallels between animal and plant development suggest that there may only be a limited number of molecular mechanisms available to create pattern during development.

Now I want to explore the practical implications of our newly acquired knowledge of animal and plant development. I first consider the applications arising from the commonalities between model invertebrate organisms such as fruit flies and vertebrates on problems relating to human health. I then consider the enormous economic impact of the ability to manipulate the development of plants. These considerable practical spin-offs of basic research should serve as potent arguments in favor of continuing our current generous level of federal support for basic research. Because important discoveries made in basic science rapidly and pervasively make their way into all of our lives, it does make Dollars and $ense to continue investing in science. After considering the practical benefits of our understanding of development, I end with a little science fiction fantasy regarding what the future may hold.

THE HUMAN GENOME PROJECT: THE EIGHTH WONDER OF THE WORLD

The objective of the Human Genome Project is to determine the complete DNA sequence of the entire human genome (i.e., of all genes). In parallel with the Human Genome Project, there have been coordinated efforts to determine the full genome sequences of model experimental organisms including bacteria, yeast, worms, the mustard plant, and the fruit fly. All except the fruit fly and human projects are now complete or are in the very final stages of completion. The fly genome project is scheduled to be finished by the time this book is published, and the human genome project is predicted to be done by 2002.

What is the value of all this DNA sequence information (which you may recall would be a book 12 stories high if typed on paper), and why do I refer to it, perhaps with some degree of hyperbole, as the eighth wonder of the world, ranking with architectural achievements such as the pyramids of Egypt or the Great Wall in China? There are two reasons for comparing the genome project to these great wonders of the world. First, it represents an enormous amount of human effort. Only a few years ago, this type of project would have been logistically impossible to accomplish. Advances in sequencing methods, gene cloning, and high-power computing have now made this project possible, but it remains a technically challenging and labor-intensive effort. Second, the human genome sequence will literally be an enduring mark of humanity lasting far beyond the current wonders of the world. With respect to the science of biology, the final century of the second millennium will most likely be remembered for the solving of the structure of DNA by Watson and Crick and for the birth of the age of genomics. These are big plaques in the Biology Hall of Fame, right up there alongside Darwin's theory of evolution.

There are many practical ramifications of the genome project. The most obvious consequence of knowing the complete DNA sequence of an organism is a dramatic acceleration in gene identification. For example, in worms, whose genome sequence is now known, it is possible to isolate a protein with some interesting property (e.g., it binds physically to another protein) and to determine which gene encodes this protein by using a computer. This positive gene identification is made by comparing the empirically determined fragmentation pattern of the protein in question (a kind of a molecular fingerprint, if you will) with a database consisting of the predicted fragmentation patterns of the proteins encoded by all worm genes. For worm proteins, this FBI-like game routinely comes up with the correct suspect, and is soon likely to live up to the famous FBI motto. Although this technical improvement on current gene-cloning methods may not seem dramatic, because it only facilitates what is already possible, it nonetheless will revolutionize the way we design experiments—what now typically takes 1–2 years to accomplish (i.e., cloning a gene for a protein of interest) will be done overnight.

Completion of the genome project could be likened to the invention of the integrated circuit or the microchip. These electronics improvements did not create any fundamentally new circuit elements, but rather created the possibility of assembling large numbers of such elements necessary for designing fast integrating processors. The practical consequences of developing cheap powerful computers cannot be underestimated. This advance has transformed our world into a cyberblitz, in which we must now endure back-talking plastic boxes on our desk tops that perform more computations per second than could be accomplished by a room full of the most sophisticated computers less than 20 years ago. The genome project will similarly enable progress in molecular biology on a massive scale. It is not too surprising that the biotech industry is paying keen attention to these developments and that one of the big gold rushes at the turn of this millennium is for genes and patents on gene function.

■ Biotech Companies ■

The fact that one can extrapolate meaningfully from results obtained in model organisms such as fruit flies or worms to principles that may be operating in humans has not been lost on entrepreneurs and sharp investors in biotechnology. Biotech companies are appearing on the scene which are devoted primarily to the idea that insights gained from analysis of biological processes in model animals such as flies or worms are likely to be of commercial value in developing drugs or treatments for human diseases and ailments. These biotech companies are betting on the very reasonable prediction that they will identify new components of medically important signaling pathways in model organisms and can then use these animals to screen for variants of relevant proteins or drugs that can interfere with specific steps in corresponding processes in humans. Entrepreneurs are hoping that such products will have significant therapeutic and commercial value. The only real unknown in this equation is whether the deep functional similarities in genes controlling development will also carry over to other genes such as those involved in human disease. Because several of the genes involved in fly and vertebrate development also cause human disease when mutated (see below), the next few years are likely to prove this gamble a good one.

Implications for Human Health—Curing Cancer, Fixing Genes

It is very likely that the single largest impact of our understanding of animal development will be on human health. This is not because most human diseases are developmental in nature. Developmental conditions can result in devastating birth defects, and many of these diseases are likely to involve defects in molecules we have discussed in this book; however, the number of such medical conditions is actually fairly limited. Rather, the impact of our knowledge of development on health derives from our greatly sharpened image of how cells acquire their identities and how they communicate with one another. For example, nearly all of the growth factor signaling pathways we have discussed in this book, such as the Dpp, Hh, and Notch pathways, have been implicated in human cancer. Because cell proliferation (i.e., multiplication by cell division) is intimately connected with development, many genes regulating developmental decisions also control cell proliferation. For example, activation of the Dpp and Notch pathways is required for outgrowth of appendages during adult fly development, and these pathways are also activated in developing vertebrate appendages. During adult life, mutations in these same signaling pathways can result in excess cell proliferation and, ultimately, in cancer. In some cases, loss of gene function, such as disruption of a component required for Dpp signaling, loss of a Hedgehog receptor, or mutation of the Notch receptor, leads to aggressive tumor formation because these signaling systems normally function to limit cell proliferation in adults. When this constraint on cell growth is lifted, cells proliferate inappropriately and take the first step to becoming cancerous. In other cases, mutations in growth factor receptors, which normally are involved in promoting cell growth, result in receptors that are active even in the absence of signal. This independence of receptor activity from normal activation by signals also causes cells to proliferate where and when they should not. Many forms of leukemia are caused by such misregulation of receptors and signaling pathways. Because of the intimate link between development and cellular growth control, advances made in understanding mechanisms guiding development will have a profound impact on treatment strategies and on rational drug design. One direct consequence of basic research on signaling pathways in model organisms like flies and worms is likely to be the identification of optimal targets for cancer drug therapy.

Another very important, although controversial, contribution of the molecular biology revolution to human disease is the looming gene replacement technology, or gene therapy. The idea of gene replacement therapy is to repair a defective disease-causing gene in an afflicted patient. For example, it is currently possible to take human blood cells from a patient suffering from β-thalassemia, a condition in which the oxygen-binding molecule hemoglobin is abnormal, fix the genetic defect in isolated blood cells in a test tube, and reintroduce these genetically repaired cells into the patient after that person has been irradiated to eliminate his or her own defective blood cells. The result of such a gene replacement treatment is that the future blood cells produced by the person will make a normal functional version of hemoglobin, and their disease will be cured for life. Treatments such as this are now at the experimental stage in clinical trials and undoubtedly will be improved over the next decade until they become as routine as in vitro fertilization is today. In a subsequent section, I consider the impact of this powerful gene-altering technology further, because there are some very significant ethical issues that arise from the ability to alter the genetic makeup of human cells, particularly if such changes are made in reproductive cells and become heritable.

Implications for Production of Food and Other Plant Products

The first significant practical impact of our understanding of development is being felt in the agricultural arena. There are several reasons for this lead in plant technology. First, there are far fewer ethical issues to consider when altering genes in plants versus animals, although it has become apparent that even these modest excursions into genetic engineering are meeting with unanticipated resistance (see box below). For example, there are commercially available tomatoes that contain crippled versions of the genes involved in ripening. These transgenic tomatoes can be harvested when they are nearly ripe and sweet and then transported to market without damage because the last steps in ripening, which soften the tomato, are significantly delayed. Once these genetically engineered tomatoes reach their destination and finally ripen, they can be sold and eaten in prime condition. Tomatoes produced by these ripening mutants differ greatly in quality from those of standard strains, which are harvested when green, transported to market, and induced to ripen with ethylene gas (the ripening signal for fruits). This "pick green and spray" method, which does not allow the fruit to stay on the plant during the critical ripening phase when it normally becomes sweet, results in tasteless tomatoes bearing almost no resemblance to those grown in one's own garden. Similar ripening mutants and related innovations are likely to lead to the generation of new and better varieties of other fruits. Another factor accelerating the use of transgenic technology in plants is that many agriculturally important plants can be propagated asexually by runners or by making grafts of a sterile plant onto a reproductively competent plant. These alternative forms of propagation make it possible to raise sterile strains of plants. Seedless navel oranges are a good example of this type of propagation.

Perhaps the most important reason that plants are experiencing the first major economic impact of developmental biology is that genes regulating development control agriculturally relevant traits such as the time to flowering, numbers of seeds produced, growth rate, shape, seed content, and crop yield. One reflection of this connection between basic research and agriculture is an exponentially growing number of patents that are being filed for technologies controlling various valued traits in the field. A matter of great practical significance regarding flowering plants is that technologies developed in model organisms such as mustard can be easily transferred to economically important crops. This fact derives from the relatively recent appearance of flowering plants during evolution (i.e., 100–150 million years ago) and the consequent similarities in the genetic control of flowering in plants that superficially appear quite different.

Considering that vertebrates and invertebrates last shared a common ancestor 600–800 million years ago and that many basic developmental mechanisms have remained unchanged during that time (see Chapters 3–6), it is not surprising that development of flowering plants, which are less than one-quarter of the age of segmented animals, is controlled by highly related genes. Another point to consider in this regard is that many cultivated plants have been created by breeding

Charles Darwin considered the abrupt appearance of flowering plants during evolution as an "abominable mystery." Although the reason for the rapid appearance and diversification of plants is still a matter of considerable debate, it has been noted that the diversity of winged insects also underwent a major evolutionary explosion as flowers entered the scene. The coincident diversification of flowering plants and flying insects has led to the hypothesis that the evolution of flowers and flying insects were coupled. Perhaps flying insects provided an efficient mechanism for pollinating flowering plants. An interesting question left unanswered by this hypothesis is: Which came first, the flower or the flier?

within only the last 10,000 years—an evolutionary blink of the eye. Perhaps the most dramatic example of the degree of morphological diversity which can be generated by agricultural breeding is that plants such as broccoli, cauliflower, cabbage, kale, and brussels sprouts all belong to the same species of plant. Although these plants look very different and have leaves of various shapes, sizes, and coloring, they can all be crossed to each other as can different breeds of dogs. Our experimental wonder the mustard plant is also closely related to this varied vegetable species. Because of the close kinship of flowering plants, technologies developed in the mustard plant can be directly transferred to many economically important crops.

One clear example of the potential for technology transfer from the lab to the field is the production of canola oil. The canola plant, which is a hybrid between two closely related mustards, is a patented variety that produces one of the healthiest plant oils, canola oil. Because of its superior qualities, canola oil is used in a wide variety of food items including cooking oils, pan sprays, fried foods such as potato chips, and many other high-consumption food items. The canola seed oil business is valued at over $12 billion per year. One of the major problems facing the canola oil business is that about half of the canola seed crop is released from pods prematurely and lost before harvest. This amounts to more than $5 billion of canola seed dropped on the ground. As mentioned in Chapter 8, Martin Yanofsky's lab has identified genes involved in releasing seeds from mature pod-type fruits. His work has shown that mutant mustard plants lacking the *shatterproof* genes or misexpressing the *FUL* gene are blocked in the developmental process of shattering (i.e., seed release). On the basis of his observation that misexpression of the *FUL* gene in the valve margin of mustard plants prevents shattering, he is currently testing whether similar misexpression of *FUL* in canola will prevent seed release in this commercially valuable relative of mustard. The value of this technology is quite significant when one considers that a potential crop savings of $5 billion per year amounts to half of the annual federal budget for all of science. It is not unreasonable to anticipate that the currently patented technologies for controlling flowering, seed release, shoot growth, and fruit ripening could soon exceed $1 trillion per year in value.

In addition to generating new beautiful varieties of flowers, the ability to control flowering is also likely to have a profound economic impact on the production of wood and paper products. It is estimated that approximately 15–20% of a tree's energy is devoted to flowering (or producing cones in conifers). Although this modest loss of energy may not seem like much, this fact has enormous economic impact on the total production of wood and paper products. There is a very simple solution to this problem, which is to generate nonflowering varieties of trees. This strategy is practical on a commercial scale, since many trees can be propagated by cuttings. For example, it is possible to take cuttings from a single tree and generate more than 1000 genetically identical tree clones within six months' time. The technology for generating nonflowering plants already exists. If one fuses the regulatory region of a floral meristem identity gene such as *AP1* to a toxic gene product, it will kill all developing flowers. This type of technology is just beginning to be used and will soon lead to significantly increased wood production. There is also a very positive side of this technology from an environmental point of view. If we can grow fewer and better trees on tree farms, it will eventually become unprofitable to ravage our national forests for wood. This will also translate into less water pollution, and perhaps to the recovery of some endangered fish populations such as salmon.

Another very important application of molecular biology to agriculture is developing plant strains that are resistant to disease and pests. Great strides have already been made by generating plants expressing bacterial proteins capable of killing a wide variety of insect pests. The best known of these bacterial insecticides is a protein called BT, which forms crystals in the gut of caterpillars and kills them. This natural insecticide is currently used widely in sprays to combat a wide range of insect pests. This cumbersome spraying technology is being replaced by inserting the *BT* gene directly into plants. Such transgenic plants will be resistant to many devastating insects. Although it is likely that BT-resistant forms of insects will eventually be selected if BT-producing plants are used indiscriminately, there are sensible ways to minimize this problem, and there are a host of other related bacterial proteins with similar properties that can be used to continue the battle. Thus, the use of this single strategy for combating insects is likely to nearly eliminate the economic impact of many insect pests on crops within the next decade. This powerful new technology not only will have a major impact on crop production, but also will greatly lessen farmers' dependence on chemical insecticides, a fact that those opposed to genetic engineering of any kind should bear in mind (see below).

Progress in the arena of pest control is also taking place at breakneck speed on the microbial front. Currently, a wide variety of bacterial, fungal, and viral pathogens are known for every crop species, causing tremendous decreases in crop yields worldwide. Plants, of course, have developed their own strategies for fighting off pests and pathogens, including the use of receptor molecules similar to the CLV1 receptor discussed in Chapter 7 to "sense" the presence of a pathogen and to shut down infection quickly. Many of these receptors, or resistance genes, have now been cloned and characterized from diverse plant species. Use of these genes to develop resistant varieties of all crops will now take place at a rapid pace, resulting in huge increases in crop yield and corresponding decreases in pesticide use.

Knowledge of plant development is likely to have a significant impact on combating many microbial pathogens such as bacteria which interact with plants and ultimately invade them by exploiting normal developmental mechanisms such as production of plant hormones. For example, bacteria that can infect plants and induce the formation of tumors produce plant hormones that trick plant cells into proliferating. A very hot field in plant biology is the study of interactions between plants and pathogens, with the aim of generating strains of plants resistant to these disease-causing agents. Because such strategies rely heavily on a detailed knowledge of normal plant development, the next generation of plant technologies is likely to arise from many of the basic discoveries in plant development being made today.

Malthus is credited, perhaps more than he deserves, for prompting both Darwin and Alfred Wallace to propose their similar theories of natural selection.

In addition to the staggering economic impact of these sophisticated biological solutions to pest control, there is a very important human factor to consider. It is predicted that the population of the earth will double within the next 50 years. At our current rate of food production, we will not be able to sustain this increased number of people. This problem of exponential population growth (human or animal) in the face of constant resources such as food was articulated by Thomas Malthus, a political economist, in *Essay on the Principle of Population* (1798).

There is significant concern that this lag between food production and population growth could result in a catastrophe of enormous magnitude. The best solution to this dilemma would obviously be to control population growth so that we can avoid the long-term consequences of having too many people suffocating our little planet. Those who are well acquainted with demographic and political realties, however, assert that this worthy goal is not realistic in the short term and that no matter what we do, a doubling of the world population is inevitable within the next century. There are only two possible outcomes of this nightmare—either there will be large-scale starvation, or we will struggle to boost agricultural yields to cope with the need for increased food production. If we take the humane course and do what is possible to boost food production, there will be an ever-increasing pressure to use pesticides and chemicals to control crop losses. This is where the real human value of genetically engineered plants is likely to be felt, since it should be possible not only to create crop strains that are resistant to pests, but also to generate varieties of crops that can grow in currently unfarmable land by controlling traits such as time to flowering and growth rates. By manipulating plants so that they can grow in salty soils, in drier conditions, or in colder climates, we should be able to increase the effective area of cultivatable land as well as crop yields. Those who are opposed to genetic engineering should think long and hard about this point.

■ A View on Genetic Engineering ■

Given the potentially great environmental benefit of using genetically engineered plants such as those producing an organic pesticide widely used today in sprays (e.g., BT), it is somewhat surprising that there has been such a strong negative reaction to genetically engineered food plants. This reaction has been particularly intense in Europe and Japan, but is also mounting in the United States. For example, major food-producing companies in Europe and Japan have stated categorically that they will not use genetically engineered plants (e.g., BT-producing soy beans, corn, or hops). In the United States, a lawsuit has recently been filed by a coalition of environmentalists and organic food growers alleging that there has been a conspiracy among seed companies to force farmers into using a small number of genetically engineered plants. Also, major baby food manufacturers have promised not to use any genetically engineered plant in their products, even a major company that is owned by a leading plant-designing biotech company! Is this reaction against BT-producing plants, or genetically altered plants in general, reasonable? Are there really legitimate hidden risks behind this new technology that scientists are keeping from the public, or is this more of an emotionally based response against the fear that a new wave of eugenics is under way?

It is my firm view that although there are real issues of concern with this new technology, as with all technologies, the scale of public outcry is greatly exaggerated. Balanced against the real issues regarding the genetic manipulation of plants should be an appreciation that the new technology offers tremendous potential benefits, not only for increasing crop yields and food quality, but also for reducing the damage to the environment that our current agricultural practices create. Can we continue to pour unlimited quantities of chemical pesticides and inorganic fertilizers into the environment? What will be the consequence of this current practice as we look toward the future? How can we save the irrigated lands that year after year accumulate increasing concentrations of toxic salts? Should we be more concerned about eating food products that have been extensively treated with toxic chemicals?

Humans have been engaged in genetic engineering of plants ever since they started breeding plants for agricultural purposes. Breeding is just a crude form of genetic engineering. To combine desired genetic traits of two plants, breeders cross plants that carry these useful traits and then select progeny for many generations which faithfully propagate these traits together. A problem, which is very well recognized among traditional plant breeders, is that in addition to combining two sought-after traits, intercrossing can result in the unintentional introduction of negatively interacting traits. These unanticipated defects (e.g., vulnerability to pests or pathogens) may not become immediately apparent, and in some cases have led to major practical disasters such as massive crop loss (see below).

Although, in principle, there could be similar problems arising from genetically engineered plants, they are much less likely to occur because through genetic engineering single genes carrying known traits are introduced without altering any other gene in otherwise hearty strains of plants which have been proven in the field. Genetic engineering is nothing more than a precise form of breeding, and therefore will be much more effective and have fewer side effects on average than traditional less well controlled breeding.

Although I believe that much of the reaction against genetically engineered plants results primarily from a lack of information and understanding by people who in principle should be among its greatest supporters (e.g., environmentalists, growers and consumers of organic foods, and those interested in feeding the hungry), there are real issues to consider regarding implementation of this new technology. Unfortunately, these important questions are being obscured by the current debate. For example, a very serious concern is whether introducing genetically engineered plants carrying only a single form of BT into the field is a good idea, because it is known that BT-resistant strains of insects can arise. The current situation with BT plants could be likened to the use of antibiotics, many of which have become nearly useless due to the appearance of multi-resistant bacterial strains. There are several ways to reduce the problem of resistance, and particularly of compound resistance. The first, which is also relevant in the case of antibiotics, is that BT strains of plants should only be used when necessary (e.g., when heavy pesticide use would otherwise be required) and that different BT strains of plants should be used and rotated through a given region. The second approach, which would take more patience on the part of seed-designing companies eager to see their crops in the field and making a profit, would be to make plants that carried different combinations of at least two distinct BT-type toxins (there are more than a hundred BT variants currently known). If crops carrying different combinations of BTs were grown in an interspersed pattern, the likelihood of a devastating outbreak of a resistant strain of pest would be greatly reduced.

Another real worry, which has been recognized but eclipsed in the furor, is that genetic engineering could lead to the worsening of an already serious reduction in the genetic diversity of commercially grown crops. Loss of genetic variability has already been occurring at an alarming pace as a result of currently accepted plant-breeding practices, although the public seems to be unaware of this potentially significant problem. Other factors such as mechanisms to ensure proprietary control over crops in the field have also contributed to the reduction in genetic diversity of crops. The great concern here is that these genetically homogeneous crops could be extremely vulnerable to the appearance of a new pathogen, which could devastate crops if all varieties of a crop carried the same genetic susceptibility to infection. A disaster of this type did indeed take place in 1970 when a strain of corn was introduced which eliminated the need for farmers to remove the corn tassels manually prior to pollination. Unbeknownst to those who developed this strain of corn, this trait also made plants particularly sensitive to infection by a fungus known as southern corn leaf blight. A particularly wet season in 1970 revealed this previously unknown vulnerability and led to widespread crop losses totaling nearly $1 billion nationwide.

Legitimate concerns with genetically manipulating crops such as those mentioned above such as should be kept in perspective, however, given that the use of BT-producing plants will greatly reduce the use of extremely harmful pesticides, and that awareness to the limitations of genetic engineering can lead to effective methods for mitigating most problematic by-products of using genetically designed crops. The central point that often seems to be missed in current diatribes against genetically engineered plants is that we really are faced with a choice. We can either go on using pesticides, which are becoming ever more toxic since insects can also become resistant to harsh chemicals, or begin using BT-producing plants in a sensible fashion. I would much rather eat BT (which we already do anyway—it is used heavily in the United States as a spray) than chemical pesticides, particularly given that vegetables and fruits sold in supermarkets often have been grown abroad and sprayed with extremely toxic pesticides that are not legal in the United States.

It is good that there is vigorous public discussion on using genetically engineered plants, and this debate should continue. Hopefully, it will become better appreciated through such dialogues, however, that the alternative to using genetically altered crops such as BT plants will be a growing reliance on harsher chemical pesticides. As we look toward the future, we must develop a sustainable form of agriculture that not only feeds our growing population, but also protects our environment for future generations. Genetic modification of crops offers tremendous opportunities to achieve these critical goals. Again, it should be emphasized that it will almost certainly be necessary to double food production in the next 50–100 years to avoid massive starvation. The recent introduction of golden rice, a strain that carries an entire enzymatic pathway required for Vitamin A biosynthesis, is a case in point. The potential of genetic engineering to help confront this problem should not be overlooked. We are at a crossroads in terms of making choices for the future path that agriculture will take. Perhaps it is worth taking a step back to evaluate not only emerging technologies, such as the use of genetic engineering technologies to modify crops, but also all current agricultural practices that we have so easily come to accept. What are the advantages and disadvantages that each approach offers? If we enter into an enlightened discussion of this important new issue, rather than react reflexively to misconceptions, we will all emerge stronger and better prepared for the future by the exercise.

THE BRAVE NEW WORLD

I want to end this chapter and the book by looking somewhat further into the future. What might be the consequences of our knowledge of molecular biology and development on our children in the next 50–100 years, and what are the longer-term implications of our being able to control fundamental developmental processes? In the near term, there are some very interesting and important issues that we all should consider as a society because they raise potentially profound ethical questions. One might compare the state of biological knowledge and practical technical potential we have today to that of physics in the first part of this century, when it become evident that, in principle, nuclear energy could be harnessed for a variety of purposes including creation of weapons of mass destruction. It is certainly the case that the ethical integrity of physicists from this time has been questioned as a result of their role in developing nuclear weapons. Development of the first atomic bomb was a complex issue, which took place in the context of a very unusual time in history. Biologists, however, should be careful not to leave a similar legacy of perceived irresponsibility, particularly as there are no potentially mitigating circumstances in the present to absolve us of the larger consequences of our actions or inaction.

Jurassic Park

With the full genome sequences of a growing number of organisms in hand, it will probably become feasible in the not-too-distant future to bring recently extinct organisms back to life. The dodo bird would be a good first project of this type because it has only recently become extinct, and tissues of these birds exist that could be used to extract relatively high-quality DNA samples. As in the fiction of *Jurassic Park*, knowing the full genome of sequence of the extinct dodo should make it possible to revive this lost species. Although a project of this nature is still within the domain of science fiction, the idea would be to take an egg of a living bird related to the dodo (e.g., a pigeon) and replace its genetic material with that of the dodo. With ever-improving genomics methods and schemes for cloning animals, this should be possible to do, at least through a series of approximations by successively replacing portions of the pigeon genome with DNA from the dodo. Although the logistics of converting the genome of one species into that of another are currently daunting, this technical barrier will most likely be surmounted in the near future as the progress in genomics continues to accelerate. I hasten to point out, however, that while I believe we will be able to revive recently lost species such as the dodo in the not-too-distant future, we should not rely on this hypothetical technology to save us from our current irresponsible path of destroying large numbers of species and their habitats. Resuscitating even one organism would be a major undertaking, even granted significant technical advances in the future, and for all we know, it might never become practical. Therefore, we should make every attempt to keep currently existing species alive and well by protecting them and their environment.

Although it will be a great day when we succeed for the first time in resuscitating one of nature's recently lost treasures, the obvious big challenge for such a technology will be to go for broke and try to recreate long-extinct creatures such as the mighty T-Rex. Outside of the ethical issue of whether or not this would be a good idea, and I would argue that author Michael Crichton greatly exaggerates the difficulty of confining such animals in Jurassic Park should they be brought back, the real problem is that it seems very unlikely that we will ever obtain sufficiently well-preserved DNA samples from dinosaur fossil remains to determine their full genome sequence. This is a very serious technical barrier for those salivating at the thought of a Jurassic Park or, more accurately, a Cretaceous Park (if we want to see a T-Rex). My guess is that it will be possible to solve this problem, at least in part, by the knowledge we acquire from systematically comparing genomic and developmental data from different species. In the relatively near future, as a result of the various genome projects, we should be able to predict many features of a new organism from its complete DNA sequence. We may even be able to predict the behavior of an animal from such DNA blueprints, if we make similar strides in understanding the nature and function of the brain. Whether such significant progress will be made on decoding information in the brain is the most uncertain element in this equation, since, in many respects, the brain is the last great frontier of biology. In any case, it should be possible from first principles to design an animal that at least looks like a T-Rex. We probably also could make our fearsome creation behave in ways consistent with our current biases of how such animals moved and hunted. Unless high-quality T-Rex DNA can be found, however, it is unlikely that we will ever be able to do more than create a living version of a dinosaur fossil, since without the actual blueprint of the animal we will always be guessing as to its nature. So, if we ever do make a Jurassic or Cretaceous Park, it is likely to be just an entertainment park that tells us little about how real dinosaurs lived and dominated the earth. Even so, it would be a lot of fun to visit such a show. Who would ever go to the zoo again?

The Island of Dr. Moreau

Another likely application of our knowledge of genomics and development within the next century is the creation of new plants and animals, either by making hybrids between species that are not possible today, or by creating new creatures from first principles. The first part of this excursion into fantasy should not be very difficult. Plant and animal breeders have made hybrid species for years, and new varieties, particularly of plants, are being generated at an ever-increasing pace using nothing fancier than selective breeding techniques. Obviously, if we can alter an organism's genome at will, the possibilities will become nearly unlimited for making hybrid species. When this becomes a reality, so will the haunting script of *The Island of Dr. Moreau*. For those deprived souls who have not seen this movie classic based on the novel by H.G. Wells (either the original 1933 version with Charles Laughton and Bela Lugosi, the 1977 version with Burt Lancaster and Michael York, or the 1996 remake with Marlon Brando and Val Kilmer), the plot involves a crazed scientist (Dr. Moreau) who is obsessed with making hybrids between humans and other species, such as members of the ape or cat family, with the hope of creating a perfect humanoid species endowed with the virtues of humans but purged of the weakness of character that we seem unable to shake despite our technological prowess. The protagonist, a lost traveler, is initially horrified, then intrigued, and finally terrified by the bizarre collection of creatures Dr. Moreau has created. For a brief moment, Dr. Moreau does seem to be on to something, but then he loses control of his semi-civilized hybrid creations, and things get very ugly as they go on a savage killing spree that ends Dr. Moreau's dream-turned-nightmare.

There are only two significant technical hurdles to overcome that currently make it impractical, as well as unethical, to realize the fantasy of Dr. Moreau. First, it is not possible to fertilize the egg of one species with the sperm of another. Second, because genes are not ordered in the same sequence in different species, and are typically not organized into the same numbers of chromosomes, the hybrid organisms created by such a cross would be sterile. The basis for this sterility is that such hybrids could not mix and match genes from mom and dad, which is a necessary step in making germ cells such as eggs or sperm. There are a few exceptions to the mating incompatibility of different species, but even in these rare instances, the progeny of the cross (e.g., ass × horse = mule) are typically sterile. It is somewhat easier to make hybrids in plants because one can make a cross between different species, take the embryo out of the seed, and culture it into a plant. Some important cultivated crops seem to be the product of such cross-species hybridizations. For example, wheat is believed to have arisen as a hybrid from three different cereal species approximately 7000 years ago in the Nile Valley.

Progress in methods for manipulating genomes and foreseeable advances likely to occur in molecular biology will probably provide the tools necessary to surmount the current biological obstacles to widescale hybridization of animals and plants. Recent progress in understanding the mechanisms by which sperm recognize eggs of their own species indicates that this recognition is remarkably simple. The sperm adheres to the egg and fuses with it when a recognition molecule carried on the surface of the sperm binds to a matching receptor from its own species present on the surface of the egg cell. Thus, in principle, all one would have to do to allow a human sperm to enter a mouse egg would be to make a strain of mice in which the egg carried the human sperm receptor. To solve the gene organization problem, which would be required for making fertile offspring, the genes of one species would have to be reordered and regrouped into chromosomes to mimic the organization of genes in the other species. If the two species belonged to the same family of animals or plants, this reorganization might be accomplished by making as few as 100–500 cut-and-paste changes in one of their genomes. This scale of genome manipulation is beyond the reach of current technology, but should be quite possible a century from now.

What could we do if we were armed with the ability to make hybrid species at will? Some crosses might be interesting from an evolutionary point of view. Perhaps we could recreate some missing links by crossing two species we believe to have descended from a common ancestor. For example, amphibians such as frogs split off from bony fishes some 300 million years ago. If we crossed a frog and a fish, we could see what a "frish" looked like and how it developed. We also might learn something about the nature of coordinated changes that need to be made in order to sculpt a new form of life from an existing species. Outside of the modest scientific insights likely to be provided by experiments of this kind, there would be practical applications for creating optimal forms of domesticated animals. For instance, hybrid animals resulting from crosses between a goat and cow or between a chicken and ostrich might be useful, respectively, for producing a tasty new type of "gowt" milk or larger "chickrich" eggs. Perhaps more generally, one could replace a limited set of genes in one organism with those of another. For example, if one replaced all the cow genes involved in milk production with those required for making human milk, commercial amounts of human breast milk could be produced that also contained human antibodies to common infectious agents. Given the nutritive and protective advantages of human milk versus synthetic baby formulas, this engineered product would most likely be very good for babies and a big hit in the marketplace.

Genetic Screening, Gene Therapy, Cloned Humans, and Targeted Evolution

Let's return briefly to the real world. It is now practical, on a very limited scale, to check the genetic makeup of a fetus by methods such as amniocentesis and decide whether to continue or end the pregnancy on the basis of whether or not the child is likely to suffer from some debilitating condition such as chromosomal abnormality or a genetic disease. This type of genetic prescreening also is being done on embryos grown in vitro. For example, a colleague and his wife are both carriers for a mutation (i.e., are $m/+$) causing Marfan's syndrome, which is a connective tissue disorder causing various problems including elongated bones. Marfan's syndrome has been in the headlines recently as it has been shown using modern forensic techniques that it afflicted Abraham Lincoln. The couple used in vitro techniques to fertilize eight eggs. The resulting embryos were grown for a period in the laboratory and a single cell from each embryo was removed and tested for the defective *Marfan* gene. It was determined that three of the eight embryos were free of the mutation (i.e., were +/+). These three eggs were then transplanted into the mother, and the result was a successful pregnancy. The other eggs were discarded. The age of eugenics is already here.

With help from the human genome project, it will soon become practical to screen for all known genetic diseases, which currently total more than 5000. There are obvious ethical and social issues that come with this technical advance, because it will also be trivial to prescreen embryos for a wide variety of genetically determined characteristics other than defects associated with disease. Thus, it will also be possible to evaluate traits such as sex; height; strength; eye, hair, and skin color; the tendency to gain weight; predisposition to alcoholism; intelligence; resistance to particular diseases; etc., prior to embryo implantation. One concern regarding the use of genetic prescreening, should this practice become widespread, is that important sources of "good" genetic variability might be closely linked to the "bad" genes being so carefully rejected. If there were a systematic attempt to eliminate these bad genes from the human population by genetic prescreening, the neighboring favorable traits might be lost before we knew they even existed. Because it seems highly unlikely that it would be technically or politically possible to cleanse our "dirty" DNA in such a total fashion, this scenario may not be a significant worry in the short term; however, it should be borne in mind that even limited selection against a bad gene might also result in loss of a "good" one. This picture is becoming frighteningly similar to that painted by Aldous Huxley in *Brave New World,* only in Huxley's work of fiction the technology required to create different strains of human beings had not yet been invented.

As described above, it also is now possible to change defective genes in cells taken from an individual afflicted with a genetic disease such as β-thalassemia, correct the genetic defect, and reintroduce these repaired cells into the affected person. In this way, the disease could be cured for the lifetime of that person. It is hard to imagine any compassionate person arguing against the merits of doing this—society at large would not be at risk and people faced with a debilitating disease could be cured. Similarly, gene therapy could be used to cure cancer, treat diabetes, and possibly halt the course of various autoimmune diseases such as multiple sclerosis or lupus erythematosus. A variety of start-up biotech companies have such laudable goals as their major focus.

In Chapter 1, we discussed Gurdon's famous frog cloning experiment (i.e., making genetically identical copies of an individual) and the recent popularized application of this method to clone a sheep named Dolly. Successful cloning of other animals has also been reported. It is very likely that cloning any mammal, including humans, will soon become technically trivial. Currently, it is routine to make specific changes in a gene in embryonic mouse cells, inject these altered cells into a mouse blastula-stage embryo, and generate living and fertile mice carrying this modified form of the gene in place of the normal gene. Recently, a human embryonic cell line equivalent to that used to generate mutant transgenic mice has been established. There you have it. We now possess the tools to change our genetic makeup and not only correct defective disease-causing cells afflicting our bodies, but also transmit these changes to our children. In addition, if we want to have a bunch of talented individuals such as Einstein and Mozart, we could make thousands or millions of copies of this great being. Is this science fiction? Not really, only modest technical advances will be necessary for making this possible within the next few years. We are building some pretty powerful tools. The question is, will we make good or bad uses of them? This is a good time to open a dialogue between scientists, politicians, and lay people about the ethical implications of human engineering. Whether we like it or not, future generations will judge us by how we handle our new-found abilities.

Regarding this last and most coveted of traits, studies in a variety of animals indicate that aging is a genetically controlled process. Many scientists who study aging do not think it unreasonable to increase human life expectancy to 150–200 years.

For a fantasy finale, let's take a trip 500–1000 years into the future along the lines of *First and Last Men* by Olaf Stapeldon. Within the next century it becomes possible to change genes, introduce them into eggs, and create children with any desired genetic trait. Of course, this technology first becomes available to the wealthiest members of society because in its early days, this technology is quite pricey. The rich are happy to pay the price, however, so that their kids will be disease-free and blessed with high intelligence, physical strength, coordination, beauty, and of course, longevity.

The resulting supertots eventually go to the best schools, as the privileged classes always have, and then become the leaders of their day. They have the distinction, however, of being the first generation of the ruling class to be truly genetically superior to those they govern, a fact which gives them a secure grasp on power felt previously only by the great rulers of history.

These superhumans are not satisfied for long. They want more—to be smarter, to see and sense new things, and to go where nobody has gone before. They develop microchips that can interface with the human nervous system.

These neurochips store massive amounts of data—whole encyclopedias of information are available at an instant. As this increased memory capacity necessitates much enlarged cortical regions of the brain to handle the enormous flow of new information, appropriate targeted changes are made to genes controlling development of the cortical regions of the brain to enlarge these areas. Over the next few generations, these bionic creatures become truly superhuman. They can think orders of magnitude faster than anyone alive today, create forms of music we could not comprehend, and become immortal. Old or malfunctioning chips are replaced with more powerful designs, and aging biological tissues are similarly replaced with more efficient and durable tissue lines. They redesign themselves and then yet another generation of more sophisticated creatures. Natural selection is definitely over. These brave new *Homo electricus* are now in full control of their destiny. When they peer back through the fog of time at the dim-witted creatures, the last of the biologically derived forms of humans, who made the first genetic changes in humans, these omnipotent beings ask themselves, how could such pathetic animals have been our ancestors? What happens next is a blur as this species vanishes in a flash deep into the future, perhaps to join similar creatures from distant worlds. The funny thing is, when they finally meet the first alien form of life, it looks just like them. I guess that there really aren't that many ways to skin a cat. Too weird? Maybe, and then again, maybe not.

A primitive form of this bionic technology exists today to aid people who have been blinded as adults to see—at least to distinguish light and dark. Similarly, cochlear implant technologies are being developed which are hoped to restore hearing to the deaf.

This final fantasy scenario is just that—my stab at a little science fiction. On the more serious side, however, many, if not all, of the technical advances featured in this wild story are likely to become reality. It is difficult to believe that future generations will never take part in their own redesign. Using the analogy of weapons of war, it is worth noting that there has never been a significant weapon that has not been used at least once in combat. Is it conceivable that we will never tamper with our genetic material? If someday we or our descendants choose such an auto-evolutionary path, the big question is how this could be accomplished in a fair and ethical fashion. Maybe the answer is that it cannot be done ethically, and that if it does happen, it will be the work of villains—at least according to us. What should we do now? Should we even bother thinking about these still abstract issues? Should we plant our feet firmly and forbid any kind of human engineering, and let the next generation worry about the problem? Should we go full speed forward and meet our destiny, uncertain and possibly horrifying as it may be? I do not pretend to have answers to any of these troubling questions. Although I believe that it is impossible to stop the progress of science, I also think this is a good time for us to step back a moment and ask ourselves, What are we doing and where do we want to go? We have landed on the naked shores of the brave new world, and we need a plan for the future we wish to create.

Glossary

ABC model A model of flower development to explain the behavior of three different classes of floral patterning mutants (i.e., A, B, and C mutants). It proposes that A, B, and C genes function in a pairwise fashion to specify the four floral organ identities: sepals (A function alone), petals (A plus B functions), stamens (B plus C functions), and carpels (C function alone).

achaete-scute (AS-C) Genes encoding transcription factors required for development of the nervous system. In *AS-C* mutants, formation of the nervous system is compromised. AS-C expression is actively excluded from the dorsal region by Dpp signaling.

Activator A transcription factor that activates expression of a gene (turns the gene on).

agamous (AG) A C-function floral homeotic gene expressed in the center of the floral meristem in cells, giving rise to stamens and carpels. In *AG* mutants, sepals are transformed into carpels.

Agricultural breeding Combining desired traits from two different plant varieties by crossing the strains and identifying progeny that carry both traits.

Amino acids The 20 subunits from which proteins are built.

Animal hemisphere The darkly pigmented half of the oocyte from which the ectoderm and mesoderm derive.

Aniridia A human disease associated with eye defects similar to those in *Small eye*-deficient mice in which the vertebrate *pax6* gene is mutant.

Antennapedia A fly gene encoding a homeobox transcription factor that defines the second thoracic segment (T2) of flies, which has wings and legs. Mutants deficient for *Antennapedia* lack wings due to the transformation of T2 into T1, which is non-wing bearing. Mutants misexpressing *Antennapedia* in the head have antennae transformed into legs.

Anterior–posterior (A/P) axis The anterior (head)–posterior (tail) axis of an animal. In the frog embryos, the anterior–posterior axis of the ectoderm forms parallel to the animal–vegetal axis of the early egg.

A/P organizer A stripe of Hh-responding cells that lies in the anterior compartment of the fly wing imaginal disc and directly abuts the anterior–posterior boundary.

apetala1 (AP1) An A-function floral homeotic gene expressed in an outer ring of floral meristem in cells, giving rise to sepals and petals. In AP1 mutants, flowers are replaced by leaves and an adjoining shoot meristem, which forms a branch. The result of this transformation is the production of a highly branched plant with leaves but no flowers.

apetala3 (AP3) A B-function floral homeotic gene expressed in a central ring of floral meristem cells that overlaps the domains of cells expressing AG and AP1 and gives rise to petals and stamens.

Apical–basal axis (A/B) The vertical axis of a plant extending from the shoot (apical end) to the root (basal end).

Apical ectodermal ridge (AER) A specialized group of cells at the junction between the dorsal and ventral surfaces of vertebrate limb buds that plays a role in patterning the dorsal–ventral axis of the limb.

Apical shoot meristem A small group of self-renewing cells located at the apical tip of a plant from which all above-ground structures of the plant derive.

apterous (ap) A fly homeobox gene required for the outgrowth of appendages and for formation of dorsal wing cells. Mutants completely lacking the function of *ap* make no wings at all.

Autoactivation The ability of a gene to activate its own expression. For example, Dpp signaling in the nonneural ectoderm of a fly embryo can autoactivate, and, if unopposed, can diffuse and spread into the neuroectoderm where it will suppress expression of the AS-C family of genes.

Auxin A plant hormone that can stimulate proliferation of plant cells and influences establishment of the apical–basal axis by favoring basal cell fates (e.g., roots).

β-Catenin A transcription factor required for dorsal–ventral patterning of the frog embryo that is concentrated in the nuclei of dorsal-most cells of the early fertilized embryo.

Bacterial clones Isolated colonies of genetically identical bacteria containing plasmids.

Bacterial toxin (BT) A protein of bacterial origin that acts as a potent insecticide by forming crystals in the gut of caterpillars and killing them.

bicoid A gene encoding the maternal morphogen (Bicoid) which is provided by the mother to pattern the anterior–posterior axis of the fly embryo.

Bithorax A homeotic gene encoding a homeobox transcription factor that is active in the third thoracic segment (T3). Bithorax mutants have two sets of wings due to the conversion of T3 to T2.

Blastoderm embryo An early embryo consisting of a hollow ball of morphologically similar cells prior to gastrulation.

Body axes The primary perpendicular axes (i.e., anterior–posterior and dorsal–ventral) of a multicellular organism such as an animal or plant.

bone morphogenetic protein-4 (BMP4) The vertebrate counterpart of the gene encoding the Dpp signal in flies. In early embryos, *BMP4* inhibits neural development and in appendages plays a role in anterior–posterior patterning.

bride-of-sevenless (boss) A gene encoding a ligand for the Sevenless receptor that is expressed in the R8 photoreceptor. *boss* mutants specifically lack the R7 photoreceptor in the eye.

Cambrian extinction A massive extinction of life forms during the early Cambrian period 550–600 million years ago. Of the many phyla that evolved in the late pre-Cambrian period, only a few survived this massive extinction to give rise to modern-day animals.

Carpel The female reproductive organ onto which the pollen is deposited to begin the life cycle of the flowering plant.

Carrier An individual having one mutant copy (m) of a given gene and one good copy (+) of that gene. A carrier appears normal because generally mutations have no effect if the organism possesses one good copy of the gene.

Catalyze To accelerate a chemical reaction by bringing together and promoting chemical interaction between two compounds.

cauliflower (CAL) A gene closely related to *AP1* that functions together with *AP1* to prevent floral meristems from developing as primary indeterminate shoot meristems.

Cell The living subunit of all organisms that is bounded by a membrane and contains the genetic information stored in the form of DNA.

Cell division The process by which a cell separates into two daughter cells. A complex organism is created from a single egg cell by 20–30 cycles of cell division.

Cell proliferation Multiplication of cells by division.

Cell wall A rigid protective layer of plant cells that prevents cell migration.

Chloroplast An intracellular structure in plants that performs photosynthesis and contains the light-absorbing chlorophyll molecule.

Chordin The vertebrate counterpart of the fly *sog* gene that is produced by cells in the neuralizing dorsal Spemann organizer. Chordin promotes neural development indirectly by blocking the neural suppressive activity of BMP4.

Chromosome A long string of genes (1,000–10,000) located in the nucleus of cells that consists of DNA (carries genetic information) and proteins bound to the DNA (transcription factors).

clavata1 (CLV1) A gene encoding the receptor protein Clavata1 (activated by the Clavata3 signal), which is required to restrict the number of proliferating meristem and is expressed in the L2 and L3 layers of the apical shoot meristem. *CLV1* mutants have greatly overgrown meristems.

clavata3 (CLV3) A gene encoding a signal, Clavata3 (activates the Clavata1 receptor), which is required to restrict the number of proliferating meristem and is expressed exclusively in the L1 layer of the apical shoot meristem. *CLV3* mutants have a great overgrowth of the shoot meristem.

Cloning (of an organism) The process of creating an exact copy of an organism.

Coding region of gene The region of the gene that contains the information necessary for synthesizing a protein product.

Compartments Domains of cells that do not intermix with each other (e.g., cells of the then anterior and posterior compartments of the fruit fly wing which are separated by the anterior–posterior boundary).

Conserved Some aspect of an organism that has remained unchanged during the course of evolution.

Cortex (L2 layer) The middle of the three germ layers partitioning the radial axis of a plant, providing rigidity and substance to the plant.

Cotyledon Seed leaves of a plant embryo. Embryos of dicot plants have two cotyledons and those of monocots have one.

Cup-shapedcotyledon1 (CUC1) A gene functioning in concert with *CUC2* to split the cotyledon primordium into two separated parts.

Cup-shapedcotyledon2 (CUC2) A gene expressed in a stripe of cells bisecting the apex of the globular embryo that is required to split the cotyledon primordium into two separated parts. Double mutants lacking functions of both *CUC1* and *CUC2* genes generate seedlings with a goblet of cotyledon material encircling the shoot meristem.

Cuticle The tough outer covering of the larvae marked by various structures, such as ventral denticles which form in the anterior portion of each segment.

Cytokinin A plant hormone that can stimulate proliferation of plant cells and influences establishment of the apical–basal axis by favoring apical cell fates (e.g., shoots).

Cytoplasm The main cellular compartment bounded by the cell membrane and containing the nucleus. Translation of RNA into protein occurs in the cytoplasm.

dachshund A gene that is essential for initiating eye development in flies. *dachshund* function is required for the *eyeless* expression. *dachshund* mutants lack eyes.

decapentaplegic (dpp) A gene encoding a secreted signal that promotes formation of the dorsal ectoderm and suppresses neural development. In *dpp* mutants, dorsal cells assume a lateral neuroectodermal identity and denticle belts encircle the embryo. *dpp* is also required to pattern the anterior–posterior axis of adult appendages.

Default state A neural state of the ectoderm in vertebrates and invertebrates.

deformed A fly homeotic gene that functions to specify a region of the fly head.

Delamination The segregation of individual neuroblasts from the ectoderm creating a three-layered embryo with a mesoderm on the inside, neuroblasts in the middle, and a skin-forming ectoderm on the outside.

Denticles An intricate pattern of hairs arrayed in rows on the ventral surface of the larval cuticle.

Determinate development A mode of meristem development (e.g., the floral meristem of mustard plants) in which cells proliferate for a limited period to generate the primordium for a structure (e.g., a flower) and then stop dividing to differentiate (e.g., into floral organs).

Dicot plants Flowering plants that have embryos with two seed leaves, or cotyledons.

Diploid cell A cell having two copies of every gene.

distalless A homeobox gene that plays an organizing role in defining the proximal–distal axis of appendages in flies and vertebrates. Flies lacking *distalless* function in legs lack the distal portions of the leg.

DNA *D*eoxyribo*n*ucleic *a*cid is a double-stranded helical molecule consisting of two complementary strands of bases that store the genetic information of all living organisms. DNA is replicated as cells divide.

DNA bases (A,C,G,T) The four subunits of DNA known as bases, which are strung together to make up a DNA strand. The bases are: adenine (A), cytosine (C), guanine (G), and thymine (T). RNA is very similar to DNA, one difference being that in place of the DNA base thymine, RNA contains the base uracil (U).

DNA-binding site A specific sequence of nucleotides where a transcription factor will bind to DNA. The transcription factor recognizes a particular sequence of bases and binds to this preferred site like a key fitting into a lock.

DNA polymerase An enzyme that carries out DNA replication by copying one strand of DNA (the template strand) to make the complementary strand. At each base on the single strand of DNA, the DNA polymerase adds a complementary base to the newly synthesized strand.

DNA replication The copying of DNA. DNA polymerase attaches itself to one end of a single strand of DNA (the template) and moves along the DNA to the other end by adding an appropriate complementary base to create the new DNA strand. DNA replication is required for cell division.

DNA strands DNA consists of two strands of bases wrapped around each other in a configuration known as a double helix. The bases on the two strands of DNA are complementary, meaning they form base pairs. In DNA, A pairs with T, and C pairs with G.

Dolly The name of a sheep that was successfully cloned at the Roslin Institute in Scotland by the process of manipulating mammalian embryos.

Donor An animal from which a part is taken (e.g., a nucleus, cell, or group of cells) to be transplanted into a recipient animal (host).

Dormancy The intervening stage between embryonic development (which occurs in the seed) and adult development (which takes place upon germination of the seedling).

Dorsal A gene encoding the maternal morphogen (Dorsal), which is a transcription factor responsible for patterning the dorsal–ventral axis. Dorsal levels are highest ventrally and lowest dorsally. In embryos lacking function of *dorsal*, all cells along the dorsal–ventral axis assume dorsal identities.

Dorsal surface of leaf The surface of the leaf forming nearest the shoot.

Dorsal–ventral (D/V) axis The dorsal (back)–ventral (belly) axis of an animal.

Double helix The three-dimensional structure of a double-stranded DNA molecule in which the two strands of bases wind around each other and are held together by complementary base pairing.

Ectoderm The outer embryonic germ layer of cells that gives rise to skin and nervous system. In frog embryos, the ectoderm is derived from the animal hemisphere of the egg.

Egg A germ cell that contains the genetic information of the female. Once fertilized by the male germ cell (sperm), an egg becomes an embryo.

Embryo A fertilized egg in its early stages of development.

Embryonic gene A gene whose function is required in the embryo. Both parents contribute embryonic genes to the embryo (e.g., the fly *hunchback* gene). In contrast to embryonic genes, only the mother contributes maternal genes to the embryo (e.g., the fly *bicoid* gene).

Embryonic mutant A mutant lacking the function of an embryonic gene.

Endoderm The inner embryonic germ layer of cells of the embryo that gives rise to gut. In frog embryos, the endoderm is derived from the vegetal hemisphere of the egg.

engrailed (en) A segment-polarity gene required for the formation of the posterior portion of each segment. In *engrailed* mutants, the posterior half of the segment is transformed into a mirror copy of the anterior portion of the segment.

Enucleated A term describing an egg after its nucleus has been removed. This type of egg is also known as the host egg in John Gurdon's classic nuclear transplantation experiment.

Enzymes Complex-shaped proteins that accelerate or catalyze chemical reactions by bringing together and promoting chemical interaction between two compounds.

Epidermis (of animal) Skin.

Epidermis (L1 layer of plant) The outer of the three germ layers partitioning the radial axis of a plant, which protects the plant from the outside world.

Essential gene A gene required for the survival of an organism. Flies have approximately 6000 essential genes.

Ethylene A hormone that promotes growth in cell width over length. Ethylene competes with gibberellic acid (GA), which, conversely, promotes growth in cell length over width.

Eugenics Improvement of the human genetic design by selective breeding for desired traits.

even-skipped (eve) A pair-rule gene encoding a transcription factor that is required for the formation of even-numbered segments. *eve* mutant lacks even-numbered segments.

Express (a gene) To transcribe a gene from DNA to RNA. Different cells express distinct subsets of genetic information (i.e., activate expression of different genes).

eyeless A homeobox encoding gene required for initiating eye development that is required to activate expression of other eye-specific genes. *eyeless* mutants have reduced eyes or no eyes at all. *eyeless* is the fly counterpart of the vertebrate *pax-6* gene.

eyes absent A gene required for initiating eye development in flies. *eyes absent* is required for *eyeless* expression, and *eyes absent* mutants lack eyes.

fibroblast growth factor (fgf) A gene encoding a secreted signaling factor (FGF) that is critical for initiating limb outgrowth and for defining the anterior–posterior polarity of the limb.

Floral meristem A small group of cells forming at the flank of the apical shoot meristem that generates the primordium for a flower and differentiates into the four floral organs (sepals, petals, stamens, and carpels).

Floral organs The concentrically organized sepals, petals, stamens, and carpels of a flower that originate from a floral meristem.

Fruit The seed-containing structure that develops from the basal portion of the carpel following fertilization of the eggs.

Fruitful (FUL) A gene expressed in the region giving rise to the primordium of the fleshy fruit valve that is required for formation of the valve.

FT A gene encoding a protein related in structure to that of TFL which promotes initiation of floral development by opposing the activity of *TFL*.

Furrow A morphologically visible crease in the imaginal disc in the eye that moves across the disc from posterior to anterior during development.

Gap gene An embryonic gene (e.g., *hunchback*) that functions to define a large block of cells along the anterior–posterior axis.

Gap mutant A mutant lacking the function of a gap gene in which a large section of the cuticle is typically missing in one restricted region of the anterior–posterior axis. Gap mutants exhibit large gaps in the cuticle in one restricted region of the anterior–posterior axis but are normal elsewhere.

Gastrulation The organized movement of cells during midembryonic development that creates a laminated embryo with distinct tissue layers.

Gene The unit of heredity composed of DNA. Genes consist of a coding region, which directs the synthesis of a protein product, and a regulatory region, which determines whether a given cell will actively express that gene (i.e., transcribe it into RNA).

Gene clone = DNA clone A single species of plasmid DNA containing an inserted piece of DNA from another organism. A gene clone is propagated in a bacterial clone carrying a single type of plasmid.

Gene cloning The isolation of a gene clone from a collection of clones, or library.

Gene therapy Replacing a defective gene with a normal copy of the gene.

Genetic code The code relating the 64 different possible triplet sequences of four bases in RNA to the 20 different amino acids in proteins.

Genetic engineering Alteration of gene(s) carried by an organism.

Genetic mosaics Animals that are heterozygous for a mutation but contain clusters of homozygous mutant cells referred to as clones.

Genetic screen = mutant screen A systematic hunt for mutations.

Genetics The subfield of biology dealing with gene function. A geneticist typically infers the normal function of a gene by studying the consequence of eliminating the activity of that gene.

Genome The complete genetic blueprint of an organism comprising all the DNA sequences of a cell or organism.

Gibberellic acid (GA) A hormone that promotes growth in cell length over width. Gibberellic acid competes with ethylene, which, conversely, promotes growth in cell width over length.

Globular embryo A spherical morphologically undifferentiated mass of embryonic cells that forms after several divisions of the fertilized plant egg.

Glutamate receptor A receptor protein that can be activated by binding the amino acid glutamate, which functions as a signal in the nervous system.

Glutamate An amino acid that is also widely used as a signal in the nervous systems of diverse animals including humans.

Heart-stage embryos The first stage of plant embryonic development when the embryo becomes visibly polarized. The primordia of cotyledons can be distinguished as the lobes of a heart.

hedgehog A segment-polarity gene encoding a secreted signal that is required for the formation of posterior structures in the segment. In *hedgehog* mutants, the naked posterior half of the segment is transformed into a mirror copy of the anterior portion of the segment, resulting in the production of denticle hairs throughout the segment.

Homeobox The DNA-binding region of a subtype of a transcription factor such as those encoded by the homeotic genes.

Homeotic gene A gene that determines regional cellular identities such as a *homeotic/Hox* gene in animal embryos, which determines segmental identities, or a floral organ identity gene in plants.

Homeotic mutant A mutant lacking the function of a homeotic gene. For example, in a homeotic animal mutant, the identity of a specific segment is transformed into that of an adjacent segment, and in a homeotic flower mutant, two adjacent floral organs are transformed into appropriate organs.

Homology A similarity in a structure or feature present in two organisms that suggests a common evolutionary origin for these organisms and hence of the structure (e.g., the human hand and the bird wing).

Homunculus theory The notion that the sperm carried inside it a complete human form in miniature, which simply grew larger during the course of development.

Hormone A circulating chemical signal in plants or animals that is typically produced by a small set of specialized cells, travels in body fluids, and triggers a response in distant target cells.

Host An animal receiving a grafted part (e.g., a nucleus, cell, or group of cells) from a donor animal.

***Hox* genes** Segment-identity genes of vertebrate embryos that are highly related in structure and function to fly homeotic genes.

Hox4 The vertebrate *Hox* gene counterpart of the fly homeotic gene, *deformed,* which is expressed in the mouse head.

Hox6 The vertebrate *Hox* gene counterpart of the fly homeotic gene, *Antennapedia,* which is expressed in the upper trunk region of the mouse embryo.

Human engineering A new, looming, and controversial technology that will make it possible to alter the heritable human genetic makeup, as well as to clone human beings.

Human Genome Project A major national initiative being carried out in concert in many laboratories to determine the complete DNA sequence of the human genome.

hunchback A gap gene encoding a transcription factor that is required for formation of the most anterior region of the embryo.

Imaginal disc A small group of embryonic cells that gives rise to an adult structure of the fly such as a leg, wing, eye, or antenna.

In situ hybridization A method for determining the pattern of gene expression in an organism or tissue. In situ hybridization permits the experimenter to determine visually which cells in a developing organism transcribe the gene from DNA to RNA.

Indeterminate development A self-regenerating mode of meristem development (e.g., the apical shoot meristem) in which cells proliferate continuously during the life of the plant to provide new cells for growth of the plant (e.g., the central stem) and for formation of secondary meristems (e.g., primordia giving rise to branches and leaves or to flowers).

Induction A change in the developmental course of a cell resulting from that cell receiving a signal from another cell.

Invagination An internalizing cell movement in which sheets of cells fold into a developing structure.

Invertebrate An animal without a backbone, such as insects or worms.

knirps A gap gene encoding a transcription factor that is required for formation of the middle–posterior (abdominal) regions of the embryo.

knolle An embryonic patterning gene required for epidermal development and for assembly and orientation of a platform in the cell upon which new cell walls are built during cell division.

Krüppel A gap gene encoding a transcription factor that is required for formation of the thoracic region of the embryo.

Larva A hatched embryo, which in the case of a fly, emerges from the egg case approximately 22 hours after fertilization. The fly larva molts twice to achieve its final size and forms a pupa; it then undergoes metamorphosis to become a fly.

Leaf primordium The default state of a secondary apical meristem which generates a blade-like structure with distinct dorsal (nearest the shoot) and ventral (farthest from the shoot) surfaces.

Leafy (LFY) A gene expressed in early developing floral meristems that is required to initiate development of the floral meristem. In *LFY* mutants, flowers are replaced by leaves and an adjoining shoot meristem which forms a branch.

Lens The cells that give rise to the transparent portion of the eye which focuses light on the retina.

Library A complete collection of bacterial clones containing fragments of the entire genome of an organism.

Limb buds Developing vertebrate appendages such as legs, wings, and arms.

MADS-box genes Genes encoding a class of transcription factors present in plants and animals. In plants, several MADS-box genes define the identities of floral organs and direct fruit development.

Margin of leaf The edge of the leaf that forms at the junction between the dorsal and ventral surfaces.

Margin of wing The edge of the wing that forms at the junction between the dorsal and ventral surfaces.

Marginal zone The equatorial zone of the frog embryo that forms within the animal hemisphere and gives rise to mesoderm and a rigid structure supporting the spinal cord called the notochord.

Master gene A gene that acts as a single regulator to specify a certain cell fate and can redirect other cells to adopt that fate.

Maternal genes Genes that are active only in the mother and/or egg and are required for development of the fertilized embryo.

Maternal information Information provided by the mother that has a role in establishing the position of the primary body axes of the embryo.

Maternal mutant A mutant in a maternal gene that lacks a function in the egg supplied solely by the mother.

Medio-lateral axis of leaf (M/L) The axis of the leaf running perpendicular to the proximal–distal axis, which is marked by structures such as veins that branch in particular locations.

Mesoderm The embryonic germ layer that gives rise to muscle, heart, and fat. In flies, the mesoderm forms from the ventral region of the embryo and in frogs, it derives from the marginal zone that runs between the ectoderm and endoderm.

Metamorphosis The transformation of a larva into an adult.

Misexpression (of a gene) The inappropriate expression of a gene in time or space.

Monocot plants A category of flowering plants whose members contain a single embryonic seed leaf, or cotyledon.

Morphogen A secreted signal that elicits different cellular responses at different concentrations. To be a morphogen a molecule must satisfy three criteria: (1) It is synthesized in a subset of cells; (2) it diffuses from its site of synthesis to become progressively less concentrated farther from the source of synthesis; and (3) cells respond to different concentrations of the morphogen by activating expression of distinct sets of genes.

Morphogenesis The process by which the developing organism attains its final shape. In animal embryos, morphogenesis is accomplished chiefly by organized cell movements during gastrulation. In a developing plant, morphogenesis is accomplished by controlling the orientation of the plane of cell division and by regulating the direction of cell expansion following division.

Mutagen A chemical compound that causes mutations.

Mutant clone analysis A genetic method in which the defects observed in genetic mosaics (i.e., individuals carrying small patches of homozygous mutant cells surrounded by normal cells) are analyzed.

Mutant An organism with a mutated gene that causes an identifiable defect.

Mutation An alteration in the base sequence of a gene.

Mutual inhibition A type of cellular communication in which two or more cells attempt to prevent each other from becoming a preferred cell type. As soon as one cell sends a stronger inhibitory signal to its neighbors than it receives, it assumes the preferred state and forces the other cells to adopt the alternative state.

Naked A segment-polarity gene required for the formation of anterior structures in the segment. In *naked* mutants, the anterior half of the segment is transformed into a mirror copy of the naked posterior portion of the segment.

Neural inducing factor/substance A secreted signal liberated by the Spemann organizer that promotes neural over epidermal development.

Neural tube A cylindrical infolding of the neural ectoderm that gives rise to the central nervous system of a vertebrate.

Neuroblast A specialized neural precursor cell that separates itself from the ventral portion of the ectoderm and divides to give rise to a particular subset of neuronal cells.

Neuroectoderm The portion of the ectoderm (outer germ layer) from which the nervous system forms. The neuroectoderm derives from the ventral ectoderm or lateral region of a fly embryo and from the dorsal ectoderm of a vertebrate embryo.

Notch A gene encoding a receptor that is required for outgrowth of the wing and formation of the wing margin in flies and for establishment of a specialized group of cells (the AER) along the margin of vertebrate limb buds. Complete loss of *Notch* function in flies results in a lack of wings, and reduced *Notch* activity results in wings with notches along the margin.

Notochord A rod-like structure that serves as a rigid support (like a backbone). The notochord is part of the mesoderm tissue layer that also gives rise to muscle, heart, and blood.

Nucleus A structure in the center of a cell, surrounded by a porous nuclear membrane, that contains the genetic material in the form of DNA. The nucleus is the site of transcription in which RNA molecules are copied from DNA.

odd-skipped A pair-rule gene encoding a transcription factor required for the formation of odd-numbered segments. *odd-skipped* mutants lack odd-numbered segments.

Oocyte An unfertilized egg. In frogs, the oocyte is a visibly polarized structure consisting of two differently pigmented hemispheres: animal and vegetal.

Optic vesicle Specialized neural tube cells that bud out from the neural tube to give rise to the vertebrate eye.

optimotor blind (omb) A very sensitive target gene of Dpp signaling in the fly wing that is expressed in a broad domain centered over the anterior–posterior organizer and contains the nested *spalt* expression domain.

Organizer A region of a developing organism that sends signals to neighboring cells to organize the formation of a morphological structure. Examples of organizers are the Spemann organizer in early frog embryos that organizes the neural axis, the ZPA, which organizes the anterior–posterior axis of vertebrate limb buds, and the narrow stripe of cells running up the center of the fly wing disc that organizes the anterior–posterior axis of the wing.

Ovary (floral) The portion of the female organ (carpel) in which the egg-containing ovules develop.

Pair-rule gene A patterning gene (e.g., *even-skipped*) required for the formation of structures in every other segment of a fly embryo.

Pair-rule mutant A mutant in a pair-rule gene that lacks cuticle derived from every other segment. Pair-rule mutants lack cuticle in alternating segments of the anterior–posterior axis.

Pathogens (agricultural) Infective organisms (e.g., bacteria, fungi, and viruses) that cause damage to crops.

Patterning The process by which cells acquire distinct identities during development.

pax6 The vertebrate counterpart of the fly *eyeless* gene that plays an essential role in initiating eye development in mice and humans.

Petals The second whorl of floral organs forming just inside the sepals that are the most visually prominent structure of the flower (e.g., the red petals of a rose).

Phantastica A snapdragon gene required for outgrowth of the leaf and for formation of the leaf margin at the junction between the dorsal and ventral surfaces of the leaf. *phantastica* mutants, like *apterous* mutants in flies, are defective for formation of the dorsal surface of leaves. In severely affected *phantastica* mutants, there is no outgrowth of the leaf at all.

Photoreceptors Light-sensitive neuronal cells in the eye that respond to light by producing an electrical impulse.

Phenylketonuria (PKU) A once-fatal human genetic disorder resulting from an inability to metabolize the amino acid phenylalanine. Parents who are both carriers for PKU have a 25% chance during each pregnancy of producing a child inheriting the mutant gene from both of them who would be afflicted by PKU.

Plasmid A small, circular, independently replicating DNA molecule propagated in bacteria that can carry foreign genes.

Plasmodesmata Large pores that connect plant cells to their neighbors and through which large molecules can diffuse directly from one plant cell to another.

Pollen grain The product of the male organ of a plant (stamen) that houses the sperm.

Pollen tube A long tube through which the sperm nucleus descends to the egg within the ovary portion of the carpel.

Primary body axes The perpendicular anterior–posterior and dorsal–ventral axes of animals established during early embryonic development.

Progressive patterning/refinement A sequence of simple patterning events during development. Each patterning event is dependent on what has happened in the previous developmental stage.

Protein A chain of amino acids that folds up to generate a complex three-dimensional form. Proteins are encoded by genes and do most of the work in a cell.

Proximal–distal (P/D) axis The proximal (e.g., shoulder)–distal (e.g., hand) axis of an animal appendage extends from the body (proximal) to the tip of the appendage (distal). In leaves of plants, the proximal–distal axis runs from the stem (proximal) to the tip (distal) of the leaf.

Radial axis The width dimension of a plant, which is subdivided into three tissue layers (outer layer = L1 = epidermis, middle layer = L2 = cortex, and inner layer = L3 = vasculature).

Receptor A molecule typically present in the membrane of a cell that mediates communication between cells by sensing a signal sent by a neighboring cell. The signal sticks to the receptor on the cell surface, triggering a response that ultimately alters gene expression in the receiving cell.

Recombinant DNA A DNA molecule formed by joining DNA fragments from different sources in a test tube.

Regulatory region of gene The region of a gene that determines when and where the gene will be active (i.e., transcribed or on) versus silent (i.e., off).

Repressor A transcription factor that prevents transcription of a gene (i.e., turns a gene off).

Retina A hollow sack of cells derived from the out-pocketing of the optic vesicle that gives rise to the portion of the eye containing the light-sensitive photoreceptor cells.

rhomboid (rho) A dorsal–ventral patterning gene required for the development of the neuroectoderm. *rho* mutants have greatly reduced ventral epidermal structures.

RNA *Ribo*nucleic *A*cid is a long, single-stranded polymer of bases similar to single-stranded DNA and essential for protein synthesis. One difference between DNA and RNA is that the base thymine in DNA is replaced with uracil in RNA.

RNA polymerase An enzyme that carries out RNA synthesis (or transcription). RNA polymerase copies one strand of RNA (the template strand) to make the complementary RNA strand using the base-pairing rules.

Saturating mutant screen A mutant screen that is extended to the point of recovering multiple independent mutations in an average gene. In a typical saturating screen, more than 95% of all genes contributing to the process being analyzed will have been mutated.

Secondary meristem A small group of cells that arise during maturation of the plant along the periphery of the apical shoot meristem and give rise to branches, leaves, or flowers.

Segmentation mutants A group of anterior–posterior patterning mutants affecting segmentation of the embryo, e.g., *engrailed*. The four basic groups of segmentation mutants are: gap mutants, pair-rule mutants, segment-polarity mutants, and homeotic mutants.

Segment-polarity gene/mutant A patterning gene required for the formation of part of each segment along the A/P axis. Mutants in segment-polarity genes exhibit defects within every segment, such as deletions and/or mirror image duplications.

Sepals The outer whorl of floral organs, resembling leaves, that encloses the flower.

sevenless A gene encoding a receptor protein for the Bride-of-Sevenless signal that is expressed in cells including the R7 photoreceptor. *sevenless* mutants lack the UV-sensitive R7 photoreceptor cell.

Shattering The process by which pod-type fruits break open and release their seeds.

***Shatterproof* genes** A pair of highly related genes expressed specifically in valve border cells, which are required for formation of valve borders. The valve borders are missing in *shatterproof* mutants.

shootmeristemless (STM) A gene encoding a homeobox-type protein that is required for determining apical shoot meristem identity in heart-stage embryos. *STM* mutants lack an apical shoot meristem and do not develop beyond the embryonic stage because they cannot elaborate a growing shoot.

short gastrulation (sog) A D/V patterning gene encoding a secreted factor that opposes the action of *dpp* in the neuroectoderm. In *sog* mutants, Dpp signaling spreads into the neuroectoderm through a combination of diffusion and autoactivation, and formation of the nervous system is compromised.

Signal A molecule produced in one cell that alters the fate of a neighboring cell.

Signaling factor A signal.

sine oculus A gene required for initiating eye development in flies that is required for the *eyeless* expression. *sine oculus* mutants lack eyes.

Small eye A mutant mouse in which the *pax6* gene (the vertebrate counterpart of the fly *eyeless* gene) has been disrupted, resulting in the development of smaller than normal eyes.

snail A dorsal–ventral patterning gene encoding a transcription factor that represses neuroectodermal gene expression in the mesoderm. *snail* mutant embryos assume a spiral morphology due to the absence of muscle.

sonic hedgehog (shh) A gene encoding a vertebrate version of the Hedgehog morphogen (Shh) that is produced by the zone of polarizing activity and has anterior–posterior organizing activity.

spalt (sal) A moderately sensitive target gene of Dpp signaling in the fly wing which is expressed in a domain centered over the anterior–posterior organizer and is contained within the wider *omb* expression domain.

Spemann organizer The organizing center of a frog embryo that induces neighboring cells to form the central nervous system.

Stamen The male reproductive organ in a flowering plant that produces pollen.

Stigma An exposed portion of the carpel onto which the pollen is deposited.

Stromatolites Organized colonies of bacterial cells that date back more than three billion years and still exist today. Stromatolites are found in isolated bodies of water that are sheltered from seaweeds and animals, and they can grow to be over a meter tall.

Suspensor An umbilical cord-like structure that connects the plant embryo to the nutrients stored within the seed.

Tay-Sachs disease A human genetic disorder that is propagated by a carrier. Parents who are both carriers for Tay-Sachs have a 25% chance during each pregnancy of producing an affected child inheriting the mutant gene from both of them.

terminal flower (TFL) A gene encoding a likely inhibitory signal that suppresses floral development in secondary meristems by repressing expression of *AP1* and *LFY.* Flowering is initiated by genes such as *FT* that oppose the action of *TFL.* In *TFL* mutants, the shoot meristem seems to be unable to restrain flower development and is consumed in the process of developing into a flower rather than continuing to serve as a regenerative growth center at the apex of the plant.

Transcription Synthesis of a single-stranded RNA copy of a DNA molecule that takes place in the nucleus of the cell. Transcription copies the coding region of a gene into a complementary RNA molecule.

Transcription factors A class of proteins that control the transcription of genes (i.e., that turn genes on or off). Transcription factors bind to the DNA in the regulatory region of a gene and determine whether RNA polymerase will be able to transcribe that gene in a given cell.

Transgene A cloned or recombinant gene that has been inserted into the genome of an organism.

Transgenic organism A genetically modified organism that contains a recombinant gene or transgene.

Translation The conversion of the base sequence of DNA and RNA into a sequence of amino acids in protein, which takes place in the cytoplasm of a cell.

Triplet A group of three RNA bases that is assigned to a specific amino acid by the genetic code.

twist A D/V patterning gene encoding a transcription factor that activates mesodermal gene expression in the mesoderm. *twist* mutant embryos have a spiral shape due to the absence of muscle.

Unicellular organism A single-celled organism such as a bacterium or yeast cell.

Valve The fleshy sectors of a fruit that get eaten in edible fruits.

Valve border The narrow stripe of cells running along the edges of the sectors of valves.

Vascular tissue (L3 layer) The inner of the three germ layers partitioning the radial axis of a plant that serves as a transport system for water and nutrients.

Vegetal hemisphere The nonpigmented half of the oocyte from which the endoderm derives.

VegT A gene encoding a transcription factor that is required for endoderm development of vegetal cells and for induction of mesoderm in adjacent cells of the animal hemisphere.

Ventral surface of leaf The surface of the leaf forming farthest from the shoot.

Vertebrate An animal with a backbone, such as a human, mouse, or frog.

vestigial (vg) A gene required for specifying wing cell fates. *vg* mutants lack wings.

Whorls The concentric rings of the four organ types in flowers. The outermost whorl consists of the sepals, the next whorl is occupied by petals, and the inner two whorls are the stamens and carpels (the male and female reproductive organs, respectively).

wuschel (WUS) A gene encoding a homeobox type protein that is required for determining apical shoot meristem identity in globular-stage embryos and activating expression of *STM*. *WUS* mutants lack an apical shoot meristem and do not develop beyond the embryonic stage because they cannot elaborate a growing shoot.

Zone of polarizing activity (ZPA) A small region in the posterior portion of each limb bud that organizes the anterior–posterior axis of the limb, at least in part, by secreting the morphogen Sonic Hedgehog.

References and Additional Reading

CHAPTER 1

Alberts B., Bray D., Lewis J., Raff M., Roberts K., and Watson J.D. 1994. *Molecular biology of the cell*, 3rd edition. Garland, New York.

Branden C. and Tooze J. 1991. *Introduction to protein structure.* Garland, New York.

Briggs R. and King T.J. 1952. Transplantation of living nuclei from blastula cells into enucleated frogs eggs. *Proc. Natl. Acad. Sci.* **38:** 455–463.

Crick F.H.C. The genetic code. III. *Sci. Am.* **215:** 55–62.

Darwin C.R. 1859. *On the origin of species.* John Murray, London.

——— 1964. (1859) Facsimile first edition of *On the origin of species.* (ed. E. Mayr). Harvard University Press, Cambridge, Massachusetts.

Gurdon J.B. 1962. The developmental capacity of nuclei taken from intestinal epithelium cells of feeding tadpoles. *J. Embryol. Exp. Morphol.* **10:** 622–641.

——— 1968. Transplanted nuclei and cell differentiation. *Sci. Am.* **219:** 24–35.

Slack J.M.W. 1987. Morphogenetic gradients–past and present. *Trends Biochem. Sci.* **12:** 200–204.

Watson J.D. and Crick F.H.C. 1953. Molecular structure of nucleic acids. A structure for deoxyribose nucleic acid. *Nature* **171:** 737–738.

——— 1953. Genetic implications of the structure of deoxyribonucleic acid. *Nature* **171:** 964–967.

Wilmut I., Schnieke A.E., McWhir J., Kind A.J., and Campbell K.H. 1997. Viable offspring derived from fetal and adult mammalian cells. *Nature* **385:** 810–813.

CHAPTER 2

Cohen S.N. 1975. The manipulation of genes. *Sci. Am.* **233:** 24–33.

History of Development at http://zygote.swarthmore.edu

Human Genome Project at http://www.nhgri.nih.gov/HGP

Okayama H. and Berg P. 1982. High-efficiency cloning of full-length cDNA. *Mol. Cell. Biol.* **2:** 161–170.

O'Neill J.W. and Bier E. 1994. Double-label in situ hybridization using biotin and digoxigenin-tagged RNA probes. *BioTechniques* **17:** 870–875.

Palmiter R.D. and Brinster R.L. 1986. Germ line transformation of mice. *Annu. Rev. Genet.* **20:** 465–499.

Rubin G.M. and Spradling A.C. 1982. Genetic transformation of *Drosophila* with transposable element vectors. *Science* **218:** 348–353.
Tautz D. and Pfeiffle C. 1989. A non-radioactive in situ hybridization method for the localization of specific RNAs in *Drosophila* reveals translational control of the segmentation gene *hunchback*. *Chromosoma* **98:** 81–85.
von Goethe J.W. 1978 (1790). *The metamorphosis of plants* (with an introduction by Rudolf Steiner), 2nd revised edition. Bio-Dynamic Literature, Wyoming, Rhode Island.

CHAPTER 3

Alan Turing Web site at http://www.turing.org.uk/turing/index.html
Alberts B., Bray D., Lewis J., Raff M., Roberts K., and Watson J.D. 1994. *Molecular biology of the cell,* 3rd edition. Garland, New York.
Anderson K.V. and Nüsslein-Volhard C. 1984. Information for the dorsal–ventral pattern of the *Drosophila* embryo is stored as maternal mRNA. *Nature* **311:** 223–227.
Anderson K.V., Jürgens G., and Nüsslein-Volhard C. 1985. Establishment of dorsal-ventral polarity in the *Drosophila* embryo: Genetic studies on the role of the Toll gene product. *Cell* **42:** 779–789.
Biehs B., Francois V., and Bier E. 1996. The *Drosophila* short gastrulation gene prevents Dpp signaling from autoactivating and suppressing neurogenesis in the neuroectoderm. *Genes Dev.* **10:** 2922–2934.
Bier E. 1997. Anti-neural inhibition: A conserved mechanism for neural induction. *Cell* **89:** 681–684.
Flybase at http://flybase.bio.indiana.edu:82
Frasch M., Warrior R., Tugwood J., and Levine M. 1988. Molecular analysis of even-skpped mutant *Drosophila* development. *Genes Dev.* **2:** 1824–1838.
Frasch M., Hoey T., Rushlow C., Doyle H., and Levine M. 1987. Characterization and localization of the even-skipped protein of *Drosophila. EMBO J.* **6:** 749–759.
Gehring W.J. 1985. Homeotic genes, the homeobox, and the genetic control of development. *Cold Spring Harbor Symp. Quant. Biol.* **50:** 243–251.
Gellon G. and McGinnis W. 1998. Shaping animal body plans in development and evolution by modulation of Hox expression patterns. *BioEssays* **20:** 116–125.
Gierer A. and H. Meinhard. 1972. A theory of biological pattern formation. *Kybernetik* **12:** 30–39.
Gray S., Szymanski P., and Levine M. 1994. Short-range repression permits multiple enhancers to function autonomously within a complex promoter. *Genes Dev.* **8:** 1829–1838.
Hartenstein V. 1993. Atlas of *Drosophila* development. Cold Spring Harbor Laboratory Press, Cold Spring Harbor, New York.
Interactive Fly http://sdb.bio.purdue.edu/fly/aimain/1aahome.htm
Levine M., Rubin G.M., and Tjian R. 1984. Human DNA sequences homologous to a protein-coding region conserved between homeotic genes of *Drososphila. Cell* **38:** 667–673.
Lewis E.B. 1978. A gene controlling segmentation in *Drosophila. Nature* **276:** 565–570.
——— 1985. Regulation of genes of the bithorax complex of *Drosophila. Cold Spring Harbor Symp. Quant. Biol.* **50:** 155–164.
——— 1995. The biothorax complex: The first fifty years. In *Les Prix Nobel.* The Nobel Foundation, Stockholm, Sweden.

McGinnis W. and Kuziora M. 1994. The molecular architects of body design. *Sci. Am.* **270:** 58–62, 64–66.
McGinnis W., Garber R.L., Wirz J., Kuroiwa A., and Gehring W.J. 1984. A homologous protein-coding sequence in *Drosophila* homeotic genes and its conservation in other metazoans. *Cell* **37:** 403–408.
Nobel Foundation Web site laureates' biographies http://www.nobel.se/laureates
Nüsslein-Volhard C. and Jürgens G. 1985. Genes affecting the segmental subdivision of the *Drosophila* embryo. *Cold Spring Harbor Symp. Quant. Biol.* **50:** 145–154.
Nüsslein-Volhard C. and Wieschaus E. 1980. Mutations affrecting segment number and polarity in *Drosophila. Nature* **287:** 795–801.
Ray R.P., Arora K., Nüsslein-Volhard D., and Gelbart W.M. 1991. The control of cell fate along the dorsal-ventral axis of the *Drosophila* embryo. *Development* **113:** 35–54.
Roth S., Stein D., and Nüsslein-Volhard C. 1989. A gradient of nuclear localization of the dorsal protein determines dorsoventral pattern in the *Drosophila* embryo. *Cell* **59:** 1189–1202.
Rushlow C.A., Han K., Manley J.L., and Levine M. 1989. The graded distribution of the dorsal morphogen is initiated by selective nuclear transport in *Drosophila. Cell* **59:** 1165–1177.
Saunders P.T., ed. 1992. *Morphogenesis: The collected works of A.M. Turing,* vol. 3. Elsevier, North-Holland, Amsterdam, The Netherlands.
Scott M.P. 2000. Development: The natural history of genes. *Cell* **100:** 27–40.
Scott M.P. and Weiner A.J. 1984. Structural relationships among genes that control development: Sequence homology between Antennapedia, Ultrabithorax, and fushi tarazu loci of *Drosophila. Proc. Natl. Acad. Sci.* **81:** 4115–4119.
Sharkey M., Graba Y., and Scott M.P. 1997. Hox genes in evolution: Protein surfaces and paralog groups. *Trends Genet.* **13:** 145–151.
Steward R. 1989. Relocalization of the dorsal protein from the cytoplasm to the nucleus correlates with its function. *Cell* **59:** 1179–1188.
Turing A.M. 1952. The chemical basis of morphogenesis. *Phil. Trans. R. Soc. Lond. Biol. Sci.* **237:** 37–72.
Wieschaus E., Jürgens G., and Nüsslein-Volhard C. 1984. Mutations affecting the pattern of the larval cuticle in *Drosophila melanogaster.* 3. Zygotic loci on the X-chromosome and the fourth chromosome. *Wilhelm Roux's Arch. Dev. Biol.* **193:** 296–307.

CHAPTER 4

Banerjee U., Renfranz P.J., Pollock J.A., and Benzer S. 1987. Molecular characterization and expression of *sevenless,* a gene involved in neuronal pattern formation in the *Drosophila* eye. *Cell* **49:** 281–291.
García-Bellido A. 1977. Inductive mechanisms in the process of wing vein formation in *Drosophila. Wilhelm Roux's Arch. Dev. Biol.* **182:** 93–106.
Hafen E., Basler K., Edstroem J.E., and Rubin G.M. 1987. *Sevenless,* a cell-specific homeotic gene of *Drosophila,* encodes a putative transmembrane receptor with a tyrosine kinase domain. *Science* **236:** 55–63.
Harris W.A., Stark W.S., and Walker J.A. 1976. Genetic dissection of the photoreceptor system in the compound eye of *Drosophila melanogaster. J. Physiol.* **256:** 415–439.
Kohler R.E. 1994. *Lords of the fly:* Drosophila *genetics and the experimental life.* University of Chicago Press, Illinois.

Lawrence P.A. 1992. *The Making of a fly: The genetics of animal design.* Blackwell Science, Oxford, United Kingdom.

Lawrence P.A. and Struhl G. 1996. Morphogens, compartments, and pattern: Lessons from *Drosophila*? *Cell* **85:** 951–961.

Lecuit T., Brook W.J., Ng M., Calleja M., Sun H., and Cohen S.M. 1996. Two distinct mechanisms for long-range patterning by Decapentaplegic in the *Drosophila* wing. *Nature* **381:** 387–393.

Morata G. and Lawrence P. 1975. Control of compartment development by the engrailed gene. *Nature* **255:** 6147–617.

Nellen D., Burke R., Struhl G., and Basler K. 1996. Direct and long range action of a Dpp morphogen gradient. *Cell* **85:** 357–368.

Perrimon N.B. 1995. Hedgehog and beyond. *Cell* **80:** 517–520.

Rubin G.M. 1989. Development of the *Drosophila* retina: Inductive events studied at single cell resolution. *Cell* **57:** 519–520.

Sturtevant M.A., Biehs B., Marin E., and Bier E. 1997. The spalt gene links the A/P compartment boundary to a linear adult structure in the *Drosophila* wing. *Development* **124:** 21–32.

Tomlinson A. 1988. Cellular interactions in the developing *Drosophila* eye. *Development* **104:** 183–193.

Tomlinson A. and Ready D.F. 1987. Neuronal differentiation in the *Drosophila* ommatidium. *Dev. Biol.* **120:** 366–376.

Zecca M., Basler K., and Struhl G. 1995. Sequential organizing activities of engrailed, hedgehog, and decapentaplegic in the *Drosophila* wing. *Development* **121:** 2265–2278.

CHAPTER 5

Beddington R.S. and Robertson E.J. 1998. Anterior patterning in mouse. *Trends Genet.* **14:** 277–284.

——— 1999. Axis development and early asymmetry in mammals. *Cell* **96:** 195–209.

Brunet L.J., McMahon J.A., McMahon A.P., and Harland R.M. 1998. Noggin, cartilage morphogenesis, and joint formation in the mammalian skeleton. *Science* **280:** 1455–1457.

Hammerschmidt M, Serbedzija G.N., and McMahon A.P. 1996. Genetic analysis of dorsoventral patern formation in the zebrafish: Requirement of a BMP-like ventralizing activity and its dorsal repressor. *Genes Dev.* **10:** 2452–2461.

Hemmati-Brivanlou A. and Melton D.A. 1997. Vertebrate embryonic cells will become nerve cells unless told otherwise. *Cell* **88:** 13–17.

Holley S.A., Jackson P.D., Sasai Y., Lu B., De Robertis E.M., Hoffmann F.M., and Ferguson E.L. 1995. A conserved system for dorsal-ventral patterning in insects and vertebrates involving sog and chordin. *Nature* **376:** 249–253.

Kimelman D. and Griffin K.J. 1998. Mesoderm induction: A postmodern view (comment). *Cell* **94:** 419–421.

Lamb T.M. and Harland R.M. 1995. Fibroblast growth factor is a direct neural inducer, which combined with noggin generates anterior-posterior neural pattern. *Development* **121:** 3627–3636.

Malicki J., Schugart K., and McGinnis W. 1990. Mouse Hox-2.2 specifies thoracic segmental identity in *Drosophila* embryos and larvae. *Cell* **63:** 961–967.

McGinnis N., Kuziora M.A., and McGinnis W. 1990. Human Hox-4.2 and *Drosophila* deformed encode similar regulatory specificities in *Drosophila* embryonic larvae. *Cell* **63:** 969–976.
McMahon J.A., Takada S., Zimmerman L.B., Fan C.M., Harland R.M., and McMahon A.P. 1998. Noggin-mediated antagonism of BMP signaling is required for growth and patterning of the neural tube and somite. *Genes Dev.* **12:** 1438–1452.
Picolla S., Sasai Y., Lu B., and de Robertis E.M. 1996. Dorsoventral patterning in *Xenopus:* Inhibition of ventral signals by direct binding of chordin to BMP-4. *Cell* **86:** 589–598.
Schmidt J., François V., Bier E., and Kimelman D. 1995 The *Drosophila short gastrulation* gene induces an ectopic axis in *Xenopus*: Evidence for conserved mechanism of dorsal-ventral patterning. *Development* **121:** 4319–4328.
Schulte-Merker S., Lee K.J., McMahon A.P. and Hammerschmidt M. 1997. The zebrafish organizer requires chordino. *Nature* **387:** 862–863.
The Mouse Anatomical Dictionary–The Jackson Laboratory at http://www.informatics.jax.org/reports/anatDictionary
The Mouse Atlas and Gene Expression Database Project at http://genex.hgu.mrc.ac.uk
The Virtual Embryo–Dynamic Devlopment at http://www.ucalgary.ca/UofC/eduweb/virtualembryo
Zimmerman L.B., De Jesus-Escobar J.M., and Harland R.M. 1996. The Spemann organizer signal noggin binds and inactivated bone morphogenetic protein 4. *Cell* **86:** 599–606.

CHAPTER 6

Desplan C. 1997. Eye development: Governed by a dictator or a junta? *Cell* **91:** 861–864.
Gehring W.J. and Ikeo K. 1999. Pax 6: Mastering eye morphogenesis and eye evolution. *Trends Genet.* **15:** 371–377.
Gould S.J. 1989. *Wonderful life: The Burgess shale and the nature of history.* Norton, New York.
Halder G., Callaerts P., and Gehring W.J. 1995. Induction of ectopic eyes by targeted expression of the eyeless gene in *Drosophila. Science* **267:** 1788–1792.
Hanson I. and Van Heyningen V. 1995. Pax6: More than meets the eye. *Trends Genet.* **11:** 268–272.
Harris W.A. 1997. Pax-6: Where to be conserved is not conservative. *Proc. Natl. Acad. Sci.* **94:** 2098–2100.
Johnson R.L. and Tabin C. 1995. The long and short of hedgehog signaling. *Cell* **81:** 313–316.
——— 1997. Molecular models for vertebrate limb development. *Cell* **90:** 979–990.
Riddle R.D., Johnson R.L., Laufer E., and Tabin C. 1993. Sonic hedgehog mediates the polarizing activity of the ZPA. *Cell* **75:** 1401–1016.
Saunders Jr., J.W., and Gasseling M.T. 1968. Ectoderm-mesenchymal interaction in the origin of wing symmetry. In *Epithelial-mesenchymal interactions* (ed. R. Fleischmajer and R.E. Billingham), pp. 159–170. Willams and Wilkins, Baltimore, Maryland.

Shubin N., Tabin C., and Carroll S. 1997. Fossils, genes, and the evolution of animal limbs. *Nature* **388:** 639–648.

Tomarev S.I., Callaerts P., Kos L., Zinovieva R., Halder G., Gehring W., and Piatigorsky J. 1997. Squid Pax-6 and eye development. *Proc. Natl. Acad. Sci.* **94:** 2098–2100.

Weatherbee S.D. and Caroll S.B. 1999. Selector genes and limb identity in arthropods and vertebrates. *Cell* **97:** 283–286.

Weatherbee S.D., Nijhout H.F., Grunert L.W., Halder G., Galant R., Selegue J., and Carroll S. 1999. Ultrabithorax function in butterfly wings and the evolution of insect wing patterns. *Curr. Biol.* **11:** 109–115.

Zeller R. and Duboule D. 1997. Dorso-ventral limb polarity and origin of the ridge: On the fringe of independence? *BioEssays* **19:** 541–546.

Zuker C.S. 1998. Specificity in signalling pathways: Assembly into multimolecular signalling complexes. *Curr. Opin. Genet. Dev.* **8:** 419–422.

CHAPTER 7

Arabinet at http://weeds.mgh.harvard.edu/atlinks.html

Chiu J., DeSalle R., Lam H.M., Meisel L., and Coruzzi G. 1999. Molecular evolution of glutamate receptors: A primitive signaling mechanism that existed before palnts and animals diverged. *Mol. Biol. Evol.* **16:** 826–838.

Chuck G., Lincoln C. and Hake S. 1996. KNAT1 induces lobed leaves with ectopic meristems when overexpressed in Arabidopsis. *Plant Cell* **8:** 1277–1289.

Clark S.E., Jacobsen S.E., Levin J.Z., and Meyerowitz E.M., 1996. The CLAVATA and SHOOT MERISTEMLESS loci competitively regulate meristem activity in *Arabidopsis. Development* **122:** 1567–1575.

Coen E.S. 1992. Flower development. *Curr. Opin. Cell Biol.* **4:** 929–933.

Coen E.S. and Meyerowitz E.M. 1991. The war of the whorls: Genetic interactions controlling flower development. *Nature* **353:** 31–37.

Fletcher J.C., Brand U., Running M.P., Simon R., and Meyerowitz E.M. 1999. Signaling of cell fate decisions by CLAVATA3 in *Arabidopsis* shoot meristems. *Science* **283:** 1911–1914.

Jurgens G. 1992. Pattern formation in the flowering plant embryo. *Curr. Opin. Genet. Dev.* **2:** 567–570.

Jurgens G., Mayer U., Busch M., Lukowitz W., and Laux T. 1995. Pattern formation in the *Arabidopsis* embryo: A genetic perspective. *Phila. Trans. R. Soc. Lond. B. Biol. Sci.* **250:** 19–25.

Lam H.M., Chiu J., Hsieh M.H., Meisel L., Oliveira I.C., Shin M., and Coruzzi G. 1998. Glumate-receptor genes in plants. *Nature* **396:** 125–126.

Laux T., Mayer K.F., Berger J., and Jurgens G. 1996. The WUSCHEL gene is required for shoot and floral meristem integrity in *Arabidopsis. Development* **122:** 887–896.

Luo D., Carpenter R., Copsey L., Vincent C., Clark J., and Coen E. 1999. Control of organ asymmetry in flowers of *Antirrhinum. Cell* **99:** 367–376.

Mayer K.F., Schoof H., Haecker A., Lenhard M, Jürgens G., and Laux T. 1998. Role of WUSCHEL in regulating stem cell fate in the *Arabidopsis* shoot meristem. *Cell* **95:** 805–815.

Mayer U. and Jürgens G. 1998. Pattern formation in plant embryogenesis: A reassessment. *Semin. Cell. Dev. Biol.* **9:** 187–193.

Skoog F. 1973. Cytokinins in regulation of plant growth. *Basic Life Sci.* **2:** 147–184.

Weigel D. and Meyerowitz E.M. 1993. Activation of floral homeotic genes in *Arabidopsis. Science* **261:** 1723–1726.
Wiegel D., Alvarez J., Smythe D.R., Yanofsky M.F., and Meyerowitz E.M. 1992. LEAFY controls floral meristem identity in *Arabidopsis. Cell* **69:** 843–859.

CHAPTER 8

Aida M., Ishida T., and Tasaka M. 1999. Shoot apical meristem and cotyledon formation during *Arabidopsis* embryogenesis: Interaction among CUP-SHAPED COTYLEDON and SHOOT MERISTEMLESS genes. *Development* **126:** 1563–1570.
Gu Q., Ferrandiz C., Yanofsky M.F., and Martienssen R. 1998. The FRUITFULL MADS-box gene mediates cell differentiation during *Arabidopsis* fruit development. *Development* **125:** 1509–1517.
Gustafson-Brown C., Savidge B., and Yanofsky M.F. 1994. Regulation of the *Arabidopsis* floral homeotic gene APETALA1. *Cell* **76:** 131–143.
Liljegren S.J., Gustafson-Brown C., Pinyopich A., Ditta G.S., and Yanofsky MF. 1999. Interactions among APETALA1, LEAFY, and TERMINAL FLOWER1 specify meristem fate. *Plant Cell* **11:** 1007–1018.
McSteen P.C., Vincent C.A., Doyle S., Carpenter R., and Coen E.S. 1998. Control of floral homeotic gene expression and organ morphogenesis in *Antirrhinum. Development* **125:** 2359–2369.
Meyerowitz E.M., Smythe D.M., and Bowman J.L. 1989. Abnormal flowers and pattern formation in floral development. *Development* **106:** 209–217.
Mozo T., Dewar K., Dunn P., Ecker J.R., Fischer S., Kloska S., Lehrach H., Marra M., Martienssen R., Meier-Ewert S., and Altmann T. 1999. A complete BAC-based physical map of the *Arabidopsis thaliana* genome. *Nat. Genet.* **22:** 271–275.
Pnueli L., Mareven D., Rounsley S.D., Yanofsky M.F., and Lifschitz E., 1994. Isolation of the tomato AGAMOUS gene TAG1 and analysis of its homeotic role in transgenic plants. *Plant Cell* **6:** 163–173.
Rounsley S.D., Ditta G.S., and Yanofsky M.F. 1995. Diverse roles for MADS box genes in *Arabidopsis* development. *Plant Cell* **7:** 1259–1269.
Sessions A. and Yanofsky M.F. 1999. Dorsoventral patterning in plants. *Genes Dev.* **13:** 1051–1054.
Sessions A., Yanofsky M.F., and Weigel D. 1998. Patterning the floral meristem. *Semin. Cell Dev. Biol.* **9:** 221–226.
Von Goethe J.W. 1978 (1790). The metamorphosis of plants (with an introduction by Rudolf Steiner, 2nd revised edition). Bio-Dynamic Literature, Wyoming, Rhode Island.
Waites R., Selvadurai H.R., Oliver I.R., and Hudson A. 1998. The PHANTASTICA gene encodes a MYB transcription factor involved in growth and dorsoventrality of lateral organs in *Antirrhinum. Cell* **93:** 779–789.
Weigel D. and Meyerowitz E.M. 1994. The ABCs of floral homeotic genes. *Cell* **78:** 203–209.
Weigel D., Alvarez J., Smyth D.R., Yanofsky M.F., and Meyerowitz E.M. 1992. LEAFY controls floral meristem identity in *Arabidopsis. Cell* **69:** 843–859.

CHAPTER 9

Crichton M. 1990. *Jurassic park.* Knopf, New York.
Island of Lost Souls. 1933. Director E.C. Kenton; writers, H.G. Wells, W. Young; stars, C. Laughton, B. Lugosi.

Island of Dr. Moreau. 1977. Director D. Taylor; writers, A. Ramus, J.H. Shaner; stars, B. Lancaster, M. York.
Island of Dr. Moreau. 1996. Director J. Frankenheimer; writers, R. Hutchinson, R. Stanley; stars, M. Brando, V. Kilmer.
Malthus R.T. 1985. *An essay on the principle of population.* Penguin English Library. (first edition of work appeared in 1798).
Online Mendelian Inheritance in Man (OMIM) at http://www3.ncbi.nlm.nih.gov/Omim/
Stapledon W.O. 1931. *Last and first men: A story of the near and far future.* Reissued 1968, Dover Publications, New York.

Credits

PHOTOGRAPHS

p.1, *Xenopus* egg (Reprinted, with permission, from Browder et al. *Developmental biology*, 3rd edition [©1991 Saunders College Publishing]).

p.1, Fruit fly egg (Reprinted, with permission, from *Embryos: Color atlas of development,* ed. J. Bard, p. 114 [©1994 Mosby–Year Book Europe Ltd.]).

p. 1, Mouse egg (Reproduced, with permission, from Aitken and Richardson *J. Reprod. Fertil. 63:* 295–307 [1981]).

p. 2, Homunculus in human sperm (Reprinted, with permission, from Moore *Heredity and development*, 2nd edition [©1972 Oxford University Press, New York]).

p. 7, Frog (Reprinted, with permission, from Gurdon *Sci. Am. 219:* 24–35 [1968]).

p. 10, John Gurdon (Photograph courtesy CSHL Library Archives).

p. 12, DNA structure (Reprinted, with permission, from Wing et al. *Nature 287:* 755–758 [©1980 Macmillan]).

p. 15, Chromosome (Reprinted, with permission, from Pelling and Allen *Chromosome Res. 1:* 221–237 [© Kluwer Academic Publishers 1993]).

p. 32, Hilde Mangold (Reprinted, with permission, from Hamburger *The heritage of experimental embryology: Hans Spemann and the organizer* [©1988 Oxford University Press, New York]).

p. 44, Johann Wolfgang von Goethe (©Austrian Archives/CORBIS).

p. 52, Gastrulating embryo (Reprinted from FlyBase, with permission, fromT.C. Kaufman [http://flybase.bio.indiana.edu:82/.bin/fbimage?show all&sort=1&startid=Fbim3. 905795]).

p. 61, Christiane Nüsslein-Volhard (Courtesy CSHL Library Archives).

p. 62, Eric Wieschaus (©The Nobel Foundation).

p. 66, Bithorax mutant (Reprinted, with permission, from Gerhart and

Kirschner *Cells, embryos, and evolution*, p. C8 [©1997 Blackwell Science, Inc.]).

p. 66, Edward B. Lewis (©The Nobel Foundation).

p. 83. Michael Levine (From http://mcb.berkeley.edu:80/faculty/GEN/levinem.html).

p. 91, Sean Carroll (Courtesy CSHL Library Archives).

p. 95, Gary Struhl (Kindly provided by Gary Struhl).

p. 98, Antonio Garcia-Bellido (Reprinted, with permission, from *Trends Genet. 9:* 103 [1993]).

p. 104, Fly eye (Reprinted, with permission, from Wolff and Ready *The development of* Drosophila melanogaster, ed. M. Bate and A. Martinez Arias, vol. 2, p. 1278 [©1993 Cold Spring Harbor Laboratory Press]).

p. 122, Bill McGinnis (Kindly provided by Bill McGinnis).

p. 125, Matt Scott (Courtesy CSHL Library Archives).

p. 142, Cliff Tabin (Kindly provided by Cliff Tabin).

p. 168, Gerd Jürgens (Kindly provided by Gerd Jürgens).

p. 184, Elliot Meyerowitz (Kindly provided by Elliot Meyerowitz).

p. 190, Marty Yanofsky (Kindly provided by Marty Yanofsky).

COLOR PLATES

Plate 1

(*A*) Kindly provided by Claude Desplan, New York University; reprinted, with permission, from Creighton T.E. *Encyclopedia of molecular biology* (©1999 Wiley, New York), pp. 276–279.

(*B,C*) Kindly provided by David Kosman and John Reinitz, Mt. Sinai School of Medicine (www.ams.sunysb.edu/~kingwai).

(*D*) Kindly provided by Bill McGinnis, University of California, San Diego.

(*E*) Provided by Henena Araujo and E. Bier.

(*F*) Reprinted, with permission, from Gray et al. *Genes Dev. 8:* 1829–2838 (1994).

(*G*) Reprinted, with permission, from Biehs et al. *Genes Dev. 10:* 2922–2934 (1996).

(*H*) Reproduced, with permission, from François et al. *Genes Dev. 8:* 2602–2616 (1994).

(*I–M*) Reprinted, with permission, from Nüsslein-Volhard and Wieschaus. *Nature 287:* 795–801 (©1980 Macmillan).

(*N*) Reproduced, with permission, from Biehs et al. *Genes Dev. 10:* 2922–2934 (1996).

Plate 2

(*A*) Reprinted, with permission, from Larabell et al. *J. Cell Biol. 136:* 1123–1136 (©1997 Rockefeller University Press).

(*B*) Reprinted, with permission, from Schmidt et al. *Development 122:* 1716 (©1996 Company of Biologists Ltd.).

(*C*) Reprinted, with permission, from Schmidt et al. *Dev. Biol. 169:* 37–50 (1995).

(*D, E*) Provided by Kimelman and E. Bier from Development Calendar, (©1996 Company of Biologists Ltd.).

(*F–H*) Reprinted, with permission, from Mayer et al. *Nature 353:* 402–407 (©1991 Macmillan).

(*I–K*) Reprinted, with permission, from Long et al. *Nature 379:* 66–69 (©1996 Macmillan).

(*L–P*) Reprinted, with permission, from Aida et al. *Development 126:* 1563–1570 (©1999 Company of Biologists Ltd.).

Plate 3

(*A,B*) Provided by M.A. Sturtevant and E. Bier.

(*C*) Reprinted, with permission, from Nellen et al. *Cell 85:* 357–368 (©1996 Cell Press).

(*D*) Provided by M.A. Sturtevant and E. Bier.

(*E*) Provided by E. Bier.

(*F*) Modified, with permission, from Sturtevant et al. *Development 124:* 21–32 (©1997 Company of Biologists Ltd.).

(*G*) Provided by E. Bier.

(*H*) Reprinted, with permission, from Halder et al. *Science 267:* 1788–1792 (©1995 American Association for the Advancement of Science).

(*I*) Kindly provided by Bill McGinnis, University of California, San Diego.

(*J*) Provided by E. Bier.

(*K*) Kindly provided by Bill McGinnis, University of California, San Diego; reprinted, with permission, from Malicki et al. *Cell 63:* 961–967 (©1990 Cell Press).

(*L–N*) Kindly provided by Cliff Tabin, Harvard University.

Plate 4

(*A*) Kindly provided by John Bowman, University of California, Davis; portion reprinted, with permission, from Bowman, *Embryonic encyclopedia of life sciences* (©1999 Macmillan).

(*B*) Kindly provided by Martin Yanofsky, University of California, San Diego.

(*C*) Kindly provided by Detlef Weigel, Salk Institute.

(*D*) Kindly provided by John Bowman, University of California, Davis; portion reprinted, with permission, from Bowman, *Embryonic encyclopedia of life sciences* (©1999 Macmillan).

(*E*) Kindly provided by John Bowman, University of California, Davis.

(*F*) Kindly provided by Martin Yanofsky, University of California, San Diego.

(*G*) Reprinted, with permission, from Weigel and Meyerowitz *Science 261:* 1723–1726 (©1993 American Association for the Advancement of Science).

(*H*) Kindly provided by John Bowman, University of California, Davis; portion reprinted, with permission, from Bowman, *Embryonic encyclopedia of life sciences* (©1999 Macmillan).

(*I*) Kindly provided by John Bowman, University of California, Davis.

(*J*) Kindly provided by Martin Yanofsky, University of California, San Diego.

(*K*) Reprinted, with permission, from Weigel and Meyerowitz *Science 261:* 1723–1726 (©1993 American Association for the Advancement of Science).

(*L*) Kindly provided by John Bowman, University of California, Davis; portion reprinted, with permission, from Bowman, *Embryonic encyclopedia of life sciences* (©1999 Macmillan).

(*M*) Kindly provided by John Bowman, University of California, Davis.

(*N,O*) Kindly provided by Martin Yanofsky, University of California, San Diego.

(*P–S*) Reprinted, with permission, from Fletcher et al. *Science 283:* 1911–1914 (©1999 American Association for the Advancement of Science).

(*T,U*) Kindly provided by Martin Yanofsky, University of California, San Diego.

(*V,W*) Reproduced, with permission, from Waites and Hudson *Development 121:* 2143–2154 (©1995 Company of Biologists Ltd.)

(*X*) Reprinted, with permission, from Chuck et al. *Plant Cell 8:* 1277–1289 (©1996 American Society of Plant Physiologists).

Index

About the Author

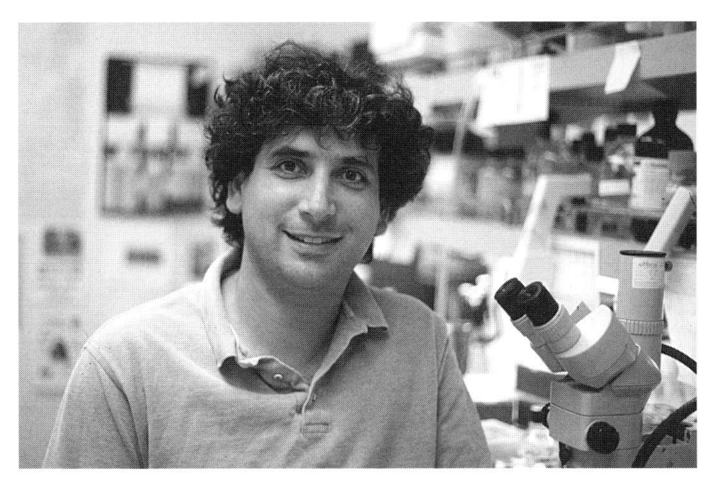

Ethan Bier (photo by Kathryn Burton)

Professor in the Department of Biology at the University of California, San Diego, Ethan Bier studies how modulation of the Dpp and EGF-receptor signaling pathways contributes to the development of the neural ectoderm of the fly embryo and of the adult wing pattern. He earned his Ph.D. at Harvard University, working with Alan Maxam, who devised the chemical method of DNA sequencing. His postdoctoral studies on neural development in flies were with Lily and Yuh Nung Jan at the University of California, San Francisco. Ethan lives with his wife Kathryn Burton and his son Benjamin in San Diego.